Shumin Han, Yuan Li, Baozhong Liu
Hydrogen Storage Alloys

Also of Interest

Series: GREEN – Alternative Energy Resources.
ISSN 2509-7237, e-ISSN 2509-7245

Green Energy and Resources for the Chemical Industry.
Centi, Perathoner, d'Hooghe (Eds.), 2018
ISBN 978-3-11-027313-7, e-ISBN 978-3-11-027729-6

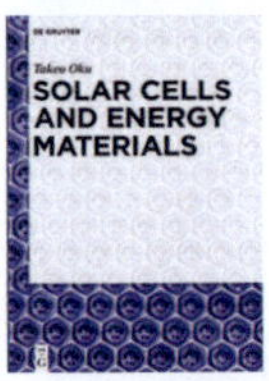

Solar Cells and Energy Materials.
Oku, 2016
ISBN 978-3-11-029848-2, e-ISBN 978-3-11-029850-5

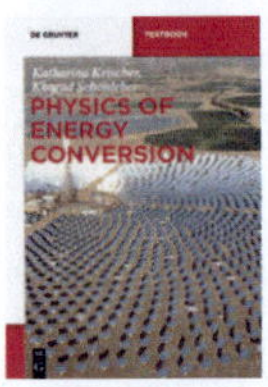

Physics of Energy Conversion.
Krischer, Schönleber (Eds.), 2015
ISBN 978-1-5015-0763-2, e-ISBN 978-1-5015-1063-2

Chemical Energy Storage.
Schlögl (Ed.), 2012
ISBN 978-3-11-026407-4, e-ISBN 978-3-11-026632-0

Shumin Han, Yuan Li, Baozhong Liu

Hydrogen Storage Alloys

With RE-Mg-Ni Based Negative Electrodes

DE GRUYTER

Authors

Professor Shumin Han
Yanshan University
State Key Laboratory of Metastable Materials Science and Technology
School of Environmental and Chemical Engineering
Qinhuangdao 066004
People's Republic of China
hanshm@ysu.edu.cn

Professor Yuan Li
Yanshan University
School of Environmental and Chemical Engineering
Qinhuangdao 066004
People's Republic of China
liyuan@ysu.edu.cn

Professor Baozhong Liu
Henan Polytechnic University
School of Materials Science and Engineering
Jiaozuo 454003
People's Republic of China
bzliu@hpu.edu.cn

ISBN 978-3-11-050116-2
e-ISBN (PDF) 978-3-11-050148-3
e-ISBN (EPUB) 978-3-11-049838-7
Set-ISBN 978-3-11-050149-0

Library of Congress Cataloging-in-Publication Data
A CIP catalog record for this book has been applied for at the Library of Congress.

Bibliographic information published by the Deutsche Nationalbibliothek
The Deutsche Nationalbibliothek lists this publication in the Deutsche Nationalbibliografie; detailed bibliographic data are available on the Internet at http://dnb.dnb.de.

Typesetting: Integra Software Services Pvt. Ltd.
Printing and binding: CPI books GmbH, Leck
Cover image: FikMik/iStock/Getty Images Plus
♾ Printed on acid-free paper
Printed in Germany

www.degruyter.com

Preface

With increasing concerns of environmental problems and depletion of fossil fuels, nick metal hydride (Ni/MH) batteries have received increasing attention as cleaner energy sources. They have been used for electric and hydride vehicles as well as other electric tools with the advantages of robust overcharge–discharge resistance, high usage safety, environmental friendly and easy recycling. Novel negative electrode materials have always been the focus during the exploitation of Ni/MH batteries. Rare earth-based AB_5-type alloys with the advantages such as easy activation and good cycling stability have been extensively commercialized. However, their discharge capacity can still hardly satisfy the requirements of high-energy density for Ni/MH batteries. Recent research found that RE–Mg–Ni-based alloys with structures of [AB_5] and [AB_2] subunits exhibit better electrochemical properties including higher discharge capacity, easier activation and better high-rate dischargeability. RE–Mg–Ni-based alloy electrodes have been applied in commercial Ni/MH batteries with promising results: long cycle life (>6,000 cycles) consumer batteries have been produced by FDK Incorporation, and large-capacity, high-power (40 cell – 1,500 Ah) stationary batteries have also been produced by Kawasaki Heavy Industries for industrial applications.

It is widely recognized that phase compositions and structures have great impacts on the electrochemical performance of RE–Mg–Ni-based alloys. Therefore, understanding the phase transformation process and its correlation with electrochemical properties of the RE–Mg–Ni-based alloys is critical for further improving their electrochemical performance. In recent years, our research group has focused on studies of the microstructures and electrochemical hydrogen storage performance, and important progresses such as the preparation and exploration of electrochemical properties of various single superlattice phases, understanding of interactions among different phases in the multiphase alloys and impact of different element composition on electrochemical performance, etc., have been achieved. An idea of publishing a book to explain the profound progresses and provide a comprehensive recognition of RE–Mg–Ni-based alloys to the researchers was naturally conceived.

Six peer-reviewed chapters are presented in this book. In Chapter 1, Dr. Jingjing Liu provides a survey on RE–Mg–Ni-based hydrogen storage alloys and describes the structure characteristics of RE–Mg–Ni-based hydrogen storage alloys. In Chapter 2, Dr. Lu Zhang reveals the preparation, electrochemical properties and gaseous hydrogen storage characteristics of the single-phase superlattice RE–Mg–Ni-based hydrogen storage alloys. In Chapter 3, Dr. Yumeng Zhao demonstrates the effect of multiphase structures on electrochemical properties of the superlattice RE–Mg–Ni-based hydrogen storage alloys. In Chapter 4, Prof. Baozhong Liu summarizes the effect of element composition on microstructures and electrochemical characteristics of the RE–Mg–Ni-based hydrogen storage alloys. Effect of different surface treatment on electrochemical characteristics of RE–Mg–Ni-based hydrogen storage alloys

DOI 10.1515/9783110501483-202

is introduced by Prof. Yuan Li in Chapter 5. In Chapter 6, Prof. Han reviews the outlook and challenges of RE–Mg–Ni-based alloys as negative electrode materials for Ni/MH batteries.

We thank the financial support by the National Natural Science Foundation of China (NOs. 20673093, 51171165, 51571173 and 21303157). We also thank all the authors and researchers for their great contribution to this book and sincerely thank Dr. Matthew A. Hood (De Gruyter group) for their conscientious cooperation in the editorial process.

Shumin Han

Contents

1 Introduction

1.1 Overview of RE–Mg–Ni-Based Hydrogen Storage Alloys

Requirements for clean energy have become increasingly urgent with the depletion of fossil energy resources and growing environmental issues. Investigations show that nearly half of the oil resources are consumed by transportation vehicles which subsequently release one-quarter of total global CO_2 [1–3]. The transformation of power sources from fossil fuels to clean energy is expected to relieve this situation. Nickel metal hydride (Ni/MH) secondary batteries are considered to be an ideal clean energy source due to the advantages of high-power output, high-rate capability, long lifetime, good reliability, non-toxicity, no memory effect and low cost [4].

The essential activity of electrochemical charge/discharge is the hydrogen absorption/desorption behaviour of hydrogen storage alloys which depends on the types of main phases, microstructure, the nature and amount of each element in the intermetallic compound, and the electrochemical process taking place. In general, the discharge pressure of metal hydride should be between 0.1 and 1 atm at room temperature to ensure the complete utilization of the stored hydrogen in the alloys [5]. The amounts of hydrogen that can be absorbed and desorbed under ambient conditions determine the electrochemical capacity of a hydride electrode, and consequently, the energy storage capacity of a battery. For applications in Ni/MH rechargeable batteries, both high hydrogen capacity and moderate plateau pressure are expected from hydrogen storage alloys. Information about these properties can be described using pressure–composition (*P–C*) isotherms [3] as shown in Figure 1.1. The *P–C* curves of each alloy sample for hydrogen absorption and desorption contain three steps: (1) a ramping up step that is typical with the α-phase at low capacity, (2) a flat plateau corresponding to a two-phase co-existing period (α to β transformation) and (3) a final increase due to the presence of metal hydride β phase. The value of hydrogen absorption/desorption equilibrium plateau pressure is used to predict the electrochemical potential of hydrogen storage alloys, and the length of plateau is used to estimate the theoretical electrochemical capacity of hydrogen storage alloys in Ni/MH batteries through the following equations [3]:

$$E = -0.9324 - 0.0291 \times \log P_{H2} \text{ (vs. Hg/HgO, 6 mol KOH)} \tag{1.1}$$

$$C(\text{mAh g}^{-1}) = 6 \times 26800 \times \frac{(\text{H/M})_5 - (\text{H/M})_{0.1}}{\text{MW}} \tag{1.2}$$

where P_{H2} represents the dehydriding equilibrium pressure, $(H/M)_5$ and $(H/M)_{0.1}$ represent the hydrogen contents of alloy according to *P–C–T* curves at the pressures of 5 and 0.1 atm, respectively, and MW is the molecular weight of the hydrogen storage alloys. The bond energy in this pressure range can be calculated by the following equation:

DOI 10.1515/9783110501483-001

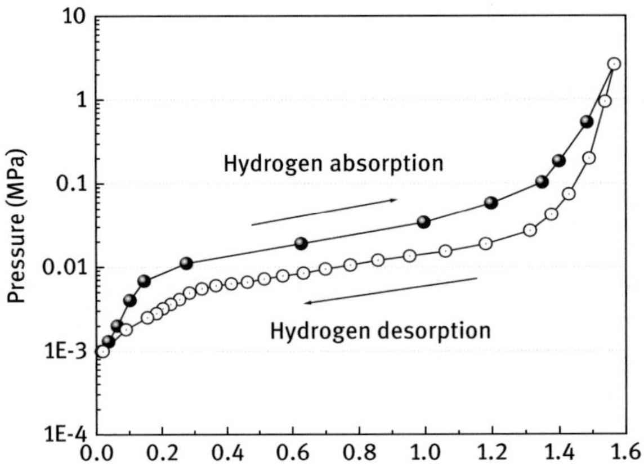

Figure 1.1: Typical pressure–composition isotherm of hydrogen storage alloys.

$$\ln P_{\mathrm{H}} = -\frac{\Delta H}{RT} + \frac{\Delta S}{R} \tag{1.3}$$

where ΔH is the change in enthalpy, ΔS is the change in entropy, R is the gas constant and T is the absolute temperature.

The electrochemical reactions of hydrogen storage alloys are very simple. Hydrogen atoms are transformed between positive and negative electrodes, as shown in Figure 1.2. During charging, hydrogen atoms from hydrolysis are absorbed into alloy bulk to form hydrides, and during discharging hydrogen atoms from the decomposed hydrides are oxidized into water.

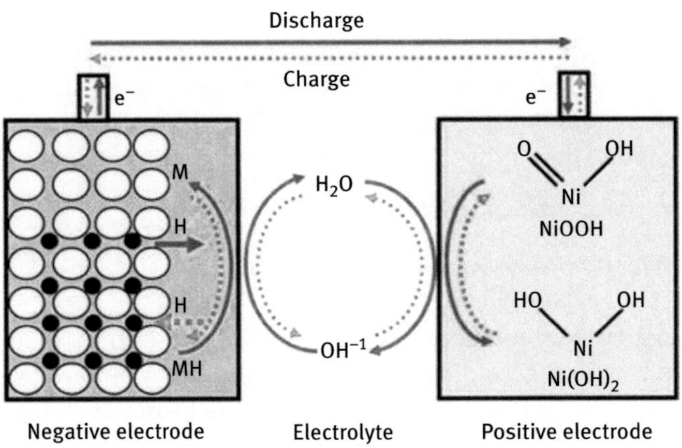

Figure 1.2: Charge/discharge diagram in Ni/MH batteries [1].

The electrochemical reactions are as follows:

$$\text{Negative}: \ M + xH_2O + xe^- \underset{\text{discharge}}{\overset{\text{charge}}{\rightleftharpoons}} MH_x + xOH^- \tag{1.4}$$

$$\text{Positive}: \ Ni(OH)_2 + OH^- \underset{\text{discharge}}{\overset{\text{charge}}{\rightleftharpoons}} NiOOH + H_2O + e^- \tag{1.5}$$

$$\text{Overall reaction}: \ xNi(OH)_2 + M \underset{\text{discharge}}{\overset{\text{charge}}{\rightleftharpoons}} xNiOOH + MH_x \tag{1.6}$$

During the exploitation of Ni/MH batteries, novel negative electrode materials have always been the focus. In general, the development of the negative electrode materials for Ni/MH batteries has experienced three generations. The representative for the first generation is rare-earth-based AB_5-type alloys. This alloy system with the advantages of easy activation and fast hydrogen desorption has been extensively commercialized. However, their discharge capacity is limited to 300–320 mAh g^{-1}, which can hardly satisfy the requirements of high-energy density for Ni/MH batteries [6]. The second generation mainly refers to AB_2-type Ti-, Zr-based Laves phase alloys. These alloys have high discharge capacity and good cycling stability, but poor activation and high-rate dischargeability due to their low Ni contents [7]. Recently the industry has put more focus on the third generation which is La–Mg–Ni-based hydrogen storage alloys. Research found that La–Mg–Ni-based alloys whose structures are the combination of both [AB_5] and [AB_2] subunits can compensate for the disadvantages of the poor discharge capacity, activation and rate capability of AB_5– and AB_2-type alloys, thus exhibiting good overall electrochemical properties including easy activation, high discharge capacity and good high-rate dischargeability [8, 9].

It was reported that the maximum discharge capacity of a La–Mg–Ni-based $La_{0.7}Mg_{0.3}Ni_{2.8}Co_{0.5}$ alloy reached as high as 410 mAh g^{-1}, about 30 % higher than that of AB_5-type alloys [10, 11]. Since then, numerous studies on La–Mg–Ni-based alloy system have been conducted to develop them as a practical electrode material for high-energy and high-power Ni/MH batteries [12–20]. And now, La–Mg–Ni-based alloy electrodes have been applied in commercial Ni/MH batteries with promising results: long cycle-life (>6,000 cycles) consumer batteries have been produced by FDK Incorporation [20], and large-capacity, high-power (40 cell – 1,500 Ah) stationary batteries have been produced by Kawasaki Heavy Industries for industrial applications [20, 21]. Meanwhile, driven by people's increasing needs for high-power, high-energy and long-cycle-life of batteries, research on La–Mg–Ni-based alloys has never been stopped.

1.2 Structure Characteristics of RE–Mg–Ni-Based Hydrogen Storage Alloys

La–Mg–Ni-based alloys were originated from the corresponding binary La–Ni-based alloys [22, 23]. The phases that are commonly seen in La–Mg–Ni-based alloys are $(La,Mg)Ni_3$, $(La,Mg)_2Ni_7$ and $(La,Mg)_5Ni_{19}$ phases with superlattice structures. Similar to La–Ni binary phases, these ternary phases are also composed of $[A_2B_4]$ and $[AB_5]$ subunits alternatively stacking along *c* axis in different patterns [24, 25], as shown in Figure 1.3. The chemical formula for this alloy system can be expressed as $m[A_2B_4]{\cdot}n[AB_5]$ $(R_{n+m}Mg_mNi_{5n+4m})$, where *m* and *n* represent the numbers of $[A_2B_4]$ and $[AB_5]$ layers, respectively in a stacking block unit (Figure 1.3). The ratio between $[A_2B_4]$ and $[AB_5]$ subunits in the alloy phases mainly includes 1:1, 1:2 and 1:3, which corresponds to $(La,Mg)Ni_3$, $(La,Mg)_2Ni_7$ and $(La,Mg)_5Ni_{19}$ phases, respectively. Each of these phases has two crystal types: hexagonal (2H-) and rhombohedra (3R-) type. The difference between the two types lies in the $[A_2B_4]$ subunits. The $[A_2B_4]$ subunits in 2H-type phases are with $MgCu_2$ Laves structure compared with $MgZn_2$ Laves structure in 3R-type phases [26]. In addition, it can be observed from Figure 1.3 that $[A_2B_4]$ and $[AB_5]$ subunits first stack into blocks, and two (2H-type, $Z = 2$) or three (3R-type, $Z = 3$) blocks further form lattice cells. This kind of structure in this unique stacking mode is called superstacking or superlattice structure. The cell volume of the phases with superstacking structure is actually the total volume of $(m + n)^*Z$ subunits.

Mg occupation of the superstacking structure also shows regularity. When Mg substitutes for La from the primary La–Ni binary phases, it exclusively enters $[A_2B_4]$ subunits instead of $[AB_5]$ subunits. For 2H-type phases, Mg substitutes for La at 4*f* site

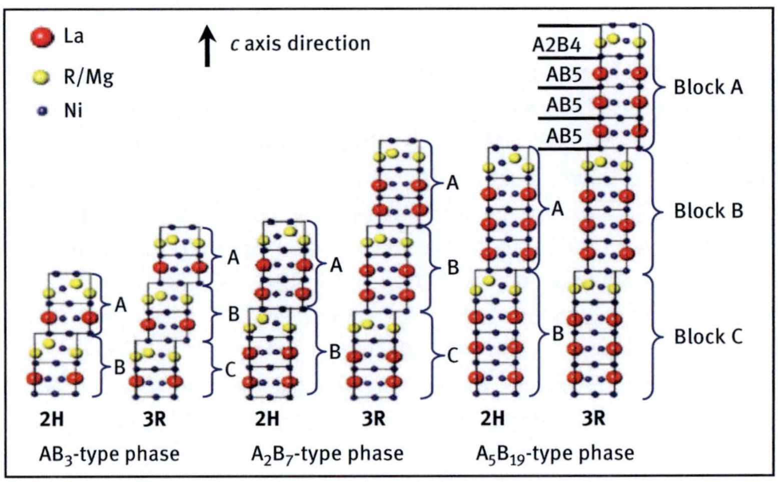

Figure 1.3: Stacking modes of AB_3-, A_2B_7- and A_5B_{19}-type phases.

of [A_2B_4] subunits, and for 3R-type phases, Mg substitutes for La at 6*c* site of [A_2B_4] subunits [26–31].

Apart from the above three common superstacking phases, rhombohedra AB_4-type phase with [A_2B_4]/[AB_5] subunit ratio of 1:4 has also been found in the $La_{0.85}Mg_{0.15}Ni_{3.8}$ alloy prepared by spark plasma sintering method [32] and the $La_{0.8}Mg_{0.2}Ni_{3.4-x}Co_{0.3}$ $(MnAl)_x$ alloys prepared by inductive melting followed by annealing [33]. Moreover, Zhang et al. [23] recently found another new type of alloy phase in ternary Ca–Mg–Ni system. This phase has rhombohedra A_5B_{13}-type structure with [A_2B_4]/[AB_5] subunit ratio of 2:1 as shown in Figure 1.4. It is the first compound reported to date with $m = 2$ in $R_{n+m}Mg_mNi_{5n+4m}$ alloy system. But its corresponding binary compound does not exist in Ca–Ni binary phase diagram.

Thanks to their special superlattice structure containing both [A_2B_4] and [AB_5] subunits, RE–Mg–Ni-based alloys exhibit both high discharge capacity of A_2B_4-type alloys and high catalytic activity of AB_5-type alloys. Therefore, these RE–Mg–Ni-based alloys with high specific capacity and good rate dischargeability are considered to be one of as the most promising next-generation negative electrode materials for high-energy and high-power Ni/MH batteries. Numerous investigations have been conducted for their practical applications [26, 30, 32–40], especially in China and Japan. This book essentially focuses on RE–Mg–Ni-based hydrogen storage alloys with high-energy densities and that are used as negative electrode materials in Ni/MH batteries. In this book, the preparation techniques, structural and electrochemical characteristics of single-phase RE–Mg–Ni-based hydrogen storage alloys, the interactions and effects of alloy phases in multiphase alloys, the effect of elemental

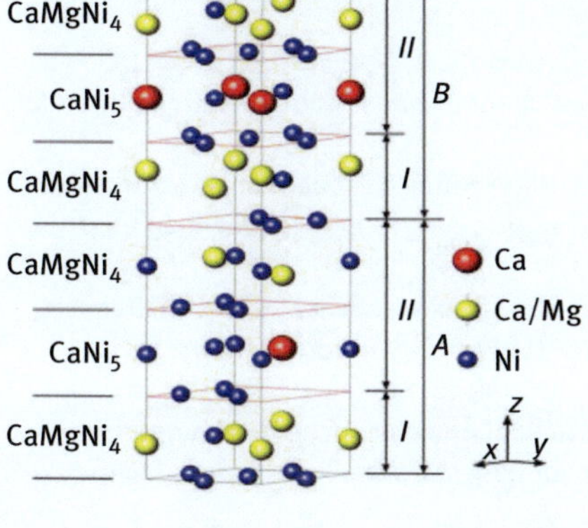

Figure 1.4: Structure of a unit cell for $Ca_3Mg_2Ni_{13}$ phase [23].

substitution and surface modification on the structural and electrochemical properties of RE–Mg–Ni-based alloys are thoroughly described [41].

References

[1] Davis SJ, Caldeira K, Matthews HD. Future CO_2 emissions and climate change from existing energy infrastructure. Science 2010;329:1130–333.

[2] Chow J, Kopp RJ, Portney PR. Energy resources and global development. Science 2003;302:1528–31.

[3] Liu YF, Pan HG, Gao MX, Wang QD. Advanced hydrogen storage alloys for Ni/MH rechargeable batteries. J Mater Chem 2011;21:4743–55.

[4] Zhu WH, Zhu Y, Tatarchuk BJ. Self–discharge characteristics and performance degradation of Ni–MH batteries for storage applications. Int J Hydrogen Energy 2014;39:19789–98.

[5] Feng F, Geng M, Northwood DO. Electrochemical behaviour of intermetallic–based metal hydrides used in Ni = metal hydride (MH) batteries: a review. Int J Hydrogen Energy 2001;26:725–34.

[6] Zhang XB, Sun DZ, Yin WY, Chai YJ, Zhao MS. Effect of La/Ce ratio on the structure and electrochemical characteristics of $La_{0.7-x}Ce_xMg_{0.3}Ni_{2.8}Co_{0.5}$ (x = 0.1–0.5) hydrogen storage alloys. Electrochim Acta 2005;50:1957–64.

[7] Pan HG, Jin Q, Gao MX, Liu YF, Li R, Lei Y. Effect of the cerium content on the structural and electrochemical properties of the $La_{0.7-x}Ce_xMg_{0.3}Ni_{2.875}Mn_{0.1}Co_{0.525}$ (x = 0–0.5) hydrogen storage alloys. J Alloys Compd 2004;373:237–45.

[8] Young K, Wong DF, Wang L, Nei J, Ouchi T, Yasuoka S. Mn in misch–metal based superlattice metal hydride alloy – part 1 structural, hydrogen storage and electrochemical properties. J Power Sources 2015;277:426–32.

[9] Zhou X, Young K, West J, Regalado J, Cherisol K. Degradation mechanisms of high–energy bipolar nickel metal hydride battery with AB_5 and A_2B_7 alloys. J Alloys Compd 2013;580:S373–7.

[10] Kohno T, Yoshida H, Kawashima F, Inaba T, Sakai I, Yamamoto M. Hydrogen storage properties of new ternary system alloys: La_2MgNi_9 La_3MgNi_9, La_4MgNi_{19}. J Alloys Compd 2000;311:L5–7.

[11] Li RF, Xu PH, Zhao Y, Wan J, Liu XF, Yu RH. The microstructures and electrochemical performances of $La_{0.6}Gd_{0.2}Mg_{0.2}Ni_{3.0}Co_{0.5-x}Al_x$ (x = 0–0.5) hydrogen storage alloys as negative electrodes for nickel/metal hydride secondary batteries. J Power Sources 2014;270:21–7.

[12] Yasuoka S, Ishida J, Kishida K, Inui H. Effects of cerium on the hydrogen absorption-desorption properties of rare earth-Mg-Ni hydrogen-absorbing alloys. J Power Sources 2017;346:56–62

[13] Li HH, Fei Y, Wang YP, Chen L, Jiang HB. Crystal structure and electrochemical performance of $La_{0.75}Ce_{0.25}Ni_{3.46}Al_{0.17}Mn_{0.04}Co_{1.33}$ alloy for high–power–type 29 Ah Ni–MH battery. J Rare Earth 2015;33:633–8.

[14] Gal L, Charbonnier V, Zhang J, Goubault L, Bernard P, Latroche M. Optimization of the La substitution by Mg in the La_2Ni_7 hydride-forming system for use as negative electrode in Ni–MH battery. Int J Hydrogen Energy 2015;40:17017–20.

[15] Zhou Z, Song Y, Cui S, Huang C, Qian W, Lin C. Effect of annealing treatment on structure and electrochemical performance of quenched $MmNi_{4.2}Co_{0.3}Mn_{0.4}Al_{0.3}Mg_{0.03}$ hydrogen storage alloy. J Alloys Compd 2010;501:47–53.

[16] Shen XQ, Chen YG, Tao MD, Wu CL, Deng G, Kang ZZ. The structure and 233 K electrochemical properties of $La_{0.8-x}Nd_xMg_{0.2}Ni_{3.1}Co_{0.25}Al_{0.15}$ (x = 0.0–0.4) hydrogen storage alloys. Int J Hydrogen Energy 2009;34:2661–9.

[17] Ma S, Gao MX, Li R, Pan HG, Lei Y. A study on the structural and electrochemical properties of $La_{0.7-x}Nd_xMg_{0.3}Ni_{2.45}Co_{0.75}Mn_{0.1}Al_{0.2}$ (x = 0.0–3.0) hydrogen storage alloys. J Alloys Compd 2008;457:457–64.

[18] Zhang P, Liu YF, Tang R, Zhu JW, Wei XD, Liu SS, Yu G. Effect of Ca on the microstructural and electrochemical properties of $La_{2.3-x}Ca_xMg_{0.7}Ni_9$ hydrogen storage alloys. Electrochim Acta 2006;51:6400–5.

[19] Tang R, Liu YN, Zhu CC, Zhu J, Yu G. Effect of Mg on the hydrogen storage characteristics of $Ml_{1-x}Mg_xNi_{2.4}Co_{0.6}$ ($x = 0–0.6$) alloys. Mater Chem Phys 2006;95:130–4.

[20] Liu YF, Pan HG, Gao MX, Zhu Y, Lei Y, Wang QD. Electrochemical studies on $La_{0.7}Mg_{0.3}Ni_{3.4-x}Co_{0.6}Mn_x$ metal hydride electrode alloys. Mater Chem Phys 2004;84:171–81.

[21] Takasaki T, Nishimura K, Saito M, Fukunaga H, Iwaki T, Sakai T. Cobalt–free nickel–metal hydride battery for industrial applications. J Alloys Compd 2013;580:S378–81.

[22] Kadir K, Sakai T, Uehara I. Synthesis and structure determination of a new series of hydrogen storage alloys; RMg_2Ni_9 (R = La, Ce, Pr, Nd, Sm and Gd) built from $MgNi_2$ Laves–type layers alternating with AB_5 layers. J Alloys Compd 1997;257:115–21.

[23] Zhang QA, Sun DL, Zhang JX, Latroche M, Ouyang LZ, Zhu M. Structure and deuterium desorption from $Ca_3Mg_2Ni_{13}$ deuteride: a neutron diffraction study. J Phys Chem C 2014;118:4626–33.

[24] Buschow KH, Van Der Goot AS. The crystal structure of rare–earth nickel compounds of the type R_2Ni_7. J Less-Common Met 1970;22:419–28.

[25] Yamamoto T, Inui H, Yamaguchi M, Sato K, Fujitani S, Yonezu I. Microstructures and hydrogen absorption/desorption properties of LaNi alloys in the composition range of La77.8 ~ 83.2 at.% Ni. Acta Mater 1997;45:5213–21.

[26] Zhang FL, Luo YC, Chen JP, Yan RX, Chen JH. La–Mg–Ni ternary hydrogen storage alloys with Ce_2Ni_7-type and Gd_2Co_7-type structure as negative electrodes for Ni/Mh batteries. J Alloys Compd 2007;430:302–7.

[27] Liu JJ, Li Y, Han D, Yang SQ, Chen XC, Zhang L, Han SM. Electrochemical performance and capacity degradation mechanism of single-phase La–Mg–Ni–based hydrogen storage alloys. J Power Sources 2015;300:77–86.

[28] Liu YF, Cao YH, Huang L, Gao MX, Pan HG. Rare earth–Mg–Ni–based hydrogen storage alloys as negative electrode materials for Ni/MH batteries. J Alloys Compd 2011;509:675–86.

[29] Kadir K, Sakai T, Uehara I. Structural investigation and hydrogen capacity of YMg_2Ni_9 and $(Y_{0.5}Ca_{0.5})(MgCa)Ni_9$: new phases in the AB_2C_9 system isostructural with $LaMg_2Ni_9$. J Alloys Compd 1999;287:264–70.

[30] Férey A, Cuevas F, Latroche M, Knosp B, Bernard P. Elaboration and characterization of magnesium–substituted La_5Ni_{19} hydride forming alloys as active materials for negative electrode in Ni–MH battery. Electrochim Acta 2009;54:1710–14.

[31] Zhang JL, Han SM, Li Y, Liu JJ, Yang SQ, Zhang L. Effects of $PuNi_3$– and Ce_2Ni_7–type phase abundance on electrochemical characteristics of La–Mg–Ni–based alloys. J Alloys Compd 2013;581:693–8.

[32] Zhang J, Villeroy B, Knosp B, Bernard P, Latroche M. Structural and chemical analyses of the new ternary La_5MgNi_{24} phase synthesized by Spark Plasma Sintering and used as negative electrode material for Ni–MH batteries. Int J Hydrogen Energy 2012;37:5225–33.

[33] Ozaki T, Kanemoto M, Kakeya T, Kitano Y, Kuzuhara M, Watada M. Stacking structures and electrode performances of rare earth–Mg–Ni–based alloys for advanced nickel–metal hydride battery. J Alloys Compd 2007;446–447:620–4.

[34] Zhang FL, Luo YC, Wang DH, Yan RX, Kang L, Chen JH. Structure and electrochemical properties of $La_{2-x}Mg_xNi_{7.0}$ ($x = 0.3–0.6$) hydrogen storage alloys. J Alloys Compd 2007;446–447:181–8.

[35] Zhao YM, Han SM, Li Y, Liu JJ, Zhang L, Yang SQ, Ke DD. Characterization and improvement of electrochemical properties of Pr_5Co_{19}-type single-phase $La_{0.84}Mg_{0.16}Ni_{3.80}$ alloy. Electrochim Acta 2015;152:265–73.

[36] Ma ZW, Zhu D, Wu CL, Zhong CL, Wang QN, Zhou WH. Effects of Mg on the structures and cycling properties of the $LaNi_{3.8}$ hydrogen storage alloy for negative electrode in Ni/MH battery. J Alloys Compd 2015;620:149–55.

[37] Zhong CL, Chao DL, Chen YG, Wang W, Zhu D, Wu CL. Microstructures and electrochemical properties of $LaNi_{3.8-x}Mn_x$ hydrogen storage alloys. Electrochim Acta 2011;58:668–73.

[38] Liu ZY, Yan XL, Wang N, Chai YJ, Hou DL. Cyclic stability and high rate discharge performance of $(La,Mg)_5Ni_{19}$ multiphase alloy. Int J Hydrogen Energy 2011;36:4370 4.

[39] Lemort L, Latroche M, Knosp B, Bernard P. Elaboration and characterization of unreported (Pr,Nd)5Ni19 hydrides. J Alloys Compd 2011;509:S823–6.

[40] Zhao YM, Han SM, Li Y, Liu JJ, Zhang L, Yang SQ, Cao J, Jia ZR. Structural phase transformation and electrochemical features of La–Mg–Ni-based AB_4-type alloys. Electrochim Acta 2016;215:142–9.

[41] Liu JJ, Han SM, Li Y, Zhang L, Zhao YM, Yang SQ, Liu BZ. Phase structures and electrochemical properties of La–Mg–Ni-based hydrogen storage alloys with superlattice structure, Int J Hydrogen Energy 2016;41:20261–75.

2 Preparation, Electrochemical Properties and Gaseous Hydrogen Storage Characteristics of the Single-Phase Superlattice RE–Mg–Ni-Based Hydrogen Storage Alloys

The knowledge of the electrochemical characteristics of single-phase La–Mg–Ni-based alloys is significant for analysing the effects of different phases in this alloy system. Among ternary La–Mg–Ni-based alloys, $AB_{3.0}$-type $LaMg_2Ni_9$ alloy with 3R-type $(La,Mg)Ni_3$ phase was first reported by Kadir et al. [1]. Its crystal structure is a stacking of a $[LaNi_5]$ subunit and a $[MgNi_2]$ subunit along *c* axis. $LaMg_2Ni_9$ alloy exhibited a rather small hydrogen storage capacity of 0.3 wt.% [2], which was explained by the inertness of the $[MgNi_2]$ unit to hydrogen absorption, similar to individual $MgNi_2$ Laves-type alloy that does not absorb hydrogen at conventional hydrogenation conditions [2, 3]. In later research, however, Liao et al. prepared a serious of $(La,Mg)Ni_3$-phase $La_{1-x}Mg_xNi_{3.0}$ alloys and found that the La_2MgNi_9 alloy exhibits good electrochemical properties, and the maximum discharge capacities reached 400 mAh g^{-1} [4]. Later on, researchers did a lot of studies on La–Mg–Ni-based alloys to improve their electrochemical properties, and remarkable progress has been made. However, most of the prepared alloys are multiphase structures. In this circumstance, it is difficult to isolate a specific phase and investigate its impact on the electrochemical properties of the entire alloy. Until 2007, Zhang et al. prepared an $La_{1.5}Mg_{0.5}Ni_7$ alloy with single $(La,Mg)_2Ni_7$ phase using inductive melting method followed by long-time heat treatment at 900°C [5]. Their method was based on the knowledge that the preferable formation temperature of $(La,Mg)_2Ni_7$ phase is around 900°C [6]. The alloy exhibited good electrochemical properties with a maximum discharge capacity (C_{max}) of 389 mAh g^{-1}, a cycling retention (S_{150}) of 82 % and a high-rate dischargeability (HRD_{900}) of 89 %. Subsequent researches on single-phase alloys tend to focus on the effect of Mg element. Denys et al. prepared single-phase $LaNi_3$ and $La_{1.5}Mg_{1.5}Ni_9$ alloys as well as La_2Ni_7 and La_3MgNi_{14} alloys using stepwise powder sintering method [7, 8], and also studied the influence of Mg on the hydrogen absorption behaviour of AB_3 and A_2B_7 phases. They found that Mg addition increased hydrogen absorption/desorption plateau pressure and enabled $[A_2B_4]$ subunits to absorb hydrogen, resulting in a nearly equal hydrogen-induced expansions on *a* and *c* axis of unit cells, thus decreasing hydrogen-induced amorphization. Férey et al. prepared single-phase La_5Ni_{19} and La_4MgNi_{19} alloys using the same method and found that Mg induced 3R-type A_5B_{19} phase in the original 2H-type A_5B_{19} phase of La_5Ni_{19} alloy. On the other hand, the addition of Mg increased the reversible hydrogen capacity and electrochemical discharge capacity [9]. Until then, the preparation of single-phase alloys was still difficult due to inadequate information of the formation conditions and preparation methods, leading to the lack of reports on the electrochemical functions and characteristics of

DOI 10.1515/9783110501483-002

specific superlattice phases. Our group recently prepared AB_3-, A_2B_7- and A_5B_{19}-type single-phase La–Mg–Ni superlattice alloys by the stepwise sintering and long-time annealing methods, and explored their formation and capacity degradation mechanism. Moreover, we also prepared AB_3-, A_2B_7- and A_5B_{19}-type single-phase superlattice RE–Mg–Ni (RE = Pr, Nd, Sm) alloys by the above methods and revealed the cycling stability improvement mechanism. We expect our findings can provide the basis and inspire new thoughts for improving the cycling performance of RE–Mg–Ni-based hydrogen storage alloys.

2.1 The Preparation and Characteristics in Electrochemical and Gaseous Hydrogen Storage of the Single-Phase Superlattice La–Mg–Ni-Based Hydrogen Storage Alloys

2.1.1 Preparation of the Single-Phase Superlattice La–Mg–Ni-Based Hydrogen Storage Alloys

The La_2MgNi_9, La_3MgNi_{14} and La_4MgNi_{19} hydrogen storage alloys with single-phase AB_3, A_2B_7 and A_5B_{19} superlattice structures were both prepared by stepwise powder sintering and induction melting methods followed by an annealing treatment.

For the stepwise powder sintering method, the La_2MgNi_9, La_3MgNi_{14} and La_4MgNi_{19} alloys were sintered from $LaMgNi_4$ and $LaNi_5$ precursors. The $LaMgNi_4$ and $LaNi_5$ precursors were firstly prepared by inductive melting La, Mg and Ni, with purity higher than 99.5 % under argon atmosphere, followed by annealing at 700°C and 1,000°C, respectively, to guarantee their composition homogeneity. The precursors exhibited single-phase structures as seen from Figure 2.1. To prepare the target alloys, the $LaMgNi_4$ and $LaNi_5$ precursors were firstly ground into powders (<300 mesh) and mixed in molar ratios of 1:1, 1:2 and 1:3 for La_2MgNi_9, La_3MgNi_{14} and La_4MgNi_{19} alloys, respectively. Extra amount of Mg powder was added to the mixture to compensate for Mg loss during the sintering process. The mixtures were firstly ball milled for an hour and then cold pressed into pellets. Each of the pellets was then wrapped into Ta foil and sealed in a nickel shell for sintering as shown in Figure 2.1. The La_2MgNi_9 alloy was sintered with temperatures increasing in four steps: 600, 700, 800 and 950°C with a holding time of 7 h at each temperature and subsequently annealed at 850°C for 96 h. The La_3MgNi_{14} alloy was sintered at 600, 700, 800 and 900°C for 1 h at each temperature, and then 980°C for 4 h, followed by an annealing treatment at 900°C for 96 h. For the La_4MgNi_{19} alloy, the sintering temperatures were 600, 700, 800 and 900°C with a 1 h holding time at each temperature and the annealing circumstance was 900°C for 115 h. During the sintering process, the heating rate was 1°C min^{-1} between each temperature and 4°C min^{-1} below 600°C for the three alloys.

For the induction melting method, an La–Ni metal block was first heated to a molten state, and an Mg metal piece was placed into the smelting furnace after

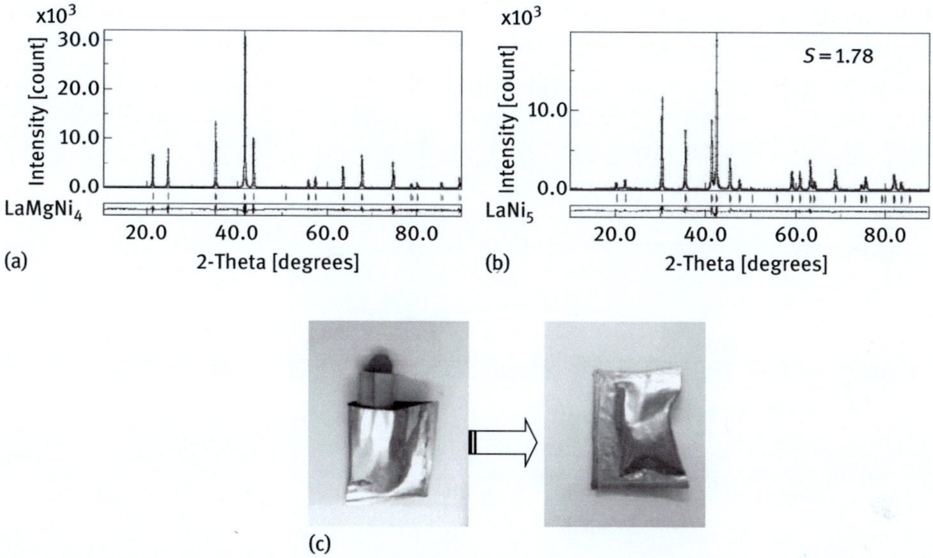

Figure 2.1: Rietveld refinements of $LaMgNi_4$ (a) and $LaNi_5$ (b) precursors. (c) Sample for sintering after wrapped and sealed by nickel shell.

temperature dropping to 1,600°C while stirring the mixture quickly and vigorously, followed by casting. Then the resulting as-cast alloys were mechanically crushed into cakes, wrapped in a nickel foil, placed in a furnace and then annealed in the medium frequency furnace under 0.04 MPa argon atmospheres. The La_2MgNi_9 and La_3MgNi_{14} alloys both were annealed at 950°C for 12 h, and that was 975°C for 24 h for the La_4MgNi_{19} alloy, during which process the heating rate was 4°C min^{-1} below 600°C and 1°C min^{-1} above 600°C, and the samples were finally quenched to room temperature in the furnace.

2.1.2 The Peritectic Reaction Process for the Formation of the Single-Phase Superlattice La–Mg–Ni-Based Hydrogen Storage Alloys

Figure 2.2 shows the thermal gravity analysis (TGA) curve of $LaMgNi_4$ and $LaNi_5$ precursors and differential thermal analysis (DTA) curve for the reactions between the precursors with temperature ramping up from 723–1,293 K. TGA curve shows that the weight of the reactant does not change during the whole heating process. The DTA curve contains a small step at 910 K and three peaks at higher temperatures. The small step around 910 K corresponds to the melting point of pure Mg element. The three peaks at higher temperatures indicate the formation of the AB_3-, A_2B_7- and A_5B_{19} single-phase alloys undergoes three major series of transformations. Noticeably, the three temperatures of the peaks are close but lower than the peritectic reaction temperatures for the binary AB_3, A_2B_7 and A_5B_{19} phases in the La–Ni binary diagram [10].

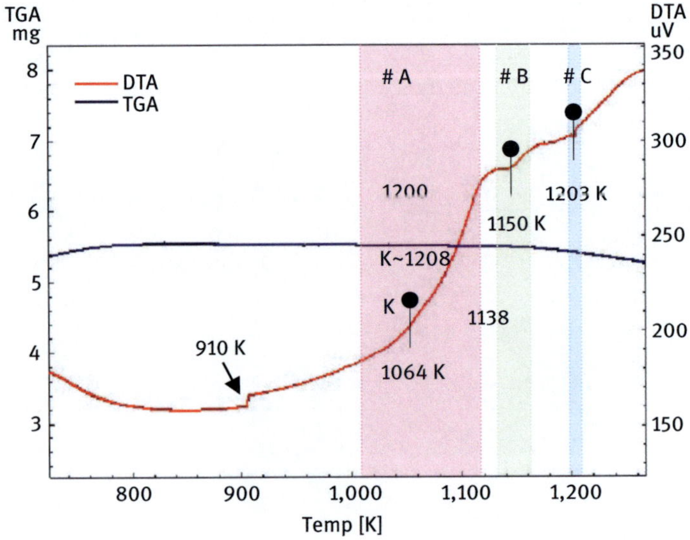

Figure 2.2: Differential scanning calorimetry (DSC) results of the cold pressing precursor mixture of $LaMgNi_4$ and $LaNi_5$.

The temperature gap is owing to the introduction of Mg element because the Mg has much lower melting point than both La and Ni, and the addition of a third element usually decreases the melting temperature of previous alloys [11, 12]. The detailed phase transformations reflected from Figure 2.2 is as follows.

In the first peak of temperature ranged from 1,010 to 1,118 K (area #A in Figure 2.2), the $LaMgNi_4$ precursor with a low melting point turns into a liquid La (Mg)-rich phase (Liquid 1; eq (2.1)), while the $LaNi_5$ precursor with a melting point of 1,623 K remains in a solid state. Based on the La–Ni binary diagram [10] and peritectic reaction theory [11], $LaNi_5$ solid phase reacts with Liquid 1 and forms $(La,Mg)Ni_3$ phase (eq (2.2)). The transformations are summarized as the following reactions:

$$LaMgNi_4 \rightarrow \text{Liquid 1} \tag{2.1}$$

$$LaNi_5 + \text{Liquid 1} \rightarrow (La,Mg)Ni_3 \tag{2.2}$$

As the temperature further increases (area #B in Figure 2.2), the $(La,Mg)Ni_3$ phase starts to liquidize to form phase Liquid 2 (eq (2.3)) which reacts with the remaining $LaNi_5$ phase to generate $(La,Mg)_2Ni_7$ phase (eq (2.4)) when the temperature increases to 1,150 K. The second transformation can be summarized as the following reactions:

$$(La,Mg)Ni_3 \rightarrow \text{Liquid 2} \tag{2.3}$$

$$LaNi_5 + \text{Liquid 2} \rightarrow (La,Mg)_2Ni_7 \tag{2.4}$$

Up to this stage, $LaNi_5$ phase is still not completely depleted, which results in an alloy contain both $(La,Mg)_2Ni_7$ and $LaNi_5$ phases.

In the last stage (area #C in Figure 2.2), as the temperature continues to rise, the $(La,Mg)_2Ni_7$ phase turns into another liquid phase (Liquid 3; eq (2.5), and the $(La,Mg)_5Ni_{19}$ phase is formed by the peritectic reaction between $LaNi_5$ phase and Liquid 3 (eq. (2.6). The reactions are summarized as follows:

$$(La,Mg)_2Ni_7 \rightarrow \text{Liquid 3} \tag{2.5}$$

$$LaNi_5 + \text{Liquid 3} \rightarrow (La,Mg)_5Ni_{19} \tag{2.6}$$

It is also noted that the peritectic reaction temperature increases with higher content of $[LaNi_5]$ subunits, which is due to the higher melting point of $LaNi_5$ phase than that of $LaMgNi_4$ ($LaNi_2$) phase.

2.1.3 The Crystal Structure Characteristic of the Single-Phase Superlattice La–Mg–Ni-Based Hydrogen Storage Alloys

Rietveld refinements of the XRD patterns of the La_2MgNi_9, La_3MgNi_{14} and La_4MgNi_{19} alloys obtained from the stepwise powder sintering and induction melting methods are shown in Figure 2.3. The La_2MgNi_9 and La_3MgNi_{14} alloys obtained from both methods exhibit a single $PuNi_3$-type (3R) $(La,Mg)Ni_3$ phase structure and a single Ce_2Ni_7-type (2H) $(La,Mg)_2Ni_7$ phase structure, respectively. The $(La,Mg)_5Ni_{19}$ phase of La_4MgNi_{19} alloy contains both Pr_5Co_{19}-type (2H) and Ce_5Co_{19}-type (3R) prepared by the powder sintering, while which is only a Pr_5Co_{19}-type (2H) structure prepared by the induction melting method. The cell parameters of the La_2MgNi_9, La_3MgNi_{14} and La_4MgNi_{19} alloys obtained from the Rietveld refinements are shown in Table 2.1. The cell volumes of the Ce_2Ni_7-type single phase of La_3MgNi_{14} alloy obtained from the two methods are close, while the cell volumes of the $PuNi_3$-type phase La_2MgNi_9 and Pr_5Co_{19}-type phase La_4MgNi_{19} alloys prepared by the induction melting method are both smaller than those of the samples prepared by the powder sintering method. The phenomenon indicates that the single-phase alloy grains obtained from the induction melting method are more refined in comparison with those from the powder sintering method. To expedient introduction, we just picked the single-phase La_2MgNi_9, La_3MgNi_{14} and La_4MgNi_{19} alloys obtained by the powder sintering method and reported their hydrogen storage and electrochemical properties in detail in this book.

Figure 2.4 shows the backscattered scanning electron microscope (SEM) images labelled with energy-dispersive spectrometer (EDS) data. It can be seen that all the alloy samples exhibit uniform structures, indicating that the elements are evenly distributed in the alloys. Noticeably, the 2H- and 3R-type $(La,Mg)_5Ni_{19}$ phases cannot be distinguished from each other under backscattering SEM due to their identical

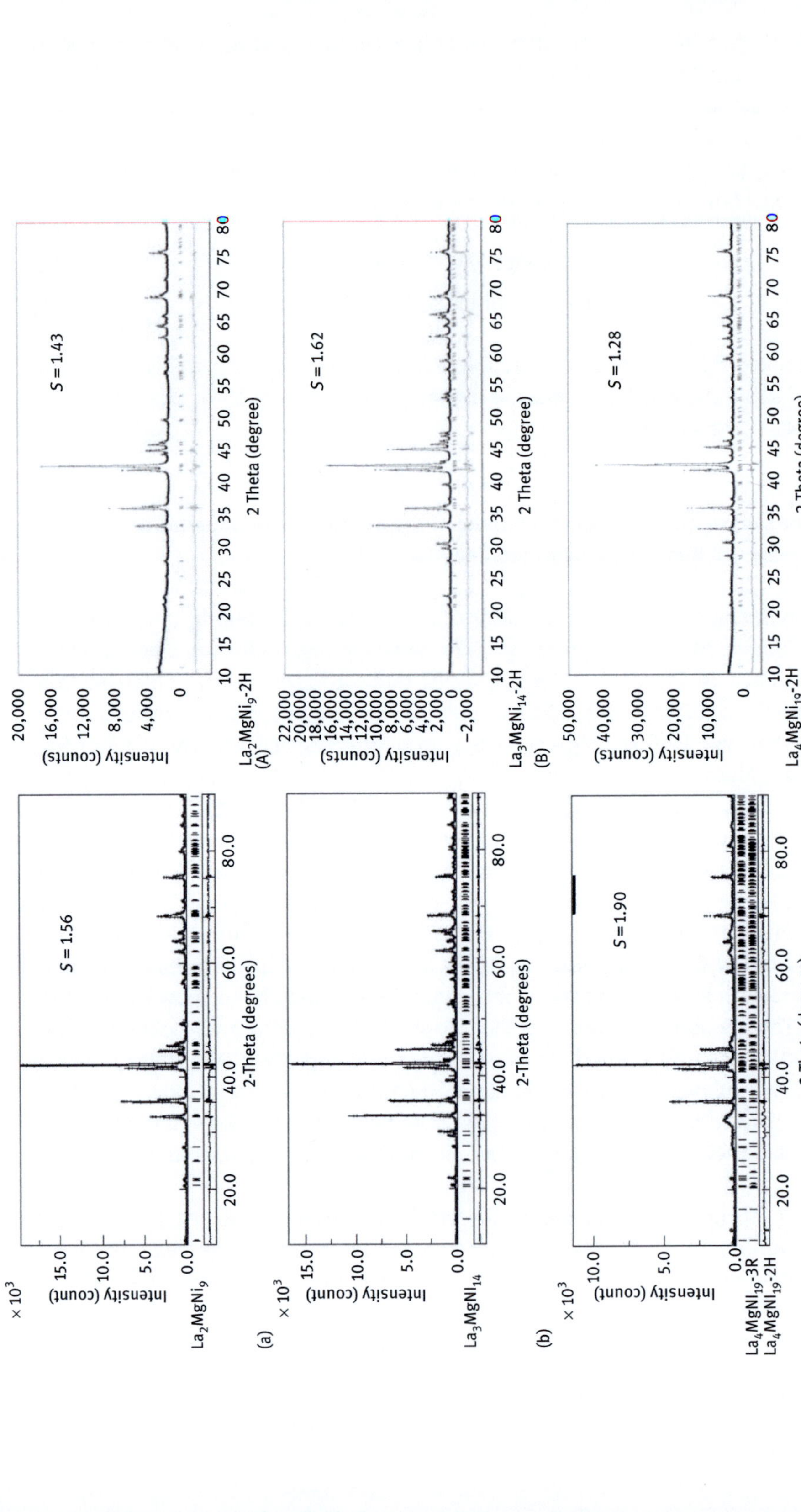

Figure 2.3: Rietveld refinements of the XRD patterns of the La_2MgNi_9, La_3MgNi_{14} and La_4MgNi_{19} alloys prepared by the stepwise powder sintering method (a–c) and the induction melting method (A–C). Vertical bars correspond to the Bragg peak positions for the constituent.

Table 2.1: Lattice parameters and phase abundances of the La_2MgNi_9, La_3MgNi_{14} and La_4MgNi_{19} alloys obtained from the stepwise powder sintering and induction melting methods.

Preparation method	Alloy	Phase structure	*a* (Å)	*c* (Å)	*V* (Å^3)
Powder sintering	La_2MgNi_9	$PuNi_3$	5.0432	24.392	537.27
	La_3MgNi_{14}	Ce_2Ni_7	5.0398	24.247	533.35
	La_4MgNi_{19}	Pr_5Co_{19}	5.0324	32.257	707.46
		Ce_5Co_{19}	5.0345	48.389	1062.2
Induction melting	La_2MgNi_9	$PuNi_3$	5.0330	24.295	532.97
	La_3MgNi_{14}	Ce_2Ni_7	5.0412	24.291	534.62
	La_4MgNi_{19}	Pr_5Co_{19}	5.0218	32.068	700.36

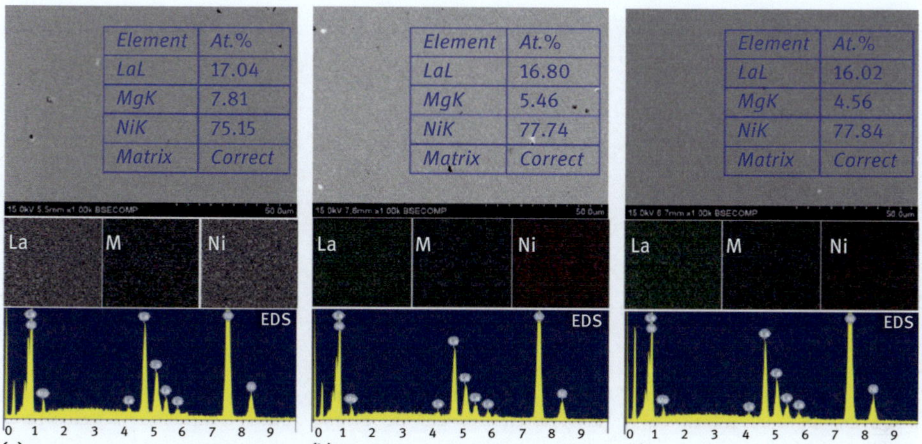

Figure 2.4: The backscattered scanning electron microscope (SEM) images and energy-dispersive spectrometer (EDS) mapping of the alloy samples.

compositions [13, 14]. The alloy compositions calculated from the EDS data are $La_{2.06}Mg_{0.94}Ni_{9.07}$, $La_{3.02}Mg_{0.98}Ni_{13.96}$ and $La_{3.89}Mg_{1.11}Ni_{18.91}$, which also agree well with the nominal compositions of the alloys.

Figure 2.5 shows the high-resolution transmission electron microscope (TEM) images and selected area electron diffraction (SAED) patterns of the alloy particle samples prepared by powder sintering method. The La_2MgNi_9 crystalline particle (Figure 2.5(a)) shows a single rhombohedra $PuNi_3$-type ($P6_3/mmc$) structure which contains [$LaNi_5$] and [$LaMgNi_4$] subunits in a ratio of 1:1, whereas the La_3MgNi_{14} crystalline particle (Figure 2.5(b)) exhibits a single hexagonal Ce_2Ni_7-type ($P6_3/mmc$) structure with [$LaNi_5$] and [$LaMgNi_4$] subunits in a ratio of 2:1. Both the 2H- and $(La,Mg)_5Ni_{19}$ phases can be seen in the La_4MgNi_{19} alloy (Figure 2.5(c) and (d)). With the XRD patterns, SEM and TEM images of the alloy samples, we confirm the

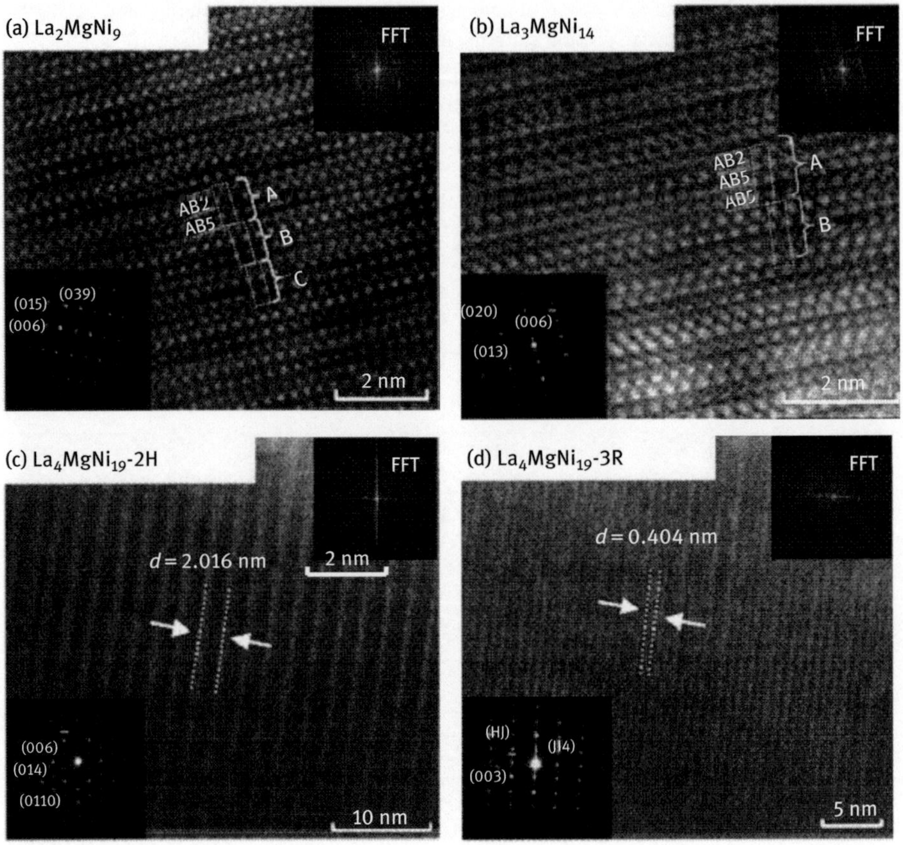

Figure 2.5: High-resolution transmission electron microscope (STEM) images and selected area electron diffraction (SAED) patterns of the alloy samples.

atomic compositions and structures of the synthesized La_2MgNi_9 alloy with single 3R-type $(La,Mg)Ni_3$ phase, La_3MgNi_{14} alloy with single 2H-type $(La,Mg)_2Ni_7$ phase and La_4MgNi_{19} alloy with both 2H- and 3R-type $(La,Mg)_5Ni_{19}$ phases. In addition, the $[AB_5]$ and $[A_2B_4]$ constituent subunits of all alloy samples are with the same compositions as $LaMgNi_4$ and $LaNi_5$.

The volume of $[LaMgNi_4]$ and $[LaNi_5]$ subunits in La_2MgNi_9, La_3MgNi_{14} and La_4MgNi_{19} alloys is shown in Figure 2.6. The $[LaMgNi_4]$ subunit volume is larger than that of $[LaNi_5]$. Similar observation was also reported in the study on the A_2B_7-type single-phase $La_{1.63}Mg_{0.37}Ni_7$ [15] and $La_{1.5}Mg_{0.5}Ni_7$ [8] alloys as well as the A_5B_{19}-type La_4MgNi_{19} alloy [13]. Compared to the La–La distance in $[La_2Ni_4]$ subunit, the La–La distance in $[LaNi_5]$ subunit is much shorter [8, 16]. It is believed that the substitution of La with Mg in the $[A_2B_4]$ subunit shortens the La–La (Mg) distance, which makes the size of the $[A_2B_4]$ subunit more match to that of the $[AB_5]$ subunit. However, in

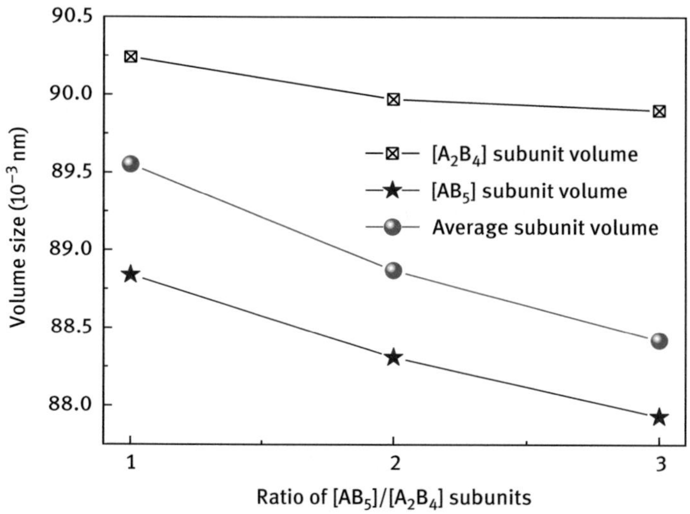

Figure 2.6: Plot of subunit volume size versus [$LaMgNi_4$]/[$LaNi_5$] subunit ratio.

this study, where 50 % of La is substituted with Mg in the [A_2B_4] subunit, the degree of shrinkage is not sufficient to make the sizes of [A_2B_4] and [AB_5] completely match, resulting in a slightly bigger volume of [A_2B_4] than that of [AB_5] [9, 15, 16]. In addition, because [A_2B_4] and [AB_5] share the same basal plane [17], their volumes may affect each other in basal plane. With increasing [AB_5]/[A_2B_4] ratio, the shrinking effect of [AB_5] subunit on its neighbouring [A_2B_4] subunit increases, while the expanding effect decreases, resulting in the decreasing trend of [A_2B_4] and [AB_5] subunit volumes with increasing [AB_5]/[A_2B_4] ratio, as seen in Figure 2.6.

2.1.4 The Gaseous Hydrogen Storage and Electrochemical Characteristics of the Single-Phase Superlattice La–Mg–Ni-Based Hydrogen Storage Alloys

2.1.4.1 The Gaseous Hydrogen Storage Characteristics of the Single-Phase Superlattice La–Mg–Ni-Based Hydrogen Storage Alloys

The *P–C–T* curves of the alloy samples measured at 30, 50 and 70°C are shown in Figure 2.7(a–c), and the results of hydrogen absorption/desorption study are summarized in Table 2.2. The *P–C–T* curves of each alloy sample for hydrogen absorption and desorption contain three steps: (1) a ramping up step that is typical with the α-phase at low capacity, (2) a flat plateau corresponding to a two-phase co-existing period (α to β transformation) and (3) a final increase due to the presence of metal hydride β-phase. The hydrogen storage capacity at room temperature for La_2MgNi_9, La_3MgNi_{14} and La_4MgNi_{19} alloys are 1.596, 1.587 and 1.569 wt.%, respectively, showing a decreasing trend with increasing [$LaNi_5$]/[$LaMgNi_4$] ratio. Only one plateau can be seen in those *P–C–T* curves, including La_4MgNi_{19} alloy with both 2H- and 3R-type

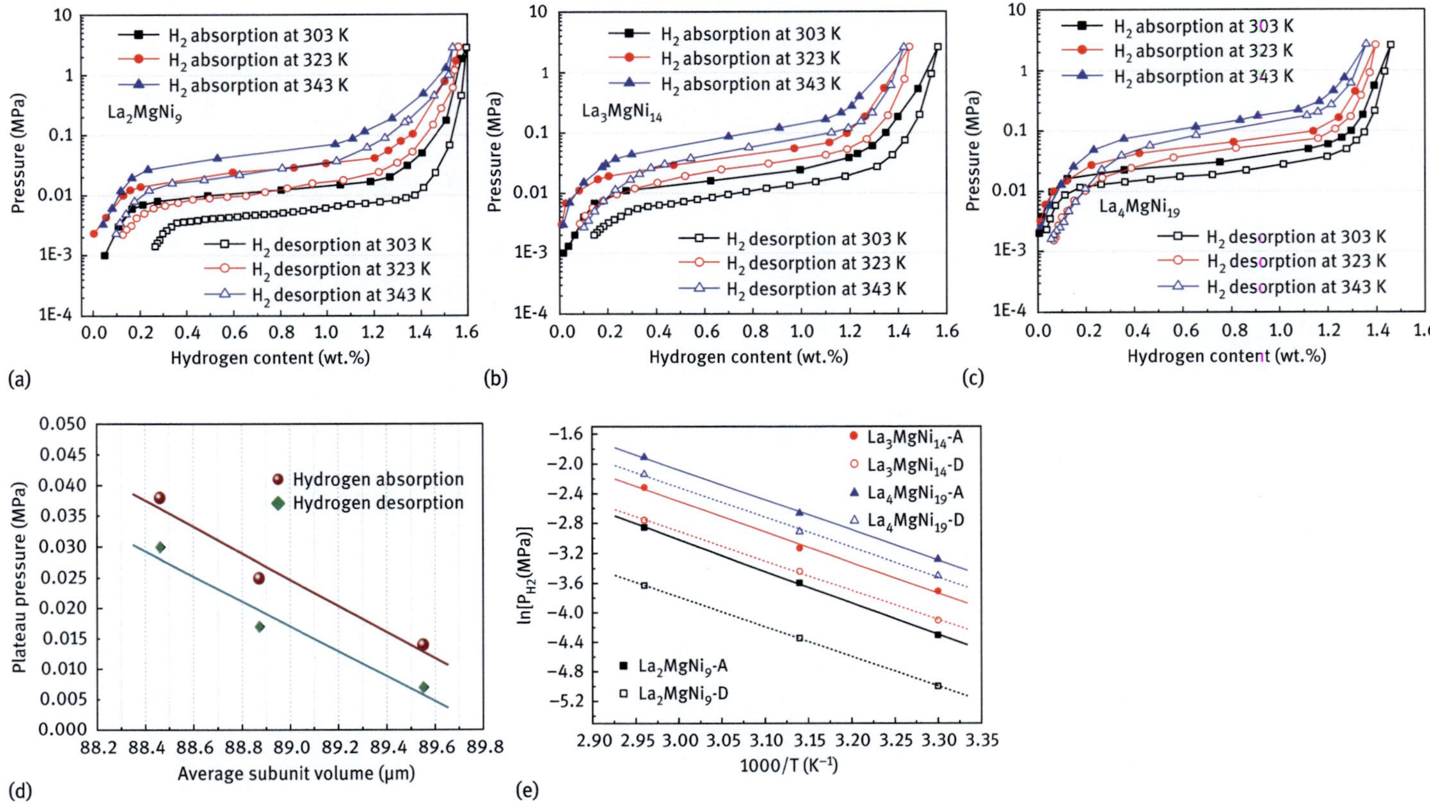

Figure 2.7: P–C–T isotherms of La_2MgNi_9 (a), La_3MgNi_{14} (b) and La_4MgNi_{19} (c) alloys at 30, 50 and 70°C. Relationships between hydrogen absorption/desorption equilibrium pressure and average subunit volume (d) and van't Hoff plots (e).

Table 2.2: Hydrogen storage properties of the alloy samples.

Alloys	H_2 storage capacity (wt.%)	P_{abs} (MPa)	P_{ads} (MPa)	$\lvert\Delta H_{abs}\rvert$ (kJ mol^{-1})	$\lvert\Delta H_{ads}\rvert$ (kJ mol^{-1})
La_2MgNi_9	1.596	0.014	0.007	35.2	33.3
La_3MgNi_{14}	1.566	0.025	0.017	33.9	33.1
La_4MgNi_{19}	1.460	0.038	0.030	33.4	32.6

phases, suggesting that 2H- and 3R-type $(La,Mg)_5Ni_{19}$ phase have similar equilibrium pressures in hydrogen absorption/desorption processes [13]. The plateau pressure of each alloy increases with increasing hydrogen absorption/desorption temperature because hydrides become less stable under high temperature. Under all three different testing temperatures, the plateau pressure of the alloy samples shows the following order: $La_2MgNi_9 < La_3MgNi_{14} < La_4MgNi_{19}$. As generally known, the stability of alloy hydrides normally decreases with decreasing crystal cell size, which leads to higher plateau pressure [18–20]. In this study, we used the alloy samples that contain different phases but are composed of the same types of constituent subunits to investigate the correlation between the average subunit volume and equilibrium pressure. Figure 2.7(d) presents the correlation between the equilibrium pressures in the hydriding/dehydriding process at 303 K and the average subunit volume of La_2MgNi_9, La_3MgNi_{14} and La_4MgNi_{19} alloys. Linear correlation between equilibrium pressure and average subunit volume is observed. Similar correlation has also been reported for $(Nd,Mg)_2Ni_7$-type alloys with different foreign elements substituting for Nd [21]. Therefore, we conclude that the linear correlation between equilibrium pressure and average subunit is applicable for practical selection of superlattice alloys with different phases but the same phase constituent subunits. The van't Hoff plots for the alloy samples are shown in Figure 2.7(e), and the absolute values of enthalpy changes ($\lvert\Delta H\rvert$) during hydrogen absorption/desorption processes are listed in Table 2.3. It can be seen that with increasing $[LaNi_5]/[LaMgNi_4]$ ratio, the $\lvert\Delta H_{abs}\rvert$ and $\Delta H_{des}\rvert$ values decrease, suggesting that the stability of the hydrides decreases with increasing $[LaNi_5]/[LaMgNi_4]$ ratio.

Table 2.3: Hydrogen storage properties of the alloy samples.

	Alloys	N_a	C_{max} (mAh g^{-1})	HRD_{1500} (%)	S_{100} (%)
Powder sintering	La_2MgNi_9	2	392	41.7	70.1
	La_3MgNi_{14}	2	386	49.6	83.0
	La_4MgNi_{19}	2	367	66.4	85.7
Induction melting	La_2MgNi_9	2	401	36.8	76.3
	La_3MgNi_{14}	2	394	37.3	84.2
	La_4MgNi_{19}	2	370	51.5	87.3

2.1.4.2 The Electrochemical Characteristics of the Single-Phase Superlattice La–Mg–Ni-Based Hydrogen Storage Alloys

The electrochemical characteristics of the La_2MgNi_9, La_3MgNi_{14} and La_4MgNi_{19} alloy electrodes are shown in Figure 2.8. All alloy samples can be fully activated within three cycles. The maximum discharge capacities of La_2MgNi_9, La_3MgNi_{14} and La_4MgNi_{19} alloys obtained by the powder sintering method are 392, 386 and 367 mAh g^{-1}, respectively. As expected, the same trend of the discharge capacity is observed for their gaseous hydrogen storage capacity. In addition, comparing the slops of the discharge capacity curves in Figure 2.8(a), the capacity degradation of the alloy electrodes is reduced with increasing $[LaNi_5]/[LaMgNi_4]$ ratios. Table 2.3 shows that the cycling stability of the alloy electrodes at the 100th charge/discharge cycle (S_{100}) increases from 70.1 % (La_2MgNi_9) to 85.7 % (La_4MgNi_{19}). The high-rate dischargeability (*HRD*) of the alloy electrodes under large current densities (300, 600, 900, 1,200 and 1,500 mA g^{-1}) is shown in Figure 2.8(b). At all discharge current densities applied, the *HRD* of the alloy electrodes shows the following order: $La_2MgNi_9 < La_3MgNi_{14} < La_4MgNi_{19}$. This is also attributed to the lower stability of alloy hydrides with increasing $[LaNi_5]/[LaMgNi_4]$ ratio, which the $(La,Mg)_5Ni_{19}$ phase with higher Ni content desorbs hydrogen faster compared to the $(La,Mg)_2Ni_7$ phase during discharge process. In general, for single-phase superlattice alloys, the *HRD* and cycling stability both increase with increasing $[LaNi_5]/[LaMgNi_4]$ ratio, while the discharge capacity shows the opposite trend. La_3MgNi_{14} alloy exhibits the best overall electrochemical properties in terms of discharge capacity, *HRD* and cycle stability.

In comparison with the above single-phase alloys prepared by the powder sintering method, the single-phase La_2MgNi_9, La_3MgNi_{14} and La_4MgNi_{19} alloys obtained by the induction melting method have higher discharge capacity and cycling stability, but lower *HRD*, as shown in Table 2.3. The phenomenon is closely related to the higher compactness of crystal structure of the induction melting alloys.

2.1.5 The Capacity Degradation Mechanism of the Single-Phase Superlattice La–Mg–Ni-Based Hydrogen Storage Alloys

Earlier, it was commonly believed that the fast capacity degradation during cycling of RE–Mg–Ni-based alloys is attributed to two main factors, the pulverization and oxidation/corrosion of the active components. Later, Liu's group detailed the capacity degradation process of the RE–Mg–Ni-based alloys, which composes three consequent stages: (1) the pulverization and Mg oxidation stage, (2) the Mg and La oxidation stage and (3) the oxidation and pulverization stage [22]. Meanwhile, they found that the pulverization of the alloy particles is caused by lattice expansion/contraction of the alloy due to hydrogen absorption/desorption process [23]. Recent studies on ternary La–Mg–Ni alloys show that the lattice strain generated,

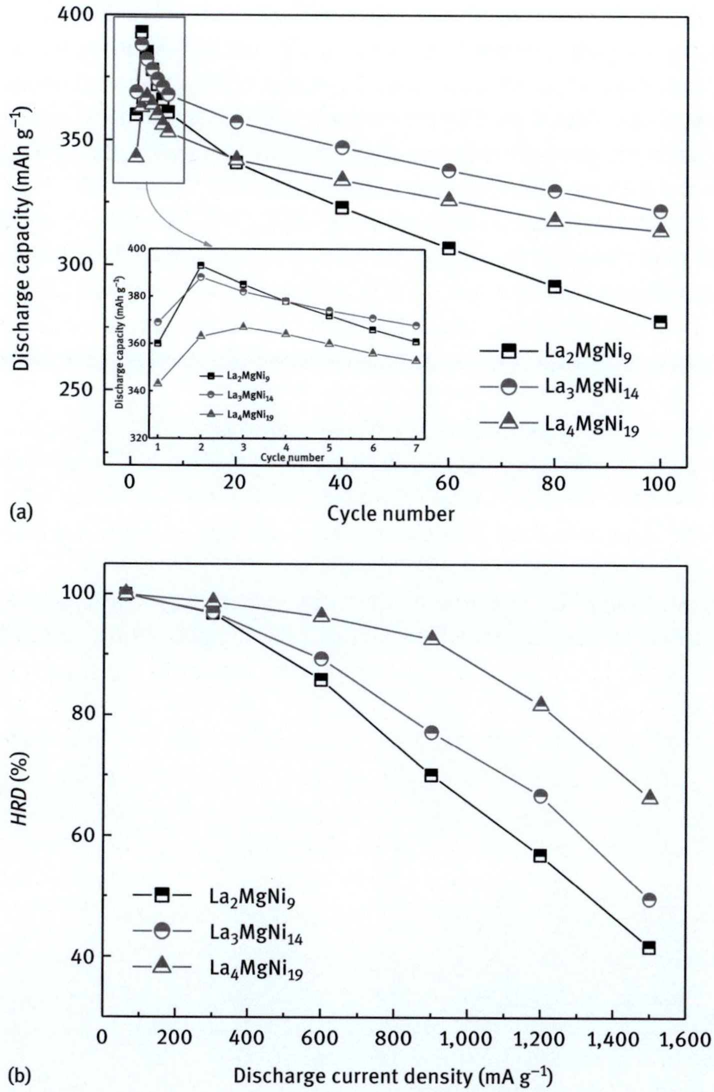

Figure 2.8: Discharging capacity of the alloy electrodes from 0 to 100 cycles (a) and *HRD* curves at different discharging current densities (b).

induced from the volume difference between $[A_2B_4]$ and $[AB_5]$ subunits the during hydrogen absorption/desorption, is believed to be the root cause leading to particle refinement of hydrogen storage alloys. Unexpectedly, the most significant strain does not occur at the full hydride stage where the cell volume expansion rate is the highest but at the medium stage. Therefore, understanding the cause of the generation and the varying trend of internal strain during the entire hydrogen absorption/desorption process for the ternary superlattice La–Mg–Ni alloys is essential to recognize their cycling degradation mechanism and to further improve their cycling stability. However, it is important to accurately study the capacity degradation process of the single phase for multiphase alloys. Because large amounts of lattice strain among the phase boundaries and changes in desynchronized cell volume due to expansion/contraction between different phases during hydrogen absorption/desorption processes both exist and can lead to aggravated pulverization of the multiphase alloys and then severe the oxidation/corrosion. For multiphase alloys, the pulverization and oxidation/corrosion processes of one phase are interconnected with other phases. This makes the determination in the capacity degradation causes for one specific phase in difficulty. Therefore, we should clear the cause of the generation and the varying trend of internal strain during the entire hydrogen absorption/desorption process for the ternary single-phase superlattice La–Mg–Ni alloys.

Figure 2.9 compares the XRD patterns of the single-phase La_2MgNi_9, La_3MgNi_{14} and La_4MgNi_{19} alloy powders after 100 charge/discharge cycles (Figure 2.9 red) with those before cycling (in blue). The alloys retain their original $(La,Mg)Ni_3$, $(La,Mg)_2Ni_7$ and $(La,Mg)_5Ni_{19}$ phase structures after 100 charge/discharge cycles, but their XRD pattern peak intensities decrease and peak widths broaden.

The HR-TEM images and SAED patterns of the alloy particles after 100 hydrogen absorption/desorption cycles are shown in Figure 2.10. It can be seen that the HR-TEM images of the cycled alloys become blurring than those of the as-prepared

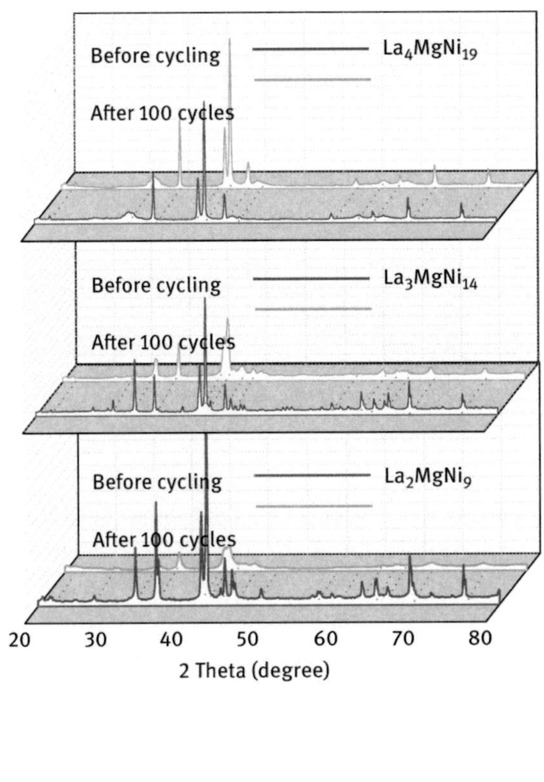

Figure 2.9: XRD patterns of the La_2MgNi_9, La_3MgNi_{14} and La_4MgNi_{19} alloy powders before and after cycling.

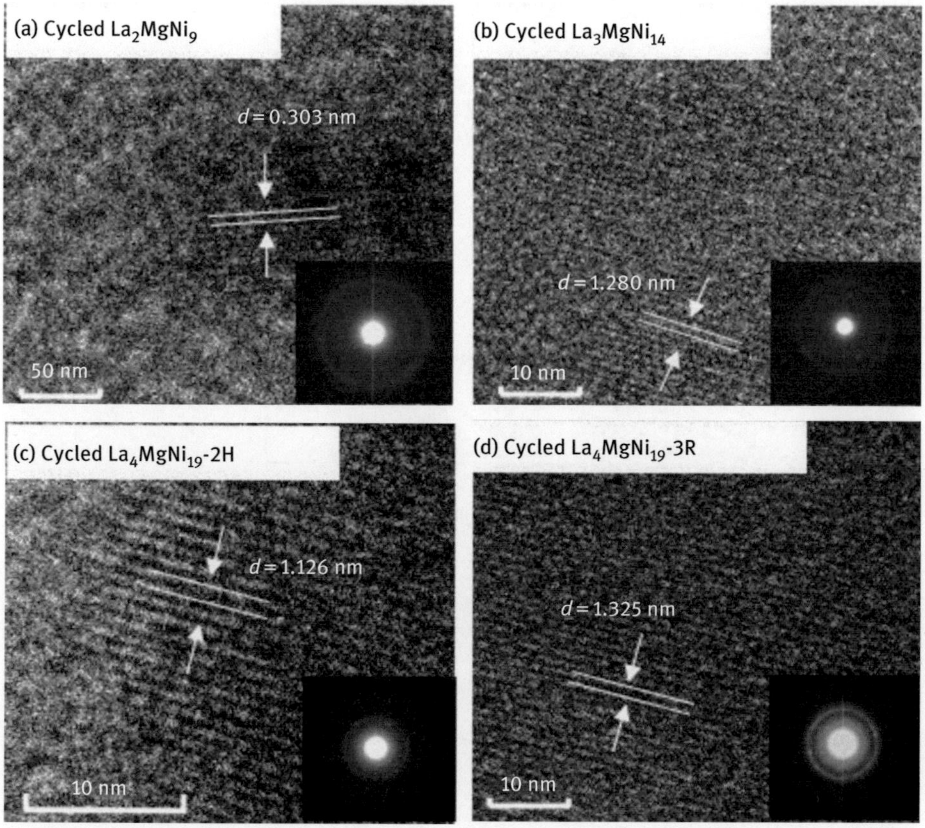

Figure 2.10: High-resolution STEM images and SAED patterns of the La_2MgNi_9, La_3MgNi_{14} and La_4MgNi_{19} alloy samples after 100 cycling.

alloys. Moreover, the SAED patterns turn to halo together with ring from dots. These observations indicate the refinement and amorphization of alloy powders after charge/discharge cycling. In addition, the particle size distribution curves before and after 100 cycles (Figure 2.11(a)) also shows a significant decrease in the size of the alloy powders after 100 charge/discharge cycles.

Theoretically, if the lattice strains were caused solely by the expansion and contraction of crystal cells during hydrogen absorption and desorption, it should keep increasing with cell expansion until it reaches the maximum hydrogen adsorption. However, the experimental data show that as the charge/discharge depth increases, the lattice strains of each alloy sample first increases, reaching to the maximum value when the charge/discharge depth is close to 50 %, and then decreases, as shown in Figure 2.11(b). This result demonstrates that the root cause of the lattice strains is not simply the hydrogen-induced expansion/contraction of cell volumes.

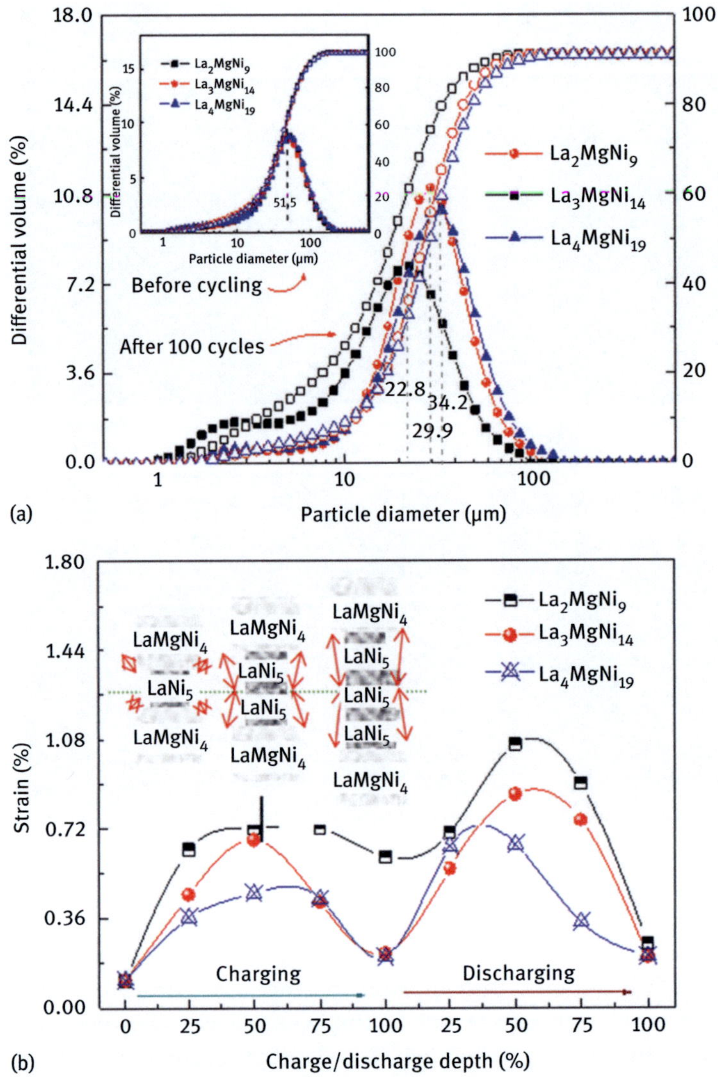

Figure 2.11: Particle size distributions of alloy powers after 100 charging/discharging cycles (a), Alloy internal strains at various charging/discharging stages within one cycle (b).

To further explore the root cause of strains inside the superlattice during hydrogen absorption/desorption, the structures of [$LaMgNi_4$] and [$LaNi_5$] constituent subunits in La_3MgNi_{14} superlattice at various charging/discharging stages were investigated. Figure 2.12(a) shows the XRD patterns of La_3MgNi_{14} at various charge/discharge depths. During the charging process where charge depth increases from 0 to 100 %, the XRD peaks shift towards lower angles, which is attributed to the lattice expansion caused

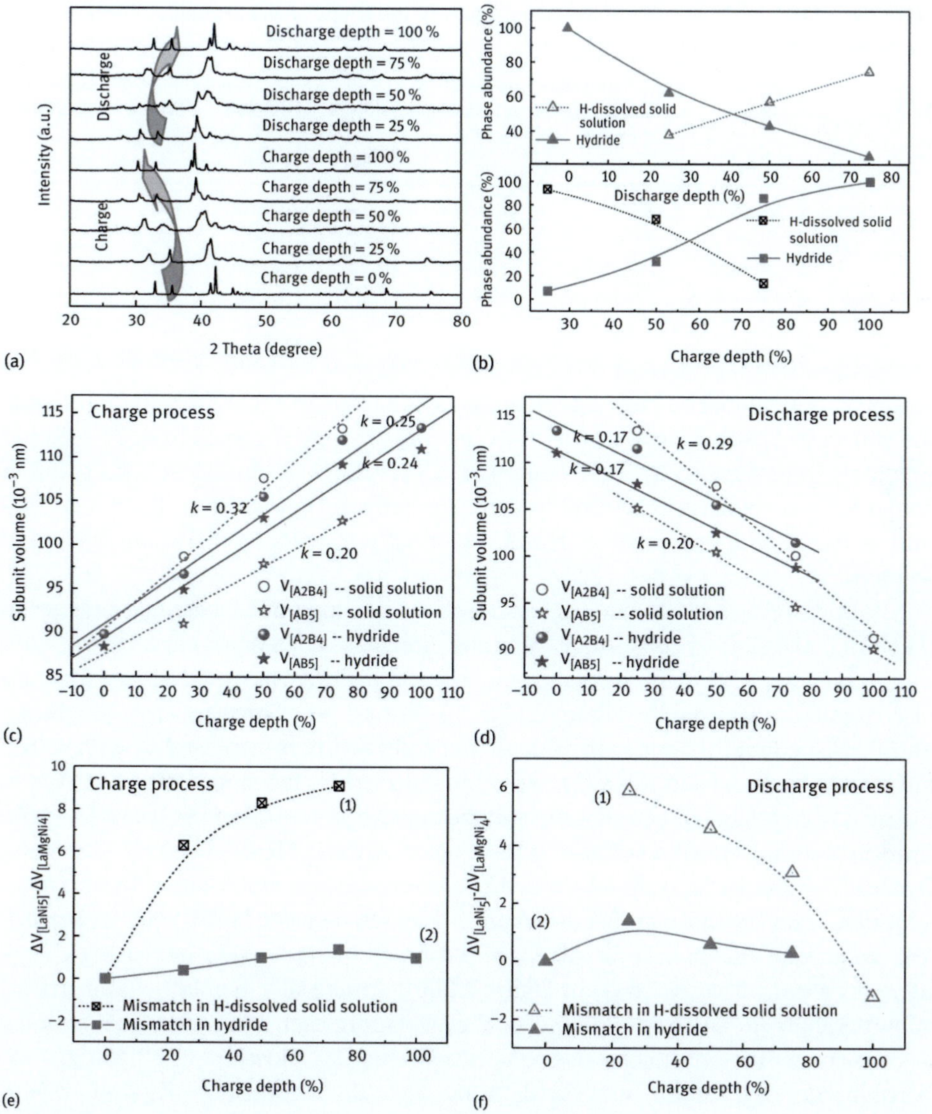

Figure 2.12: XRD patterns of the alloys at various charge/discharge depths (a), phase abundances of hydrogen dissolved solution and hydride at various charge/discharge depths (b), subunit volume during various charging (c) and discharging (d) stages and difference of the volume size between [$LaMgNi_4$] and [$LaNi_5$] subunits in hydrogen dissolved solution and hydride at various charge (e) and discharge (f) depths.

by dissolved hydrogen in the host lattice, whereas the XRD peaks shift towards higher angles due to lattice shrinkage during discharging process. During charge/discharge, H-dissolved solid solution and hydride in the alloys transform from/to each other; thus, the alloys contain both H-dissolved solid solution and hydride. The concentration of the H-dissolved solid solution and hydride at different charging/discharging

Table 2.4: Lattice constants and phase abundances of the single-phase Pr_2MgNi_9, Pr_3MgNi_{14} and Pr_4MgNi_{19} alloys.

Alloy	Phase	Space group	Lattice parameters		Volume	Phase abundance (wt.%)
			a (nm)	*c* (nm)	*V* (nm^3)	
Pr_2MgNi_9	$PuNi_3$-type	*R*-3*m*	0.4987	2.4196	0.5212	100
Pr_3MgNi_{14}	Gd_2Co_7-type	*R*-3*m*	0.4989	3.6120	0.7786	58.2
	Ce_2Ni_7-type	$P6_3/mmc$	0.4968	2.4301	0.5194	41.8
Pr_4MgNi_{19}	Pr_5Co_{19}-type	$P6_3/mmc$	0.4993	3.2197	0.6951	100

stages was calculated by Rietveld refinements and listed in Table 2.4, and their trends are shown in Figure 2.12(b). During the charging process, H-dissolved solid solution gradually transforms to hydride, leading to a decrease in H-dissolved solid solution content and an increase in hydride content. Similarly, as the discharge depth increases during the discharging process, hydride gradually transforms to H-dissolved solid solution.

Both charging and discharging processes are accompanied by volume changes in [$LaMgNi_4$] and [$LaNi_5$] constituent subunits. Volumes of [$LaMgNi_4$] and [$LaNi_5$] subunit in the form of H-dissolved solid solution and hydride were plotted as a function of charge/discharge depth (Figure 2.12(c–d)). During the charge process, [$LaMgNi_4$] and [$LaNi_5$] subunit volumes in H-dissolved solid solution increase but with different hydrogen expansion rates (k), which is reflected by the slopes of the curves in Figure 2.12(c). It can be seen that the volume expansion rate (k) of the [$LaMgNi_4$] subunits in H-dissolved solid solution is 0.32, whereas that of [$LaNi_5$] is 0.20. Therefore, [$LaMgNi_4$] subunits in H-dissolved solid solution become much larger than [$LaNi_5$] subunits, resulting in a significant volume mismatch between [$LaNi_5$] and [$LaMgNi_4$]. This volume mismatch in H-dissolved solid solution increases with increasing charge depth as shown from curve (1) in Figure 2.12(e). Structurally, [$LaMgNi_4$] and [$LaNi_5$] subunits share the same basal plane, so they constrain each other when their volume sizes become different, giving rise to lattice strains [24]. However, the hydrogen expansion rates of [$LaMgNi_4$] and [$LaNi_5$] subunits in the hydride are very similar, with the values being 0.25 and 0.24, respectively. Thus, nearly no size mismatch occurs between [$LaMgNi_4$] and [$LaNi_5$].

In general, the volume of [$LaMgNi_4$] subunit in H-dissolved solid solution becomes progressively larger than that of [$LaNi_5$] during charging process, causing increasing volume mismatch between [$LaMgNi_4$] and [$LaNi_5$]. However, because the concentration of H-dissolved solution decreases with charge depth, the internal strains in the alloys first increase and reach the maximum at the charging depth close to 50 %, then decrease (Figure 2.11(b)).

In the discharge process, the volumes of [$LaNi_5$] and [$LaMgNi_4$] subunits in both H-dissolved solid solution and hydride decrease. At the initial discharge stage (25 %

discharge depth), the volume of [$LaMgNi_4$] subunits in H-dissolved solid solution is much larger than that of [$LaNi_5$] due to the previous charging process. As the discharge goes on, the volumes of [$LaNi_5$] and [$LaMgNi_4$] subunits in H-dissolved solid solution get closer, as shown in Figure 2.12(d) and curve (1) in Figure 2.12(f). Similar to what happens in the charge process, the shrinkage of [$LaNi_5$] and [$LaMgNi_4$] subunits in the hydride occurs at similar rates, and their volume difference is at a very low level throughout the discharge process, as shown in curve (2) in Figure 2.12(f). Consequently, the internal strains in the alloy also first increase and then decrease during the discharge process.

Comparing the strains of the alloy samples at different charge/discharge depths (Figure 2.11(b)), we observe very similar strains for three alloy samples before charging, which is attributed to their similar preparation process and stacking structures. However, La_2MgNi_9 alloy exhibits much higher strains than that of the La_3MgNi_{14} and La_4MgNi_{19} alloys during the charging/discharging processes, and the strains in La_4MgNi_{19} alloy are always the lowest at any charging/discharging stage. This is in accordance with the peak height decrease and peak width widening of its XRD patterns after 100 cycles (Figure 2.9). Considering that the strains originate from mutual constriction of the basal plane shared by [$LaMgNi_4$] and [$LaNi_5$] subunits, the degree of the strains in the alloys should be dependent on the volume mismatch between [$LaMgNi_4$] and [$LaNi_5$] subunits. As shown in the insets of Figure 2.11(b), there is one, two and three [$LaNi_5$] subunit in (La,Mg)Ni_3 phase, A_2B_7 phase (La_2MgNi_9) and (La_3MgNi_{14}), respectively. When [$LaMgNi_4$] subunits become larger due to more hydrogen absorption, the deformation of the two subunits will occur, and the deformation decreases with longer buffer distance. Consequently, the strains at each hydrogen absorption/desorption stage all decrease with increasing [$LaNi_5$]/[$LaMgNi_4$] ratios. As mentioned previously, inter-lattice strains are the driving force of pulverization. Therefore, pulverizations of La_2MgNi_9, La_3MgNi_{14} and La_4MgNi_{19} alloys after 100 charge/discharge cycles decrease with increasing [$LaNi_5$]/[$LaMgNi_4$] ratios (Figure 2.11(a)). In addition, the oxygen contents of the alloy electrodes after 100 charge/discharge cycles in KOH electrolyte are 4.23, 3.53 and 2.94 mass%, respectively. As expected, larger particles have better oxidation/corrosion resistance to alkaline solution. Overall, the decreased pulverization and oxidation/corrosion resistance leads to enhanced electrochemical cycling stability in the alloys with higher [$LaNi_5$]/[$LaMgNi_4$] ratios.

In conclusion, the discrete expansion/contraction of [$LaNi_5$] and [$LaMgNi_4$] subunits during hydrogen absorption/desorption leads to the volume mismatch between [$LaNi_5$] and [$LaMgNi_4$], which leads to alloy pulverization. Higher [$LaNi_5$]/[$LaMgNi_4$] subunit ratios can improve the accommodation capability for the volume mismatch between [$LaNi_5$] and [$LaMgNi_4$] subunits during charge/discharge; thus, increasing the cycling stability by reducing the pulverization and oxidation.

2.2 The Preparation and Characteristics in Electrochemical and Gaseous Hydrogen Storage of the Single-Phase Superlattice Pr–Mg–Ni-Based Hydrogen Storage Alloys

In the past decades, people usually used elemental substitution method to improve the poor cycling stability of La–Mg–Ni-based alloys, that is, the replacement of La by other rare earth elements, such as Ce, Pr, Nd and Sm, or the replacement of Ni by transition metal elements, such as Co, Al, Mn and Fe. Thereinto, rare earth element Pr or Nd with lower chemical activity is found to be helpful for the formation of superlattice structures and significant extension of the cycle life of the alloys. Li et al. found that Pr or Nd partial substitution can increase the La_2Ni_7 phase abundance and effectively improve the cycle life and *HRD* properties by studying $La_{0.60}R_{0.20}Mg_{0.20}(NiCoMnAl)_{3.5}$ (R = La, Ce, Pr, Nd) alloys [25].

Recently, researches on Pr or Nd as the only RE element in RE–Mg–Ni-based alloys have also been conducted. The ternary Pr–Mg–Ni and Nd–Mg–Ni alloys are interesting. For example, the ternary Pr–Mg–Ni-based alloys were found to possess good hydrogen storage capacity and show outstanding solid–gas cyclic stability compared with the La–Mg–Ni-based alloys [26–28]. Lemort et al. reported that the maximum capacities of the $Pr_{1.5}Mg_{0.5}Ni_7$ and $Pr_{3.75}Mg_{1.25}Ni_{19}$ alloys reached 1.73 and 1.77 wt.% at room temperature under 10 MPa, respectively [26]. Iwase et al. found that the cyclic property of the Pr_4MgNi_{19} alloy showed a retention rate of 90 % of the initial cycle after 1,000 hydrogenation/dehydrogenation cycles [29]. Nonetheless, little is known for the electrochemical characteristics of the ternary Pr–Mg–Ni alloys.

Being similar to the La–Mg–Ni-based alloys, the Pr–Mg–Ni-based alloys with different structure have similar compositions and formational temperatures as shown in Section 3.1.1. Therefore, the realization of the controllable preparation for the Pr–Mg–Ni-based alloys with specific structures is a difficult work. While studying the formation mechanism for the RE–Mg–Ni is important, it provides a theoretical foundation for their preparation. Therefore, in this section, we have tried to prepare AB_3, A_2B_7 and A_5B_{19}-type single-phase superlattice Pr–Mg–Ni-based alloys by strictly controlling the zoning sintering and annealing processes, and introduce the formation mechanism and the electrochemical cyclic stability in terms of the structural characteristics of their constituent subunits in ternary Pr–Mg–Ni alloys in detail.

2.2.1 The Preparation of the Single-Phase Superlattice Pr–Mg–Ni-Based Hydrogen Storage Alloys

Single-phase Pr_2MgNi_9, Pr_3MgNi_{14}, Pr_4MgNi_{19} alloys were prepared by strictly controlling the zoning sintering and annealing process with $PrMgNi_4$ and $PrNi_5$ as precursors. The prepared process of the samples was as follows. Firstly, the initial alloys were crushed into −300 mesh size. Secondly, the $PrMgNi_4$ and $PrNi_5$ powders

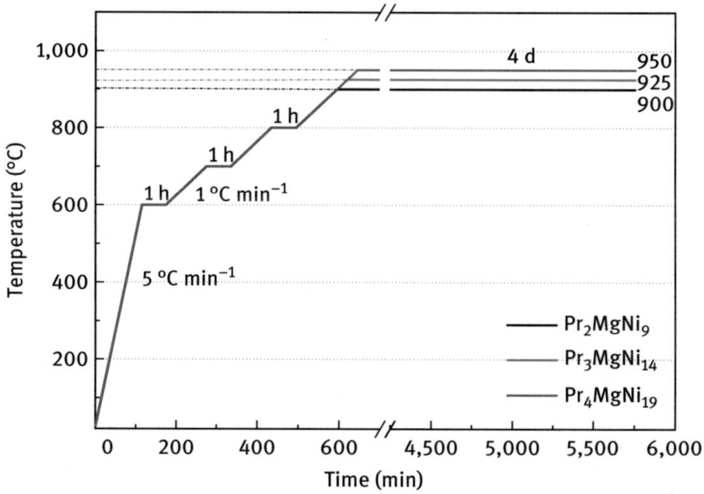

Figure 2.13: Zoning sintering and annealing processes of the three single-phase Pr_2MgNi_9, Pr_3MgNi_{14} and Pr_4MgNi_{19} alloys.

were mixed with different molar ratios, and the molar ratio $PrMgNi_4/PrNi_5$ was denoted as x, in which $x = 1.3$, 4.0 and 2.5, respectively. To compensate for the evaporative loss of Mg during the melting and sintering process, $PrMgNi_4$ was excessively added. Then, the mixtures were pressed into a pellet (3 g), which were packaged in nickel films. Finally, the nickel film was placed in a furnace for sintering. The sintering process included four heating and heat preservation stages, and the details for zoning sintering and annealing process are illustrated in Figure 2.13. In the process of heating up, the heating rate is 5°C min^{-1} below 600°C and the heating rate is 1°C min^{-1} above 600°C. In order to prevent the impurity phases arising, the sintering process is carried out under argon atmosphere.

2.2.2 The Crystal Structure of the Single-Phase Superlattice Pr–Mg–Ni-Based Hydrogen Storage Alloys

Figure 2.14 shows the low angle part (2–25°) of the XRD patterns for Pr_2MgNi_9, Pr_3MgNi_{14} and Pr_4MgNi_{19} alloys. A single reflection peak at $2\theta = 10.86°$ is observed in Figure 2.14(a), which corresponds to the lattice plane (003) of $PuNi_3$-type structure with $d = 0.8$ nm. Figure 2.14(b) contains two reflection peaks at the $2\theta = 7.3°$ and $2\theta = 14.4°$, and are determined to the lattice plane (003) and (006) of Gd_2Co_7-type structure, respectively, and also (002) and (004) of Ce_2Ni_7-type structure. The two reflection peaks in Figure 2.14(c) are assigned to the Pr_5Co_{19}-type structure. The Rietveld refinement of the alloy samples is presented in Figure 2.14(d–f), and the results are listed in Table 2.4. Pr_2MgNi_9 and Pr_4MgNi_{19} alloys are of single-phase

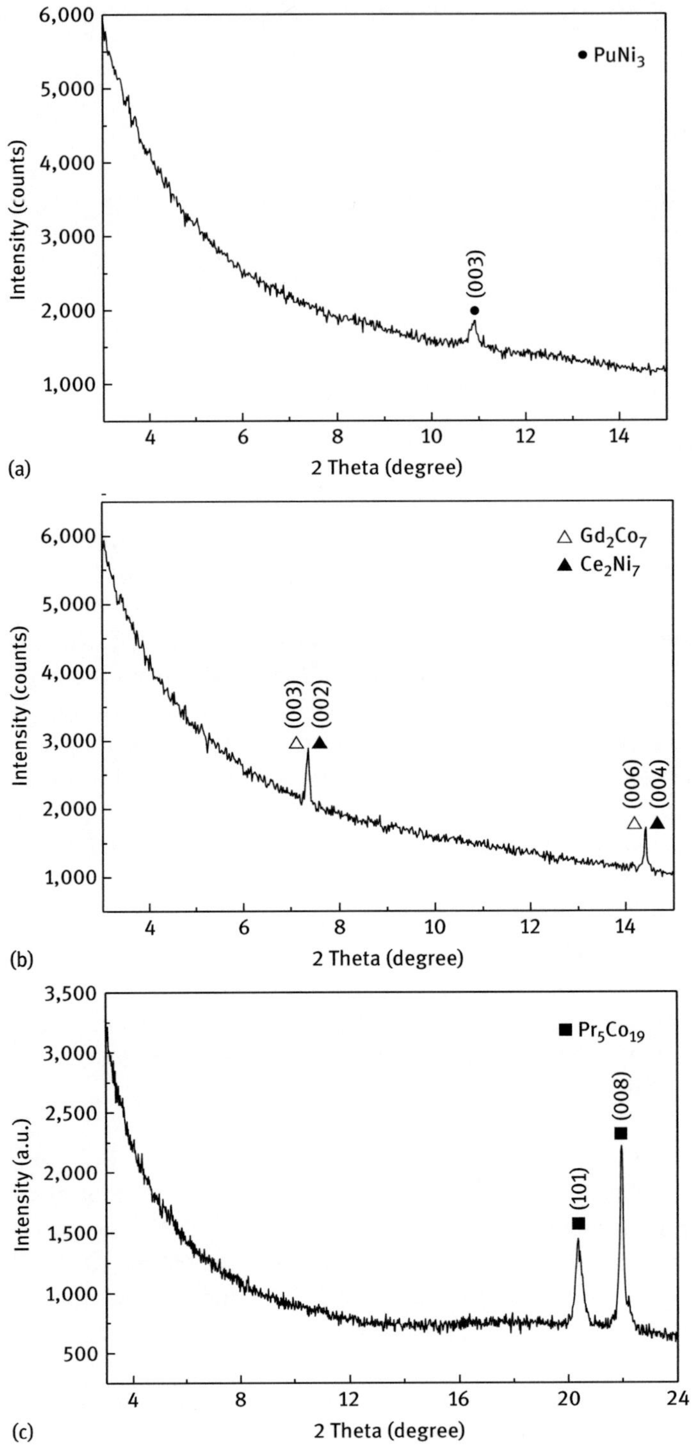

Figure 2.14: Low angle region of XRD patterns in 2θ range between 3°–25°: (a) Pr_2MgNi_9; (b) Pr_3MgNi_{14}; (c) Pr_4MgNi_{19}.

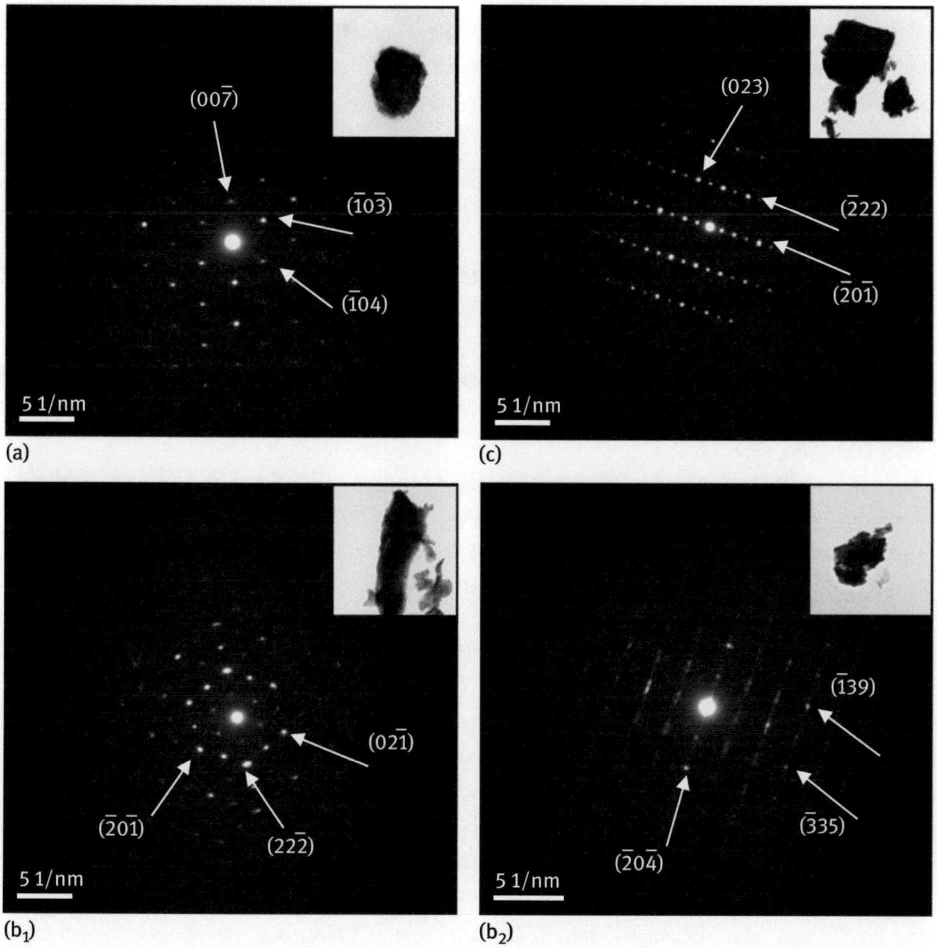

Figure 2.15: SAED and magnified morphology images of the single-phase $PuNi_3$-type Pr_2MgNi_9 (a), Gd_2Co_7-type Pr_3MgNi_{14} (b_1), Ce_2Ni_7-type Pr_3MgNi_{14} (b_2) and (c) Pr_5Co_{19}-type Pr_4MgNi_{19} alloys.

$PuNi_3$- and Pr_5Co_{19}-type phases, with the phase content of 100 wt.%. Pr_3MgNi_{14} contains Gd_2Co_7- and Ce_2Ni_7-type $(Pr,Mg)_2Ni_7$ phases, the phase abundance for the two structures are 58.2 and 41.8 wt.%, respectively.

The SAED patterns and the magnified morphology images of Pr_2MgNi_9, Pr_3MgNi_{14} and Pr_4MgNi_{19} alloys are shown in Figure 2.15. Generally, the electron diffraction spectrum with different crystal structure will have different characteristics of reciprocal lattice. According to the distance between the diffraction spot and the centre spot, it can be confirmed that the Pr_2MgNi_9 alloy possesses a $PuNi_3$-type (*R-3m*) phase structure, and the Pr_3MgNi_{14} alloy contains both Ce_2Ni_7- ($P6_3/mmc$) and Gd_2Co_7-type (*R-3m*) phase structures and the Pr_4MgNi_{19} alloy exhibits a Pr_5Co_{19}-type ($P6_3/mmc$) phase structure.

2.2.3 The Formation Mechanism of the Single-Phase Superlattice Pr–Mg–Ni-Based Hydrogen Storage Alloys

According to the Pr–Ni binary diagram [20], it can be considered that the $PrNi_3$ and Pr_2Ni_7 binary alloys can be formed through the peritectic reactions with $PrNi_5$ and $PrNi_2$ precursors during the sintering process. Being similar to the La–Mg–Ni-based alloys, the Pr–Mg–Ni-based ternary alloys are derived from the Pr–Ni binary alloys, and peritectic reactions will take place with $PrNi_5$ and $PrMgNi_4$ precursors. Based on the peritectic theory [5], the $PrMgNi_4$ phase with lower melting point partially melts firstly, while the $PrNi_5$ phase with a much higher melting point still remains in a solid state. At 900°C, the molten $PrMgNi_4$ phase gradually diffuses towards the $PrNi_5$ solid phase and reacts to generate the AB_3-type Pr_2MgNi_9 phase with the molar ratio $PrMgNi_4/PrNi_5$ of 1.3.

When sintering the two precursors with the molar ratio of 4.0, the molten $PrMgNi_4$ phase first interacts with a $PrNi_5$ solid phase and forms the AB_3-type phase when the temperature increases up to 900°C. As the temperature is above 900°C, the AB_3-type phase fully turns into a liquid and continues to react with the remaining $PrNi_5$ solid phase to form the A_2B_7-type phase at 925°C.

As for the molar ratio $PrMgNi_4/PrNi_5$ of 2.5, similar reactions take place between 900 and 925°C, and the AB_3 and A_2B_7-type phases generate at their corresponding peritectic temperature 900 and 925°C, respectively. When the temperature is higher than 925°C, the generated A_2B_7-type phase melts into a liquid phase, and meanwhile, the $PrNi_5$ solid phase is sufficient, the liquid phase reacts with the residual $PrNi_5$ solid phase and forms the A_5B_{19}-type phase at 950°C.

The formation mechanism can be summarized as follows:

$$PrNi_5 + PrMgNi_4\ (\text{liquid}) \xrightarrow{900^\circ C} Pr_2MgNi_9 \tag{2.7}$$

$$PrNi_5 + Pr_2MgNi_9\ (\text{liquid}) \xrightarrow{925^\circ C} Pr_3MgNi_{14} \tag{2.8}$$

$$PrNi_5 + Pr_3MgNi_{14}\ (\text{liquid}) \xrightarrow{950^\circ C} Pr_4MgNi_{19} \tag{2.9}$$

2.2.4 The Electrochemical Characteristics of the Single-Phase Superlattice Pr–Mg–Ni-Based Hydrogen Storage Alloys

2.2.4.1 Electrochemical Pressure–Composition (*P–C*) Isotherm of the Single-Phase Superlattice Pr–Mg–Ni-Based Hydrogen Storage Alloys

High hydrogen capacity and moderate hydride stability are important parameters for hydrogen storage alloy. Herein, the two properties can be obtained from the *P–C* isotherm where the length of plateau is used to calculate the theoretical electrochemical capacity, and the height of the hydrogen absorption/desorption equilibrium plateau pressure is used to estimate the stability of hydride, respectively.

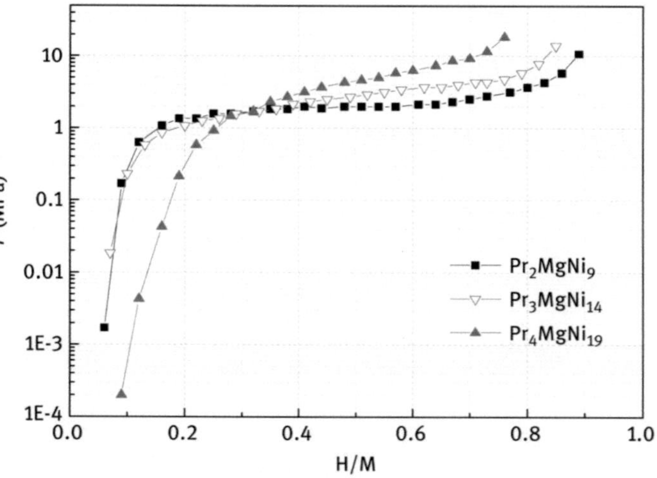

Figure 2.16: *P–C* isotherms of the single-phase Pr_2MgNi_9, Pr_3MgNi_{14} and Pr_4MgNi_{19} alloy electrodes.

The electrochemical *P–C* isotherms for the three single-phase alloy electrodes at room temperature were measured and drawn in Figure 2.16. It can be observed only a plateau region is visible for each of the alloy electrode, and the length of the plateau for the three single-phase alloy electrodes shows the following order: $Pr_2MgNi_9 > Pr_3MgNi_{14} > Pr_4MgNi_{19}$, which is in accordance with the trend of the electrochemical hydrogenation. Besides, the equilibrium plateau pressure elevates as increasing the $[PrNi_5]/[PrMgNi_4]$ subunit ratio, which indicates that the corresponding hydride becomes stable.

2.2.4.2 Activation Property and Maximum Discharge Capacity of the Single-Phase Superlattice Pr–Mg–Ni-Based Hydrogen Storage Alloys

Table 2.5 summarizes the electrochemical properties including the activation property (N_a), maximum discharge capacity (C_{max}) and capacity retention (S_n) within 250 cycles. The electrode activation periods of the single-phase Pr–Mg–Ni alloys are apparently longer than that of the La–Mg–Ni alloys, which reaches nine cycles. The maximum discharge capacities (C_{max}) of the three single-phase alloy electrodes follow the next order at 0.2 C rate of charge/discharge measurement: $Pr_2MgNi_9 > Pr_3MgNi_{14} > Pr_4MgNi_{19}$, which are 343, 333, and 316 mAh g^{-1}, respectively. It can be seen that the C_{max} of the alloy electrodes gradually decreases with the increasing of $[PrNi_5]/[PrMgNi_4]$ subunit ratio. Here are the reasons:

Combined with the Rietveld results, it is known that the substituted Mg atoms only occupy Pr2 (6c) site in $[A_2B_4]$ slabs for these three single phases, and no other Pr sites or Ni sites are occupied by Mg atoms in $[AB_5]$ slabs [27–30]. Electronegativity reflects the electron-attracting ability of an atom, and the atom with larger electronegativity has stronger bonding electron electron-attracting ability when the chemical bond

Table 2.5: Electrochemical properties of the three single-phase Pr–Mg–Ni alloy electrodes.

Alloy samples	N_a	C_{max} (mAh g^{-1})	S_{100}^{b} (%)	S_{200}^{b} (%)	S_{250}^{b} (%)
Pr_2MgNi_9	8	343	73.7	51.9	42.0
Pr_3MgNi_{14}	9	333	77.5	68.9	65.4
Pr_4MgNi_{19}	9	316	80.3	71.5	68.4

[a]The capacity cycling retention at the 100th, 200th and 250th cycle, respectively.

forms. The electronegativity of Mg element (1.3) is larger than Pr element (1.1), which represents that Mg atom may combine with more hydrogen atoms than Pr atom does when hydrogen atom enters into the lattice. Moreover, according to the schematic of the super-stacking structure for La–Mg–Ni-based alloys [31], [A_2B_4] subunits account for 1/2, 1/3 and 1/4 in AB_3-, A_2B_7- and A_5B_{19}-type phases, respectively. Denys et al. studied the saturated $La_2MgNi_9D_{13.1}$ hydride and found the stacking subunits of [$LaNi_5$] and [$LaMgNi_4$] were both occupied by D atoms forming the $LaNi_5D_{5.2}$ and $LaMgNi_4D_{7.9}$ [32], from which we can find that the hydrogen storage site in [A_2B_4] subunit is more than that in [AB_5] subunit of AB_3-type phase. Based on the above points of view, we can draw the conclusion that the maximum discharge capacity gradually decreases with increasing the subunit ratio.

2.2.4.3 Cycling Stability and the Capacity Degradation Mechanism of the Single-Phase Superlattice Pr–Mg–Ni-Based Hydrogen Storage Alloys

The cycle life of the Pr_4MgNi_{19} alloy electrode is higher than that of Pr_2MgNi_9 and Pr_3MgNi_{14} alloy electrodes during charge/discharge progress. The Pr_4MgNi_{19} alloy electrode shows a capacity retention rate of 68.4 % at the 250th charge/discharge cycle, which is superior to those of Pr_2MgNi_9 (42.0 %) and Pr_3MgNi_{14} (65.4 %). Comparing the slops of the discharge capacity curves in Figure 2.17, it is found that the discharge capacity of Pr_2MgNi_9 alloy electrode decreases steeply with a capacity loss rate of 23.9 % between 150 and 250 cycles, while for Pr_3MgNi_{14} and Pr_4MgNi_{19} alloy electrodes, the discharge capacities decrease slowly and show a liner relationship with the cycle number, which are 12 and 12.1 %, respectively. In addition, for Pr_4MgNi_{19} alloy electrode, the decrease of the discharge capacity levels out between 150 and 200 cycles, the capacity retention rate decreases from 73.8 % at the 150th cycle to 71.6 % at the 200th cycle. In general, for single-phase Pr–Mg–Ni alloy electrodes, the cyclic stability has improved with the increasing of [$PrNi_5$]/[$PrMgNi_4$] subunit ratio.

In order to investigate the mechanism of the capacity degradation of the single-phase superlattice Pr–Mg–Ni-based hydrogen storage alloys, the SEM micrographs and corresponding XRD patterns of the alloy particle before activation and after charge/discharge cycles are analysed in detail. During the charge/discharge testing process, the alloy electrode was charged at 36 mA g^{-1} for 5 h and discharged at the same current density to a cut-off voltage of 1.0 V.

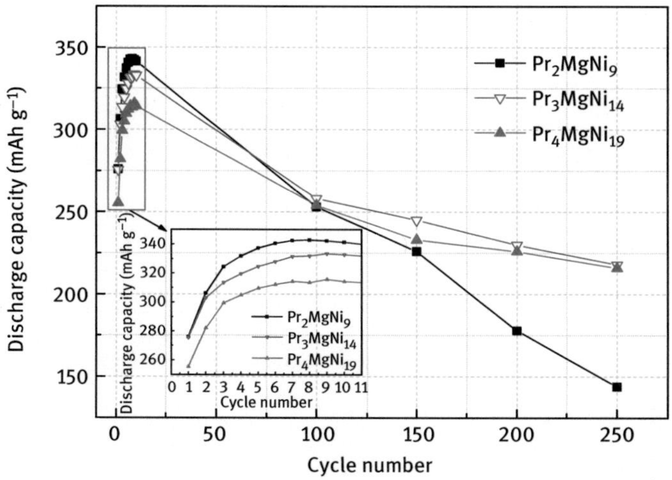

Figure 2.17: Discharge capacities of the single-phase Pr_2MgNi_9, Pr_3MgNi_{14} and Pr_4MgNi_{19} alloy electrodes versus cycle number.

Figure 2.18 illustrates the XRD patterns of the single-phase Pr_2MgNi_9, Pr_3MgNi_{14} and Pr_4MgNi_{19} alloys before activation and after 50, 100 charge/discharge cycles. It can be seen that the intensity and width of the major diffraction peaks of the alloys gradually become weaker and broader with the increasing of the cycle numbers, respectively. The full width at half maximum (FWHM) can reflect the crystal grain size of the alloys, of which the larger FWHM values represent smaller crystal grains [33, 34]. The FWHM values of the alloys after 100 cycles (analysed by Jade 6.0 software) are listed in Table 2.6. It is easily observed the FWHM value of Pr_4MgNi_{19} (0.219) alloy is smaller than Pr_2MgNi_9 (0.274) and Pr_3MgNi_{14} (0.259) alloys, indicating less pulverization and amorphization of the alloy particle after cycling. Besides, the new phases $Mg(OH)_2$ and $Pr(OH)_3$ hydroxide appear after cycling and their phase abundances gradually increase with the increasing cycle number, and which the corresponding phase abundances are listed in Table 2.6. It is noticeable that the phase abundances of the hydroxides both increase by increasing $[PrNi_5]/[PrMgNi_4]$ and the oxidation rate of each alloy electrode within 50 cycles is higher. Comparing the intensity of the diffraction peak of hydroxide at the same cycle number, it is found that the oxidation degree of Pr_4MgNi_{19} alloy electrode is smaller than the other two alloys.

The radius of hydrogen atom is larger than that of the lattice interstice of the alloys. Therefore, when hydrogen atoms go into/out the interstitial sites of the lattice, it will inevitably engender the expansion and contraction of the cell volume, which cause the generation of lattice stress in the alloy [33]. In order to release such stress, microscopic crack will appear on the surface of the alloy particle and the pulverization of the alloy electrode will be accelerated with the appearance of the cracks. Figure 2.19 presents the SEM micrographs of working electrode before activation and after 50, 100 charge/discharge cycles. SEM images can intuitively reflect the pulverization progress

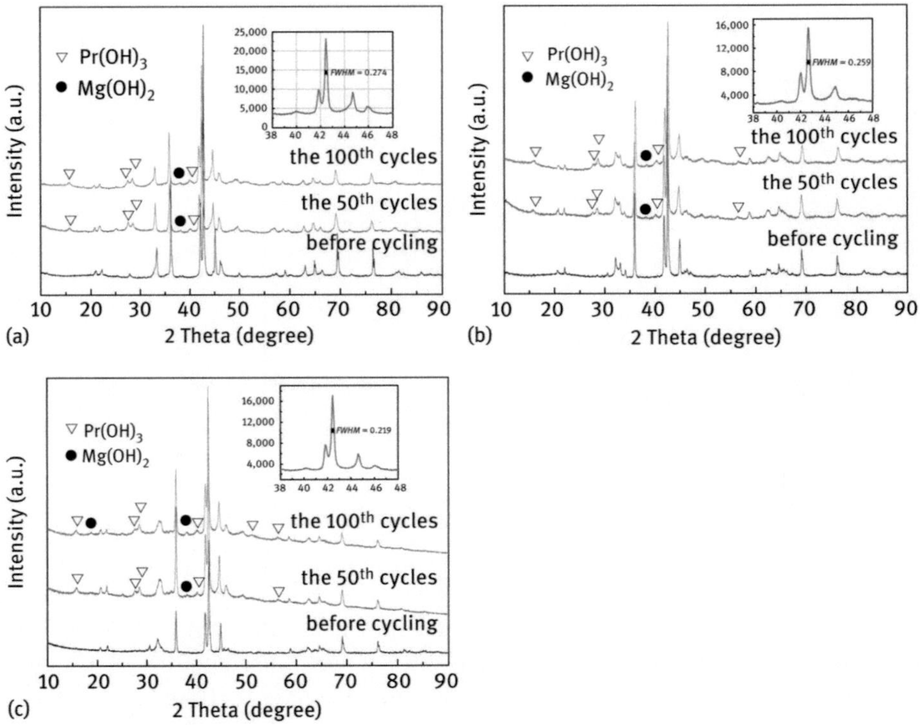

Figure 2.18: XRD patterns of the alloys before cycling and after the 50th and 100th cycles: (a) Pr_2MgNi_9; (b) Pr_3MgNi_{14}; (c) Pr_4MgNi_{19}.

Table 2.6: The phase abundances of the hydroxides for the three single-phase Pr–Mg–Ni alloy electrodes after cycling.

Alloy samples	FWHM values	$Pr(OH)_3$ content (wt.%)		$Mg(OH)_2$ content (wt.%)	
		50th	100th	50th	100th
Pr_2MgNi_9	0.274	10.3	16.7	3.1	3.7
Pr_3MgNi_{14}	0.259	7.7	12.9	1.6	2.1
Pr_4MgNi_{19}	0.219	4.3	8.7	1.4	1.9

of the alloy particles. By observing the SEM micrographs, it can be found that the surface of the three single-phase alloy particle is smooth and flat before cycling, but as the cycling numbers increases, different degree of cracks appear on the alloy surface as shown in Figure 2.19(a_1–c_2). Comparing the micrographs of the alloy electrodes at the same cycle number, it is found that with the increasing of [$PrNi_5$]/[$PrMgNi_4$] subunit ratio, the crack is less apparent. Studies found that the mismatch between [AB_5] and [A_2B_4] subunits after cycling is the main reason [35].

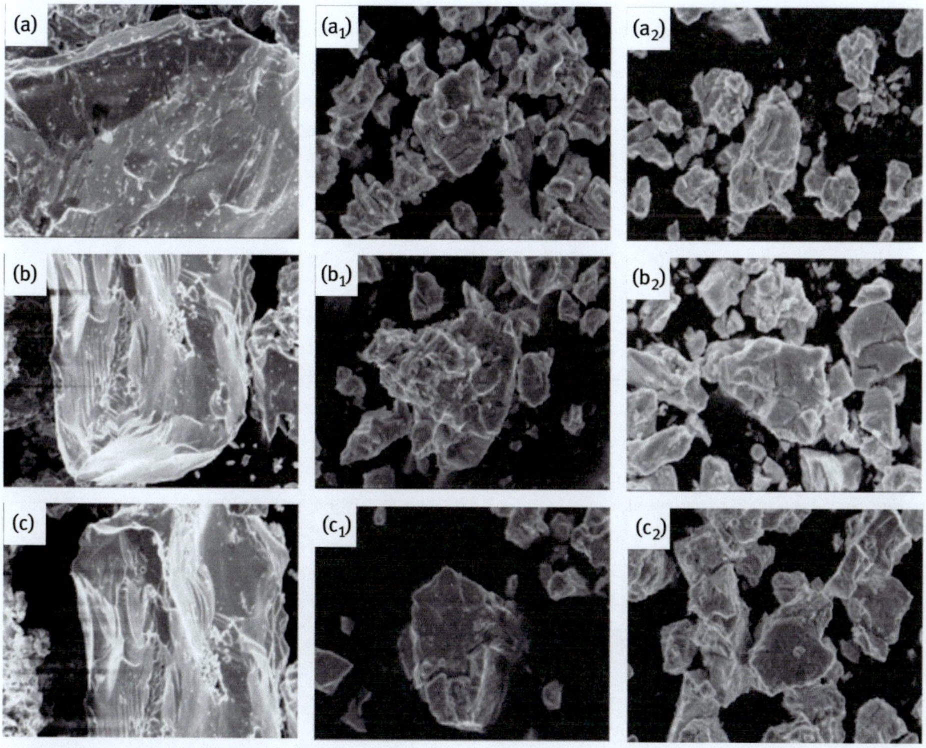

Figure 2.19: SEM images of the alloys before cycling and after the 50th and 100th cycles: (a–a_2) Pr_2MgNi_9; (b–b_2) Pr_3MgNi_{14}; (c–c_2) Pr_4MgNi_{19}.

Based on the refinement results of the atom coordinates and cell parameters, the volume of the [AB_5] and [A_2B_4] slabs, the volume expansion rate of each subunit, the difference in the volume of [$PrNi_5$] and [$PrMgNi_4$] subunits before and after the 50th and 100th cycles, and the mismatch degree which is defined as the relative values on the difference of the subunits volume between the two subunits after cycling are calculated and shown in Table 2.7. The volume expansion of [$PrMgNi_4$] subunit is larger than that of the [$PrNi_5$] subunit during charge/discharge process in each of the alloy electrode. In addition, the calculation of volume expansion in subunits shows that Pr_4MgNi_{19} alloy has a lower mismatch degree during charge/discharge process, the mismatch degree at the 50th and 100th are (12.4 %) and (22.0 %), respectively. For Pr_2MgNi_9 alloy, the mismatch degree is up to (22.9 %) and (48.6 %) at the corresponding cycles, respectively. Based on the above calculation, it can be found that the cyclic stability of the Pr_4MgNi_{19} alloy is superior to those of the Pr_2MgNi_9 and Pr_3MgNi_{14} alloys.

In order to further compare the amorphous character of the three single-phase alloy electrodes after cycling, the magnified images and SAED patterns of the alloy particle after 100 charge/discharge cycles are analysed in Figure 2.20. It can be observed that after 100 charge/discharge cycles, the SAED patterns for Pr_2MgNi_9 alloy

Table 2.7: Comparison of crystallographic parameters of the three single-phase Pr–Mg–Ni alloys.

Alloy	Cycle	$V_{[A2B4]}$ (Å^3)	$V_{[AB5]}$ (Å^3)	$\Delta V_{[A2B4]}/V_{[A2B4]}$ (%)	$\Delta V_{[AB5]}/V_{[AB5]}$ (%)	$V_{[A2B4]}-V_{[AB5]}$	$\Delta V_{[A2B4]}-V_{[AB5]}]$ (%)
Pr_2MgNi_9	Original	86.41	84.97	–	–	1.44	–
	50th	88.22	86.35	+2.09	+1.62	1.87	29.9
	100th	89.52	87.38	+3.60	+2.84	2.14	48.6
Pr_3MgNi_{14}	Original	87.10	85.40	–	–	1.70	–
	50th	88.66	86.61	+1.79	+1.41	1.99	17.2
	100th	90.01	87.80	+3.34	+2.81	2.21	30.0
Pr_4MgNi_{19}	Original	88.12	86.26	–	–	1.86	–
	50th	89.55	87.46	+1.62	+1.39	2.09	12.4
	100th	90.75	88.48	+2.98	+2.57	2.27	22.0

Figure 2.20: TEM images of the Pr_2MgNi_9, Pr_3MgNi_{14} and Pr_4MgNi_{19} alloys after 100 charge/discharge cycles: (a) $PuNi_3$-type Pr_2MgNi_9; (b_1) Gd_2Co_7-type Pr_3MgNi_{14}; (b_2) Ce_2Ni_7-type Pr_3MgNi_{14}; (c) Pr_5Co_{19}-type Pr_4MgNi_{19}.

become hazy and the halo together with ring from the central spots forms, displaying an obvious characteristic of amorphous structure. While for Pr_3MgNi_{14} alloy, diffraction spots can still be identified as both for the Gd_2Co_7- and Ce_2Ni_7-type structures, which indicates that the Pr_3MgNi_{14} alloy is partially amorphous. The SAED pattern for Pr_4MgNi_{19} alloy is clearly visible, and it means that the crystal structure of the alloy is integrated. According to the above analysis and discussion, it can be seen that the Pr_5Co_{19}-type single-phase Pr_4MgNi_{19} alloy maintains a better crystalline structure after cycling, which is consistent with the FWHM values. This implies that higher $[PrNi_5]/[PrMgNi_4]$ subunit ratio is beneficial to improving the electrochemical cycling stability.

2.2.4.4 *HRD* and Electrochemical Kinetics Properties of the Single-Phase Superlattice Pr–Mg–Ni-Based Hydrogen Storage Alloys

Figure 2.21 shows the *HRD* curves of the three single-phase Pr–Mg–Ni alloy electrodes and the discharge capacities at the current density of 1,200 mA g^{-1} (4 C) are summarized in Table 2.8. The *HRD* of the alloy electrode increases with rising $[PrNi_5]/[PrMgNi_4]$ subunit ratio from 1 to 3, and the *HRD* at 1,200 mA g^{-1} increases from 49.5 % (Pr_2MgNi_9) to 55.1 % (Pr_4MgNi_{19}). As Willems and Buschow reported, the *HRD* is determined by two factors, that is, the charge-transfer reaction at the alloy/electrolyte interface and the hydrogen diffusion rate from the interior to the surface of the alloy particles [36]. In order to clarify the main factor of the kinetic characteristic, linear polarization and potential step method are performed, from which we can obtain the exchange current density (I_0), the polarization resistance (R_p) and hydrogen diffusion coefficient (D).

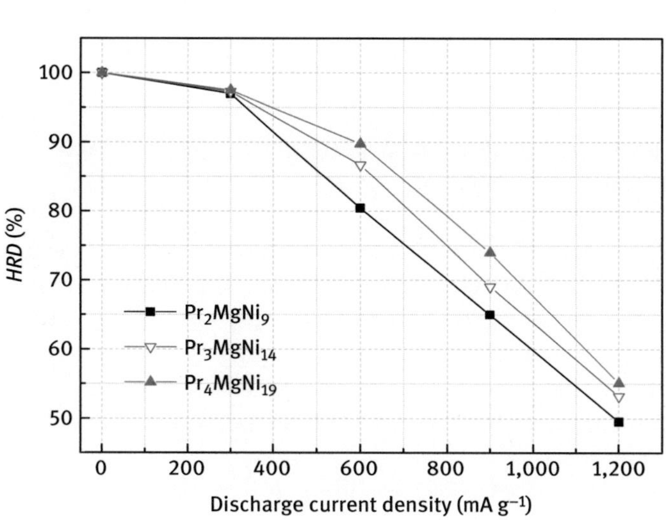

Figure 2.21: High rate dischargeability (*HRD*) curves of the Pr_2MgNi_9, Pr_3MgNi_{14} and Pr_4MgNi_{19} alloy electrodes at different discharge current densities.

2.3 The Preparation and Characteristics in Electrochemical and Gaseous Hydrogen Storage of the $PuNi_3$-Type Single-Phase Superlattice Nd–Mg–Ni-Based Hydrogen Storage Alloys

Form Sections 2.1 and 2.2 we can know that the AB_3-type super-stacking structure alloys exhibit better discharge capacity than those of the RE–Mg–Ni alloys with A_2B_7-type or A_5B_{19}-type super-stacking structure. However, the cyclic stability and low *HRD* of the AB_3-type alloys are poor. Nd partly replaces La and is found an effective way to improve the cycle life and *HRD* of RE–Mg–Ni-based MH alloys due to the lower chemical activity and the smaller atomic radius of Nd [37, 38]. Therefore, in order to obtain the AB_3-type super-stacking structure with the improved cycle life and *HRD*, using Nd to totally replace La is an effective method. Therefore, we introduce a $PuNi_3$-type single-phase Nd_2MgNi_9 alloy prepared by the induction melting method followed by an annealing treatment in this chapter and make a comparison with the $PuNi_3$-type single-phase La_2MgNi_9 alloy.

2.3.1 The Preparation of the $PuNi_3$-Type Single-Phase Superlattice Nd–Mg–Ni-Based Hydrogen Storage Alloys

Nd_2MgNi_9 were prepared by inductive melting as in the case of the La_2MgNi_9 alloy in Section 2.1.1. The Nd_2MgNi_9 annealed under argon atmosphere in tubular furnace as follows: the as-cast Nd_2MgNi_9 alloy was first heated to 600°C with a heating rate of 5°C min^{-1}, and then increased the temperature to 1,025°C with the heating rate of 1°C min^{-1} and maintained at 1,025°C for 24 h under Ar gas atmosphere. In order to compensate Mg evaporation, an appropriate excess of Mg_2Ni alloy was added during the annealing procedure. La_2MgNi_9 alloy was prepared as shown in Section 2.1.1.

2.3.2 The Crystal Structure of the Single-Phase Superlattice Nd–Mg–Ni-Based Hydrogen Storage Alloys

Figure 2.22 shows the XRD patterns of the as-cast and annealed samples. It is found that both the as-cast La_2MgNi_9 and Nd_2MgNi_9 alloys possess multiphase structures. Among them, the as-cast La_2MgNi_9 alloy comprises of $PuNi_3$-type, Gd_2Co_7-type, Ce_5Co_{19}-type and $CaCu_5$-type phases and the as-cast Nd_2MgNi_9 alloy comprises of $PuNi_3$-type, Gd_2Co_7-type and Ce_2Ni_7-type phases. When the La_2MgNi_9 and Nd_2MgNi_9 alloys annealed at 950°C for 12 h and 1,025°C for 24 h, respectively, both of the alloys only comprise of a $PuNi_3$-type phase.

Figure 2.23 shows the low-angle region XRD profiles in the 2θ between 3° and 24° for annealed La_2MgNi_9 and Nd_2MgNi_9 alloys, respectively. According to the $PuNi_3$-type (PDF 41-1129) structure, the peaks in Figure 2.23(a) at 2θ = 10.75°, 20.53°, 21.45° and

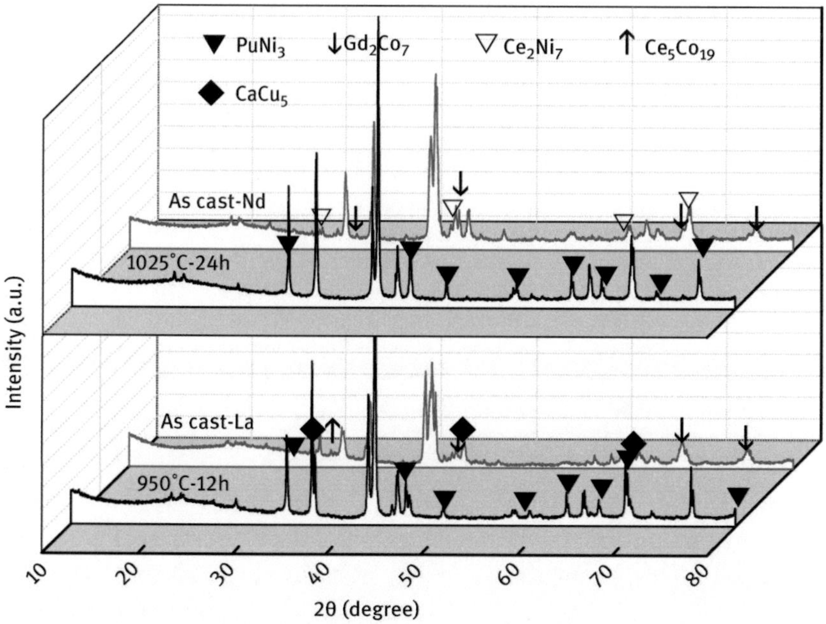

Figure 2.22: XRD patterns of the as-cast and annealed RE_2MgNi_9 (RE = La and Nd) alloys in the 2θ range of 10–80°.

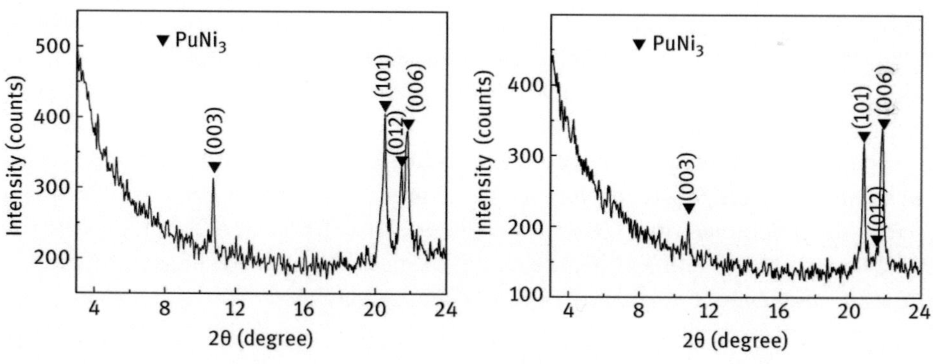

Figure 2.23: Low-angle region of XRD profiles in 2θ range between 3° and 24° for the La_2MgNi_9 alloy (a) and the Nd_2MgNi_9 alloy (b).

21.75° are the (003), (101), (012) and (006) lattice planes of the La_2MgNi_9 alloy, respectively, corresponding to the d = 0.823, 0.432, 0.414 and 0.408 nm of the $PuNi_3$-type structure (a space group $R\text{-}3m$), and the peaks in Figure 2.23(b) at 2θ = 10.82°, 20.76°, 21.51° and 21.81° are the (003), (101), (012) and (006) lattice planes of the Nd_2MgNi_9 alloy, respectively, corresponding to the d = 0.810, 0.425, 0.408 and 0.405 nm. The

Table 2.8: Lattice parameters and phase abundances of alloys from XRD and Rietveld analysis.

Process	Phase	Lattice parameters (Å)		Volume	Phase abundance
		a	*c*	(Å^3)	(wt.%)
As-cast	$PuNi_3$	5.0441	24.451	538.76	43
–La_2MgNi_9	Gd_2Go_7	5.0342	36.291	796.51	32
	Ce_5Go_{19}	5.0306	48.373	1060.2	10
	$CaCu_5$	5.0365	3.9980	87.831	15
950°C × 12 h –La_2MgNi_9	$PuNi_3$	5.0402	24.356	535.83	100
As-cast	$PuNi_3$	4.9926	24.358	525.8	53
–Nd_2MgNi_9	Gd_2Go_7	5.0299	37.061	812.06	39
	Ce_2Ni_7	5.1311	23.789	542.41	8
1,025°C × 24 h –Nd_2MgNi_9	$PuNi_3$	5.0004	24.391	528.17	100

results show that both of the annealed La_2MgNi_9 and Nd_2MgNi_9 alloys are $PuNi_3$-type single-phase structure and the diffraction peaks of the single-phase $PuNi_3$-type Nd_2MgNi_9 alloy are at higher angles compared with the single-phase $PuNi_3$-type La_2MgNi_9 alloy.

Figure 2.24 shows the Rietveld refinements of the XRD profiles of the annealed La_2MgNi_9 and Nd_2MgNi_9 alloys. Table 2.8 summarizes the Rietveld refinement results of the alloys. For the as-cast La_2MgNi_9 alloy, the contents of $PuNi_3$-type, Gd_2Co_7-type, Ce_5Co_{19}-type and $CaCu_5$-type phases are 43, 32, 10 and 15 wt.%, respectively. After annealing at 950°C for 12 h, most of the above phases in the annealed La_2MgNi_9 alloy disappear and only the $PuNi_3$-type phase exists. Meanwhile, by annealed treatment at 1,025°C for 24 h, the $PuNi_3$-type (53 wt.%), Gd_2Co_7-type (39 wt.%) and Ce_2Ni_7-type (8 wt.%) phases of the as-cast Nd_2MgNi_9 alloy turn to a single-phase $PuNi_3$-type alloy.

Figure 2.25 shows the SEM images of the alloys. It can be found that the annealed La_2MgNi_9 and Nd_2MgNi_9 alloys have single phases, which single areas of area A_1 and A_2 with the stoichiometry of 3.02 and 3.03 (Figure 2.25(b)) and areas A_1 and A_2 with the stoichiometry of 2.99 and 3.01 (Figure 2.25(d)) suggest the AB_3-type phases, coinciding with the inductively coupled plasma (ICP) chemical analysis.

2.3.3 The Electrochemical Characteristics of the Single-Phase Superlattice Nd–Mg–Ni-Based Hydrogen Storage Alloys

2.3.3.1 Activation Property and Maximum Discharge Capacity of the Single-Phase Superlattice Nd–Mg–Ni-Based Hydrogen Storage Alloys

Figure 2.26(a) shows the relationship among maximum discharge capacity, cyclic stability and cycle number of the annealed Nd_2MgNi_9 and La_2MgNi_9 alloy electrodes. It can be seen that both of the La_2MgNi_9 and Nd_2MgNi_9 alloys exhibit excellent activation property, which can be completely activated at the third and fourth cycles,

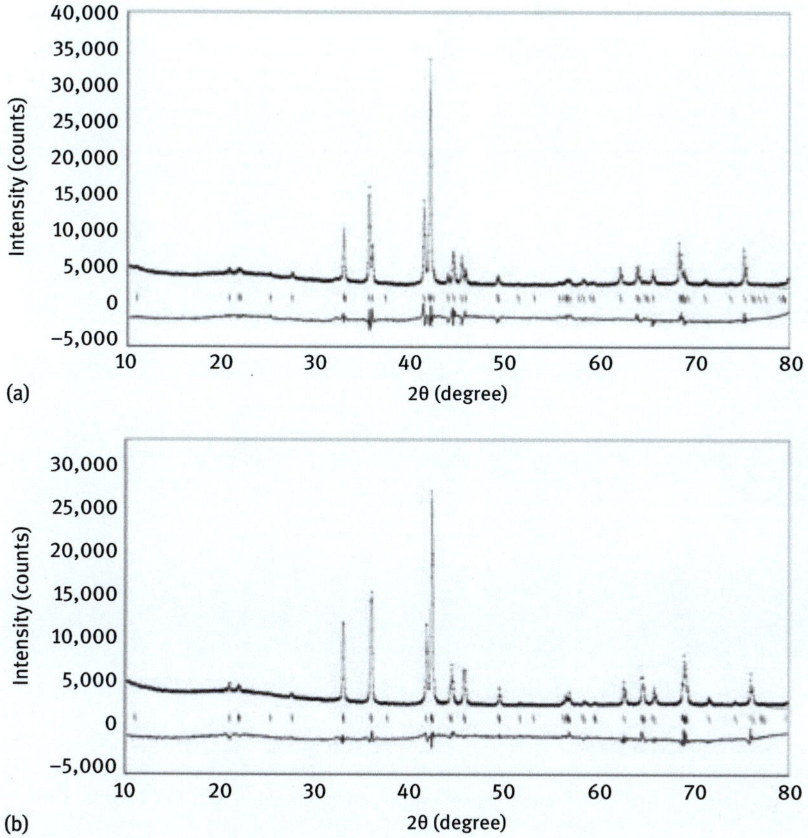

Figure 2.24: Rietveld refinement of XRD patterns of the alloy: (a) La_2MgNi_9; (b) Nd_2MgNi_9.

respectively. After activation, the maximum discharge capacity of the La_2MgNi_9 alloy is about 399.4 mAh g^{-1} at 0.2°C rate of charge/discharge measurement. However, the C_{max} of the Nd_2MgNi_9 alloy is lower than that of the former alloy. As reported by Li et al. [25] and Zhang et al. [39], partial substitution of La by other rare elements was harmful to the discharge capacity of the alloy due to that reduction of the cell volume of the alloy and sites for hydrogen atoms in the alloy bulk. In this study, the cell volume of $PuNi_3$-type structure of the La_2MgNi_9 and Nd_2MgNi_9 alloys are 535.83(3) $Å^3$ and 528.17(4) $Å^3$, respectively, which results in the increasing sites of the La_2MgNi_9 alloy for hydrogen atoms to access into and then possessing higher maximum discharge capacity than that of Nd_2MgNi_9 alloy. Figure 2.26(b) shows the hydrogen absorption curves of the La_2MgNi_9 and Nd_2MgNi_9 alloys, whose maximum hydrogen storage capacities are 1.38 and 1.45 wt.% at different cycles after activation process. During the activation process, the La_2MgNi_9 and Nd_2MgNi_9 alloys can be completely activated after the first hydrogenation/dehydrogenation cycle.

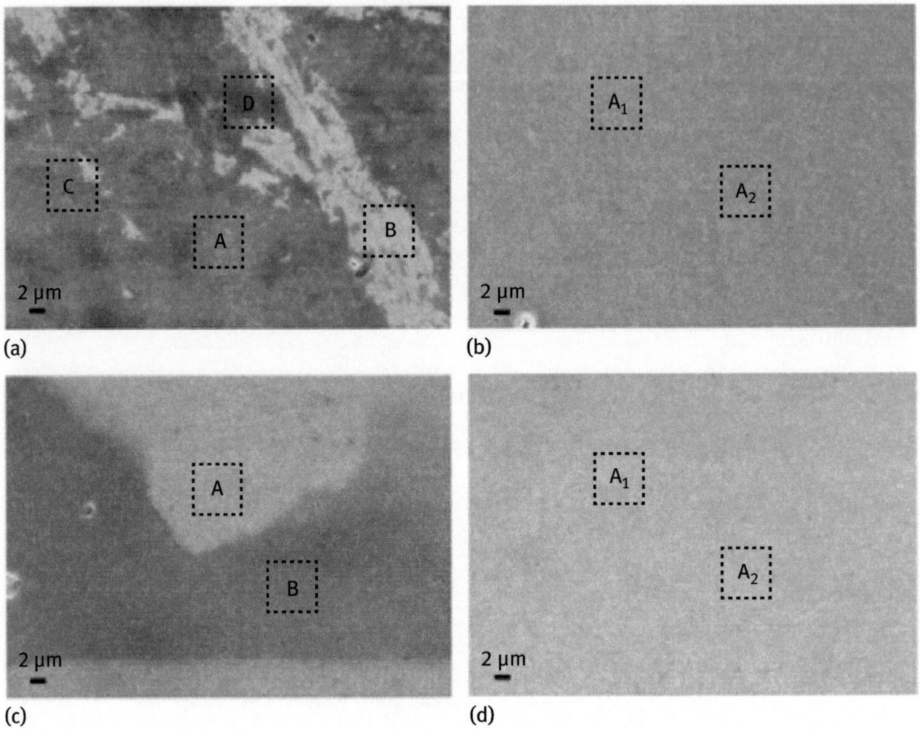

Figure 2.25: SEM images: (a) as-cast $La_{0.67}Mg_{0.33}Ni_{3.0}$ alloy, the La_2MgNi_9 alloy annealed at 950°C for 12 h (b), as-cast $Nd_{0.67}Mg_{0.33}Ni_{2.95}$ alloy (c) and the Nd_2MgNi_9 alloy annealed at 1,025°C for 24 h (d).

2.3.3.2 Kinetic Properties of the Single-Phase Superlattice Nd–Mg–Ni-Based Hydrogen Storage Alloys

The discharge curves at different discharge current densities of the Nd_2MgNi_9 and La_2MgNi_9 alloy electrodes are plotted in Figure 2.26(c). All of the discharge capacities at different discharge current densities of Nd_2MgNi_9 and La_2MgNi_9 alloy electrodes decrease with the increase of the discharge rates (from 0.2 to 4C). Compared with La_2MgNi_9 alloy electrode, the Nd_2MgNi_9 alloy electrode possesses higher discharge voltage plateau and discharge capacity at 0.2, 1, 2, 3 and 4C, respectively. The study by Zhang et al. found that Nd_2MgNi_9 alloy had the higher dehydriding pressures and lower enthalpy changes than those of La_2MgNi_9 alloy [40]; therefore, the hydrogen atoms were relatively easy to break loose from the phase structure. Then, Nd_2MgNi_9 alloy exhibited excellent *HRD* properties.

The *HRD* is a composite index representing the kinetic properties of alloy electrodes and is mainly determined by two factors: the charge transfer on the alloy electrode surface and hydrogen diffusion in alloy electrode body. *HRD* can be defined as the following equation:

$$HRD = C_d/C_{max} \times 100\% \tag{2.10}$$

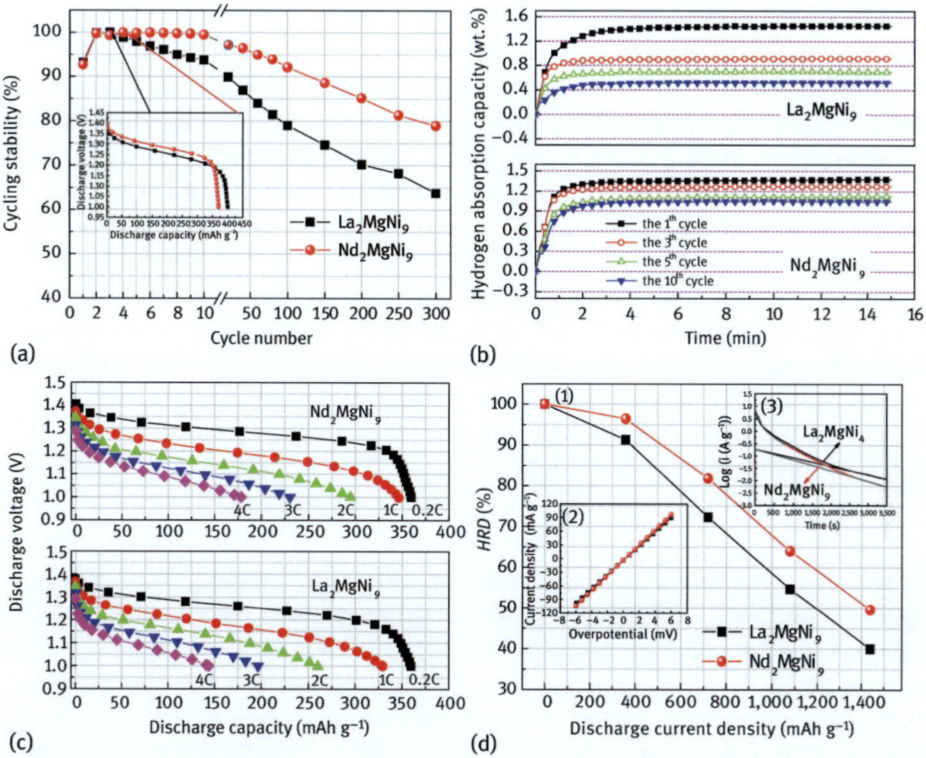

Figure 2.26: (a) Cycling stability versus cycle number of the La_2MgNi_9 and Nd_2MgNi_9 alloys electrodes; (b) hydrogen absorption curves of single-phase $PuNi_3$-type La_2MgNi_9 and Nd_2MgNi_9 alloys; (c) the discharge curves of the La_2MgNi_9 and Nd_2MgNi_9 alloys electrodes at different discharge current densities; (d) the kinetic properties curves of the La_2MgNi_9 and Nd_2MgNi_9 alloys electrodes: (1) the high-rate dischargeability, (2) linear polarization curves at 50 % DOD and 25°C and (3) semi-logarithmic curves of anodic current versus time responses at +500 mV.

where C_d is the discharge capacity of an alloy electrode discharged at a current density I_d and C_{max} is the maximum discharge capacity of the same alloy electrode discharged at 72 mA g^{-1} and I_d is selected to different discharge current density, namely 360, 720, 1,080 and 1,440 mA g^{-1}. Compared with the *HRD* at different discharge current densities of the La_2MgNi_9 alloy electrode, the Nd_2MgNi_9 alloy electrode exhibits higher *HRD*. The HRD_{1440} increases from 40.0 % to 49.6 %.

In order to further elucidate the observed variation in the *HRD* and the implications for the kinetic properties of the alloy electrodes, the linear polarization was evaluated and potential step experiments were conducted.

Figure 2.26(d)-1 shows the linear polarization curves of the alloy samples. The exchange current density I_0 is used to characterize charge transfer rate on the alloy surface. When the over-potential (η) is changed within a small range, the exchange current density I_0 is calculated by the following equation:

$$I_0 = RTI/(F\eta)(\eta \to 0) \tag{2.11}$$

where R (J mol^{-1} K^{-1}), T (K), I (mA), F (C mol^{-1}) and η (mV) are the gas constant, the absolute temperature, the applied current, the Faraday constant and the total over potential, respectively. It is found that the I_0 value of the single-phase $PuNi_3$-type Nd_2MgNi_9 alloy electrode is slightly higher than that of the La_2MgNi_9 alloy electrode, which indicates the charge transfer on the former alloy surface is faster than that of the latter one.

The semi-logarithmic plots of anodic current versus time responses of the alloy electrodes are illustrated in Figure 2.26(d)-2. The hydrogen diffusion coefficient D (cm^2 s^{-1}), used to characterize hydrogen diffusion rate in the alloy body, is estimated through fitting the slope of linear region of potential-step curves given in Figure 2.26(d)-2 via the following equation:

$$\log i = \log(6FD(C_0 - C_s)/(da^2)) - \pi^2 Dt/(2.303a^2) \tag{2.12}$$

where i (A g^{-1}), D (cm^2 s^{-1}), C_0 (mol cm^{-3}), C_s (mol cm^{-3}), a (μm), d (g cm^{-3}) and t (s) are the diffusion current density, the hydrogen diffusion coefficient, the initial hydrogen concentration in alloy body, the hydrogen concentration on the surface of alloy particles, the alloy particle radius, the alloy density and the discharge time, respectively. Compared with the hydrogen diffusion coefficient (D) of the La_2MgNi_9 alloy electrode, the D value of the Nd_2MgNi_9 alloy electrode is higher, increasing from 1.362×10^{-10} cm^2 s^{-1} to 1.711×10^{-10} cm^2 s^{-1}. Combined with the results of the *HRD*, linear polarization and semi-logarithmic plots measurements, the Nd_2MgNi_9 alloy electrode exhibits excellent kinetic properties than those of the La_2MgNi_9 alloy electrode.

2.3.3.3 Cyclic Stability of the Single-Phase Superlattice Nd–Mg–Ni-Based Hydrogen Storage Alloys

As shown in Figure 2.26(a), the cycle life of the Nd_2MgNi_9 alloy electrode is higher than that of La_2MgNi_9 alloy electrode during the charge/discharge progress, of which the Nd_2MgNi_9 alloy electrode can be up 79 % after 300 charge/discharge cycles, 200 cycles more than that of the La_2MgNi_9 alloy electrode achieved the same cycle life. Meanwhile, hydrogen absorption/desorption properties of the La_2MgNi_9 alloy powder are also poorer than those of the Nd_2MgNi_9 alloy powder. The hydrogen capacity the former is reduced from 1.46 wt.% for the initial cycle to 0.52 wt.% for the 10 cycles, and the retention rate is reduced to 36 %. However, the hydrogen capacity of the latter only decreases from 1.38 wt.% (the initial cycle) to 1.04 wt.% (after 10 cycles) and the retention can still be 76 %; the above results show that the Nd_2MgNi_9 alloy possesses better cyclic stabilities.

Figure 2.27 shows the XRD patterns of the $PuNi_3$-type single-phase La_2MgNi_9 and Nd_2MgNi_9 alloys after 1st, 3rd, 5th and 10th absorption/desorption cycles, respectively.

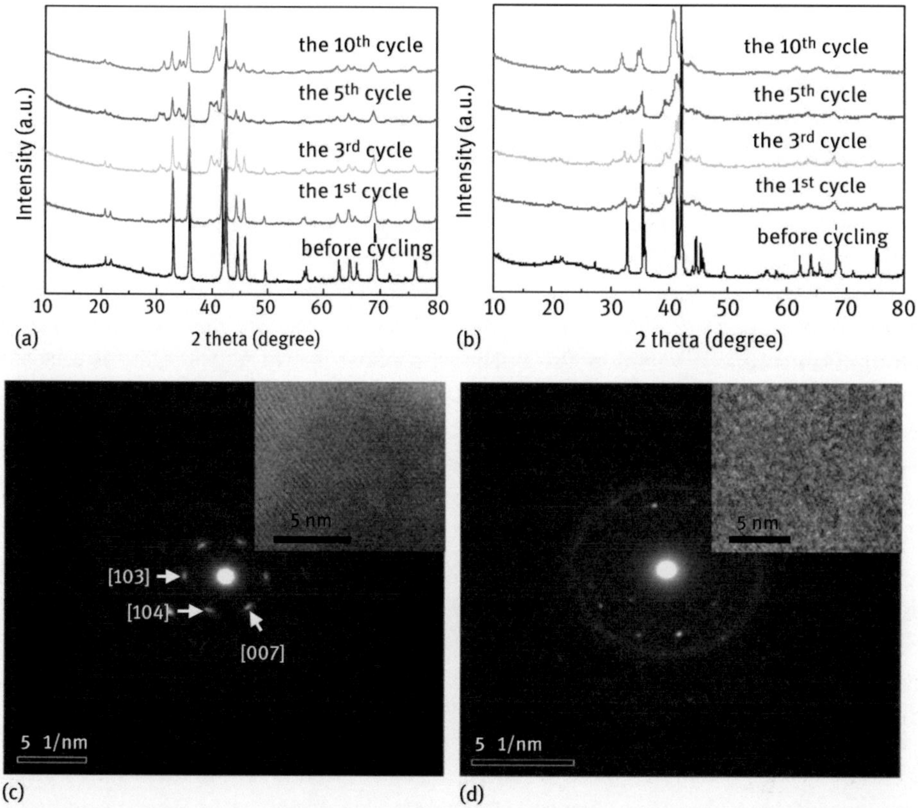

Figure 2.27: XRD patterns of the annealed RE_2MgNi_9 (RE = La and Nd) alloys in the 2θ range of 10–80°: (a) the La_2MgNi_9 alloy before cycling and after the 1st, 3rd, 5th and 10th hydrogenation/dehydrogenation cycles, (b) the Nd_2MgNi_9 alloy after the 1st, 3rd, 5th and 10th hydrogenation/dehydrogenation cycles; the selected area electron diffraction (SAED) pattern of RE_2MgNi_9 alloys after the first hydrogenation/dehydrogenation cycle (inset: high-resolution TEM (HRTEM) image): (c) Nd_2MgNi_9 alloy; (d) La_2MgNi_9 alloy.

After hydrogenation/dehydrogenation cycling, it can be seen that the Nd_2MgNi_9 alloy powders keep better peak patterns and higher intensity of diffraction peaks than those of the La_2MgNi_9 alloy powders.

In order to further comprehend the crystalline state of $PuNi_3$-type phase after the first absorption/desorption cycle, SAED and HRTEM analysis in TEM were also performed on the RE_2MgNi_9 (RE = La and Nd) alloys. Diffraction spots can still be identified as a $PuNi_3$-type structure (Figure 2.27(c)). However, the halo-liked pattern proves the existence of amorphous in the La_2MgNi_9 alloy (Figure 2.27(d)), demonstrating that the $PuNi_3$-type phase of La_2MgNi_9 alloy observably transformed into amorphous phase. Based on the results by XRD and TEM characterization, it is suggested that the single-phase $PuNi_3$-type Nd_2MgNi_9 alloy maintains a better crystalline structure than that of the La_2MgNi_9 alloy during the absorption/desorption cycles.

According to the refinement results of the atom coordinates and cell parameters, the detailed deformations of lattice parameters, the unit-cell volume and the volume of the $[AB_5]$ and $[A_2B_4]$ slabs of the $PuNi_3$-type structures after first absorption/desorption cycle are calculated and shown in Table 2.9. Volume deformation rates $\Delta V/V$ (representing a variable compared to original alloy) of the $[AB_5]$ and $[A_2B_4]$ slabs in the $PuNi_3$-type structure of the La_2MgNi_9 alloy after first hydrogenation/dehydrogenation cycle are +4.10 % and +5.45 %, respectively, and those of the Nd_2MgNi_9 alloy are +1.47 % and +0.90 %, respectively. Besides, the cell volume deformation ratio $\Delta V/V$ (%) of the $PuNi_3$-type phase in Nd_2MgNi_9 alloy is 1.19 %, which is obviously smaller than that of La_2MgNi_9 alloy (4.74 %), indicating that the Nd_2MgNi_9 alloy undergoes a smaller cell volume deformation after hydrogenation/dehydrogenation cycling, and a subsequent lower pulverization of the alloy particles [41]. Meanwhile, Iwase et al. reported that the lattice strain occurred in the hydriding process and is almost saturated at the initial cycle [28]. In this study, it is found that the lattice strain (0.3180(3) %) of Nd_2MgNi_9 alloy is lower than that (0.5241(1) %) of the La_2MgNi_9 alloy after hydrogenation/dehydrogenation, proving that the Nd_2MgNi_9 alloy possesses a higher structural stability than La_2MgNi_9 alloy. Young et al. reported that the c/a ratio was important for the prediction of the pulverization rate of AB_2 and AB_5 alloys and more elongation along the c axis during the hydride from a smaller lattice constant may promote an easier pulverization [42]. The $\Delta c/a$ ratio of the $PuNi_3$-type phase in Nd_2MgNi_9 alloy is smaller than that of the $PuNi_3$-type phase in La_2MgNi_9 alloy after first absorption/desorption cycle. It can be deduced that the decrease of $\Delta c/a$ maybe resulted in better pulverization resistance of the alloy (Table 2.9).

Table 2.9: Comparison of crystallographic parameters of the $PuNi_3$-type phase in La_2MgNi_9 and Nd_2MgNi_9 alloys.

Lattice parameter	La_2MgNi_9		Nd_2MgNi_9	
	Original	The first cycle	Original	The first cycle
a (Å)	5.0402	5.1058	5.0004	5.0069
c (Å)	24.356	24.858	24.391	24.500
Volume (Å^3)	535.83	561.21	528.17	531.91
$V_{[A2B4]}$ (Å^3)	93.79	97.63	90.42	91.75
$V_{[AB5]}$ (Å^3)	84.81	89.43	85.65	86.42
$\Delta c/a$(%)	–	0.751	–	0.317
$\Delta V/V$ (%)	–	+4.74	–	+1.19
$\Delta V_{[A2B4]}/V_{[A2B4]}$ (%)	–	+4.10	–	+1.47
$\Delta V_{[AB5]}/V_{[AB5]}$ (%)	–	+5.45	–	+0.90
Strain (%)	–	0.5241	–	0.3180

Figure 2.28 shows SEM micrographs of the La_2MgNi_9 and Nd_2MgNi_9 alloys before and after gas solid hydrogenation/dehydrogenation cycling. SEM observations indicate the pulverization progress of the alloy particles and the passivation process of the electrode surface [22]. It is found that the alloy particles are pulverized and, subsequently, the particle size becomes smaller with increasing number of cycles due to the cell volume expansion and contraction during hydrogenation/dehydrogenation progress [41]. After the 10th cycle, a large number of cracks are produced in the La_2MgNi_9 alloy particle surface, which is more serious than that of Nd_2MgNi_9 alloy particle, indicating that the La_2MgNi_9 alloy undergoes more serious pulverization.

As is known to all, pulverization of the alloy particles and oxidization/corrosion of the active components are commonly the key factors causing the capacity decays. In order to further study the different effects of the factors on the cyclic stabilities of the alloy electrodes, we conducted XRD, particle size distribution and SEM analyses after the charge/discharge cycling. Besides, the corrosion behaviours of alloy electrodes after the charge/discharge cycling were investigated by potentiodynamic polarization method. Due to the hydrogen atoms entering and leaving the interstitial sites of the lattice resulting in the expansion/contraction of the phase cells during the charge/discharge progress, the rising lattice stresses are inevitable, and it is the real driving force that leads to the pulverization of the La–Mg–Ni-based alloys [43]. The particle sizes and the strains (analysed by Jade 6.0 software) of the $PuNi_3$-type phase alloys after charge/discharge cycling are listed in Table 2.10.

It can be noticed that the particle sizes of the Nd_2MgNi_9 alloy are bigger than those of the La_2MgNi_9 alloy after the 15th, 30th, 60th and 100th charge/discharge cycles (Figure 2.29(a–b)), respectively, and, meanwhile, the strains of the former are smaller than those of the latter, which indicate that the Nd_2MgNi_9 alloy shows weaker pulverization degree than that of the La_2MgNi_9 alloy. In addition, the corrosion potential and current density values of the Nd_2MgNi_9 and La_2MgNi_9 alloy electrodes after activation, 15th, 30th, 50th and 100th charge/discharge cycling are obtained and shown in Table 2.11. The Tafel curves of the Nd_2MgNi_9 and La_2MgNi_9 alloy electrodes after different charge/discharge cycles are presented in Figure 2.29(c–d). The corrosion potential is a thermodynamic parameter which only characterizes the corrosion reaction tendency, and it is not necessarily related to the corrosion rate. The corrosion behaviour of alloy electrodes is mainly determined by the current density. The smaller value of corrosion current density represents the higher corrosion resistance of alloy electrode. With increasing cyclic numbers, it can be seen that current density values of the Nd_2MgNi_9 alloy electrode first increase from 96.2 mA cm^{-2} to a maximum (142.7 mA cm^{-2}) and then decrease to 127.4 mA cm^{-2} and that of La_2MgNi_9 alloy electrode increase from 128.3 mA cm^{-2} to a maximum (147.6 mA cm^{-2}) and then decrease to 111.5 mA cm^{-2}. The maximum current density values of the $PuNi_3$-type single-phase Nd_2MgNi_9 (142.7 mA cm^{-2}) and La_2MgNi_9 (147.6 mA cm^{-2}) alloys appear at the 50th and 30th cycles, respectively, and the maximum current density values of the former are smaller than that of the latter. The result indicates that the appearances of cracks on

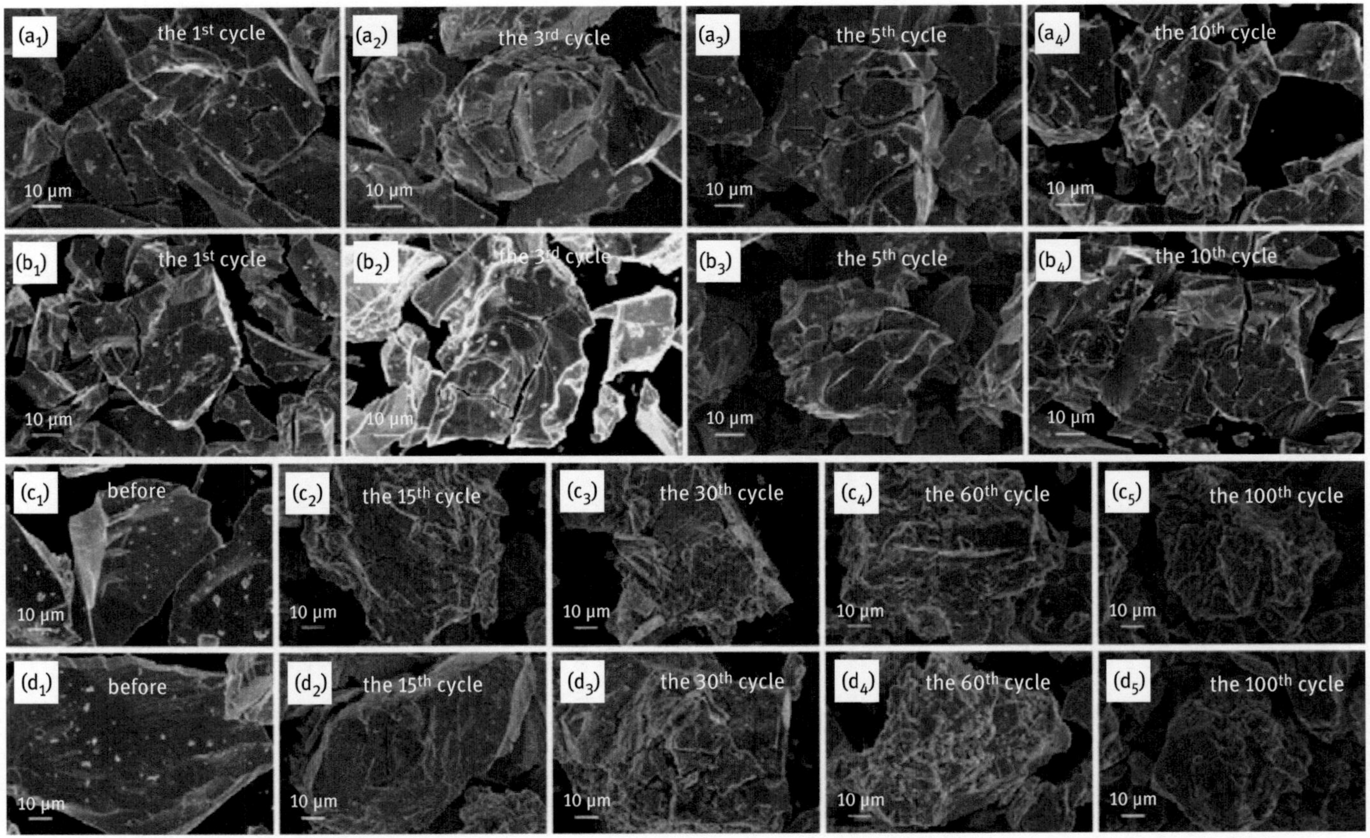

Figure 2.28: SEM images of the RE_2MgNi_9 alloys: after the 1st, 3rd, 5th and 10th hydrogenation/dehydrogenation cycles of La_2MgNi_9 alloy powder ($a_{1–4}$) and Nd_2MgNi_9 alloy powder ($b_{1–4}$); before activation and after the 15th, 30th, 60th and 100th cycles during the charge/discharge progress of La_2MgNi_9 alloy electrode ($c_{1–5}$) and Nd_2MgNi_9 alloy electrode ($d_{1–5}$).

Table 2.10: Particle sizes and strains of the La_2MgNi_9 and Nd_2MgNi_9 alloys for 30th, 60th and 100th cycles.

Alloy	Particle size (μm)			Strain (%)		
	30th cycle	60th cycle	100th cycle	30th cycle	60th cycle	100th cycle
La_2MgNi_9	26.14	25.63	24.81	0.3980	0.4046	0.5404
Nd_2MgNi_9	28.06	27.95	27.88	0.3297	0.3352	0.3633

Table 2.11: Corrosion potential and current density of the La_2MgNi_9 and Nd_2MgNi_9 alloys electrodes.

Cycle	La_2MgNi_9		Nd_2MgNi_9	
	Potential (V)	Current density (mA cm^{-2})	Potential (V)	Current density (mA cm^{-2})
5th cycle	−0.913	128.3	−0.921	96.2
10th cycle	−0.913	134.2	−0.929	110.1
15th cycle	−0.926	139.1	−0.956	123.9
30th cycle	−0.939	147.6	−0.953	127.9
50th cycle	−0.931	124.1	−0.925	142.7
100th cycle	−0.920	111.5	−0.922	127.4

the alloy surface and the exposed fresh surface areas of the La_2MgNi_9 alloy are earlier than those of the Nd_2MgNi_9 alloy and the corrosion resistance of the Nd_2MgNi_9 alloy electrode is higher than that of the La_2MgNi_9 alloy electrode.

In addition, the XRD patterns of the La_2MgNi_9 and Nd_2MgNi_9 alloys electrodes before cycling and after the 15th, 30th, 60th and 100th charge/discharge cycles are presented in Figure 2.29(e–f). Figure 2.29(e) shows that the diffraction peak intensities of the La_2MgNi_9 alloy are significantly reduced with increasing the number of cycling, and the proportions of La_2O_3, $La(OH)_3$ and $Mg(OH)_2$ phases are correspondingly increased. Conversely, Figure 2.29(f) shows that the Nd_2MgNi_9 alloy retains the intensities of the main diffraction peaks, and that the diffraction peak intensities representative of oxide and hydroxides of Nd and Mg are slightly increased. Moreover, compared with the La_2MgNi_9 alloy electrode, the intensities of oxide and hydroxides peaks of the pure $PuNi_3$-type Nd_2MgNi_9 alloy are weaker. The Nd_2MgNi_9 alloy possessing better anti-corrosion ability is due to its natural atomic property. In the view of their electronic shell structure, the electrons in 4f shell of Nd are constrained, while the electrons in 5d and 6s of La are more active, therefore La relatively tends to be oxidized to La^{3+} in the KOH electrolyte, increasing the degree of oxidation compared to Nd element [40].

Figure 2.28(c_1–c_5) and (d_1–d_5) present the SEM images of the electrode materials of the La_2MgNi_9 and Nd_2MgNi_9 alloys, respectively, before activation and after 30th, 60th and 100th charge/discharge cycles. It is observed from the above-discussed

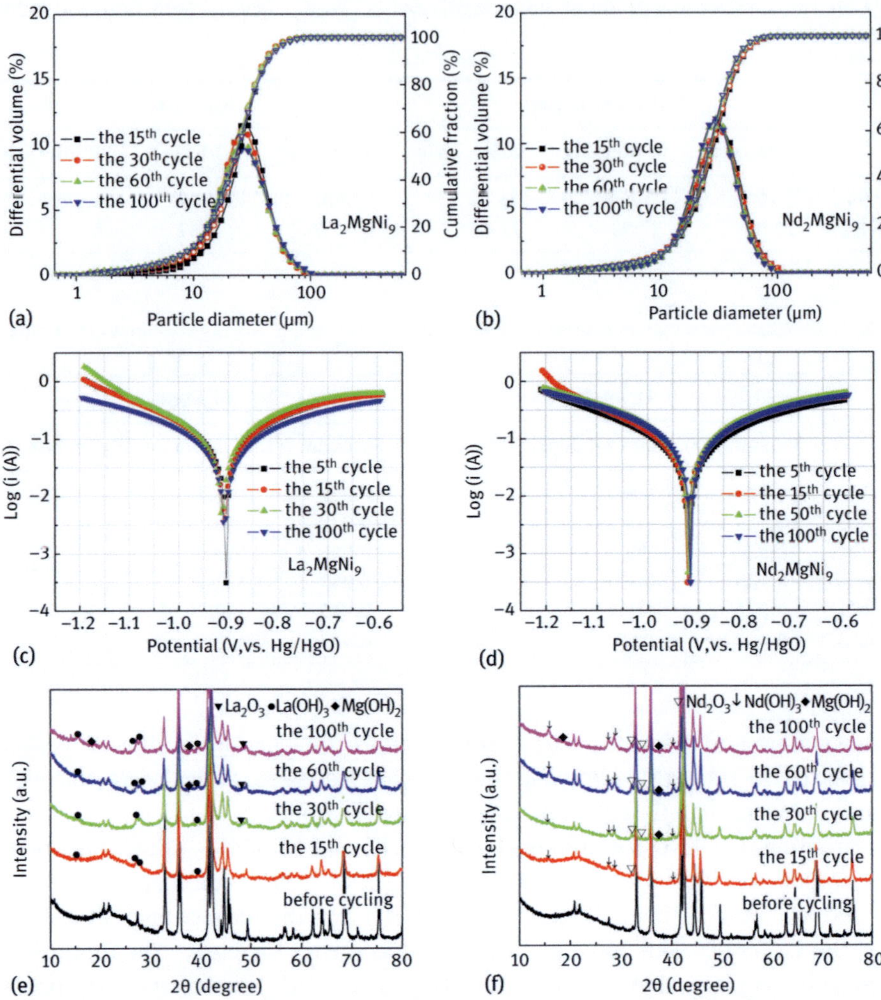

Figure 2.29: Particle size distribution after 15th, 30th, 60th and 100th charge/discharge cycles: La_2MgNi_9 alloy electrode (a) and Nd_2MgNi_9 alloy electrode (b); the Tafel curves of the alloys electrodes after the 5th, 15th, 50th and 100th cycles: La_2MgNi_9 alloy electrode (c) and Nd_2MgNi_9 alloy electrode (d); XRD patterns of the alloys before cycling and after the 15th, 30th, 60th and 100th cycles: La_2MgNi_9 alloy electrode (e) and Nd_2MgNi_9 alloy electrode (f).

micrographs that the $PuNi_3$-type Nd_2MgNi_9 alloy electrode undergoes reduced pulverization and oxidation relative to the same structural La_2MgNi_9 alloy electrode during the charge/discharge progress. A better structural stability of the Nd_2MgNi_9 alloy reduces the number of exposed surfaces of the alloy to be eroded, and, consequently, reduces the loss of discharge capacity.

It is generally accepted that the main factors influencing the cyclic properties of RE–Mg–Ni alloy are the corrosion and pulverization. Based on the above potentiodynamic polarization, XRD, particle size distribution and SEM analyses, we conclude that the pure $PuNi_3$-type Nd_2MgNi_9 alloy electrode has better cyclic properties than La_2MgNi_9 alloy electrode. This is attributed to the enhanced structural stability $PuNi_3$-type phases of the alloy, which is conducive to prevention of oxidation and pulverization with the corresponding benefits for cyclic stability.

2.4 The Preparation and Characteristics in Electrochemical and Gaseous Hydrogen Storage of the Gd_2Co_7-Type Single-Phase Superlattice Sm–Mg–Ni-Based Hydrogen Storage Alloys

Recently, new-type Sm–Mg–Ni alloys are found more interesting in comparison with other superlattice RE–Mg–Ni alloys and show excellent cycling stability after hydrogenation/dehydrogenation. Chen et al. found that the hydrogen-induced amorphization and phase decomposition phenomenon did not occur for AB_2-type $SmMgNi_4$ alloy, which was different from the other AB_2-type RE–Mg–Ni alloys, and 84 % capacity remained after 50 hydrogenation/dehydrogenation cycles at 298 K [44]. Zhang et al. investigated the structural parameters and hydrogen absorption/desorption properties of Sm_2MgNi_9, Sm_3MgNi_{14} and Sm_4MgNi_{19} (as well as $SmMgNi_4$ and $SmNi_5$ as references) alloys, and reported that the superlattice Sm-Mg-Ni alloys not only possessed good hydrogen storage capacity, which were 1.03, 1.04 and 0.93 H/M under 10 MPa at 298 K, but showed superior cycling stability, whose capacity retention rate of the Sm_4MgNi_{19} was ca. 97 % without occurrence of the amorphization and phase decomposition [35]. Hence, the Sm–Mg–Ni alloys are of great research value as a novel intermetallic hydrogen storage material, and understanding of their detailed hydrogen absorption/desorption process is basic and important for further studies on the Sm–Mg–Ni alloys.

A_2B_7-type superlattice alloys exhibit good hydrogen storage properties, and some single-phase La–Mg–Ni alloys with Ce_2Ni_7-type superlattice structure have been prepared. The Gd_2Co_7-type La–Mg–Ni alloys have never been prepared. This may be because its structure is unstable in comparison with the Ce_2Ni_7-type structure and can easily transform to other phases. In this chapter, we prepared a Gd_2Co_7-type Sm–Mg–Ni alloy by powder sintering method and studied the hydrogen absorption and desorption properties. We also reveal the special hydrogen absorption/desorption process and its effect on hydrogen storage properties of the Gd_2Co_7-type Sm–Mg–Ni alloy.

2.4.1 The Preparation of the Gd_2Co_7-Type Single-Phase Superlattice Sm–Mg–Ni-Based Hydrogen Storage Alloys

A Sm–Mg–Ni alloy was prepared by stepwise sintering technique using precursors: SmNi, $MgNi_2$ and Ni powders (99.9 %). Starting precursors of SmNi and $MgNi_2$ were prepared by induction melting method in argon atmosphere. Each precursor was remelted three times to ensure homogeneity, and 5 wt.% of Sm and 20 wt.% of Mg were excessively added to compensate for the losses in the process. All of the precursors were ball milled for 1 h and sieved to below 300 mesh sizes, and then the powder mixtures were cold pressed to a tablet (3 g) under 10 MPa, which subsequently was wrapped in a nickel foil and sintered via a step heating process. To compensate for Mg volatility losses during the sintering process, $MgNi_2$ was excessively added. Heat treatment involved stepwise rising temperature from room temperature to 873, 973 and 1,073 K under 0.04 MPa argon atmospheres in the quartz tube (each temperature kept for 1 h) and then to 1,223 K with a sintering time of 96 h. Afterwards, the pellet was cooled to room temperature with the furnace. The as-sintered sample was finally polished to remove the oxide layer and mechanically crushed to powders in a glove box under argon atmosphere, where the particle size for structural analysis is below 400 meshes, and for property testing is between 200 and 400 meshes.

The chemical composition of the sintered sample was finally examined by ICP, which indicated a chemical composition of $Sm_{1.6}Mg_{0.4}Ni_7$.

2.4.2 The Crystal Structure of the Gd_2Co_7-Type Single-Phase Superlattice Sm–Mg–Ni-Based Hydrogen Storage Alloys

Figure 2.30(a) shows the XRD profile of $Sm_{1.6}Mg_{0.4}Ni_7$ alloy, whose Gd_2Co_7-type structure is determined from its characteristic peaks at 32.24° and 34.22°. In regards to the same reflection peak at 33.24° of Ce_2Ni_7-type and $PuNi_3$-type structures, XRD profile of the $Sm_{1.6}Mg_{0.4}Ni_7$ sample in 2θ region between 3° and 15° was performed to identify the impure phase structure. Except for the peaks of the Gd_2Co_7-type structure of (003 and 006), the reflections of the Ce_2Ni_7-type structure (002 and 004) are observed, and reflection of the $PuNi_3$-type structure (003) does not appear. Therefore, the characteristic peak at 33.24° is part of the Ce_2Ni_7-type structure. And the alloy is composed of Gd_2Co_7- and Ce_2Ni_7-type structures.

Figure 2.30(b) shows the Rietveld refinement of the XRD profile of $Sm_{1.6}Mg_{0.4}Ni_7$ alloy. The result shows that the content of Gd_2Co_7-type phase is 96.0 wt.% and that of the Ce_2Ni_7-type phase is 4.0 wt.%.

Figure 2.31 shows the high-resolution TEM (HRTEM) with the corresponding SAED pattern after fast Fourier transformation (FFT) and the inverse FFT image of the $Sm_{1.6}Mg_{0.4}Ni_7$ alloy particle. The [0003] lattice plane of the Gd_2Co_7-type structure can be determined from Figure 2.31(a) and the stacking mode of the $[A_2B_4]$ and $[AB_5]$ subunit in the ratio of 1:2 is clear in Figure 2.31(b). It should be noted that we do not capture the Ce_2Ni_7-type structure in the TEM analysis, which may be due to the phase abundance of the Ce_2Ni_7-type phase in the alloy is so few in comparison with that of the Gd_2Co_7-type structure. And because of the same reason, the influence of the Ce_2Ni_7-type phase on the hydrogen storage properties of $Sm_{1.6}Mg_{0.4}Ni_7$ alloy is deemed as that of the Gd_2Co_7-type phase hereafter in regards to the similar structure and the hydrogen storage properties of the allotrope.

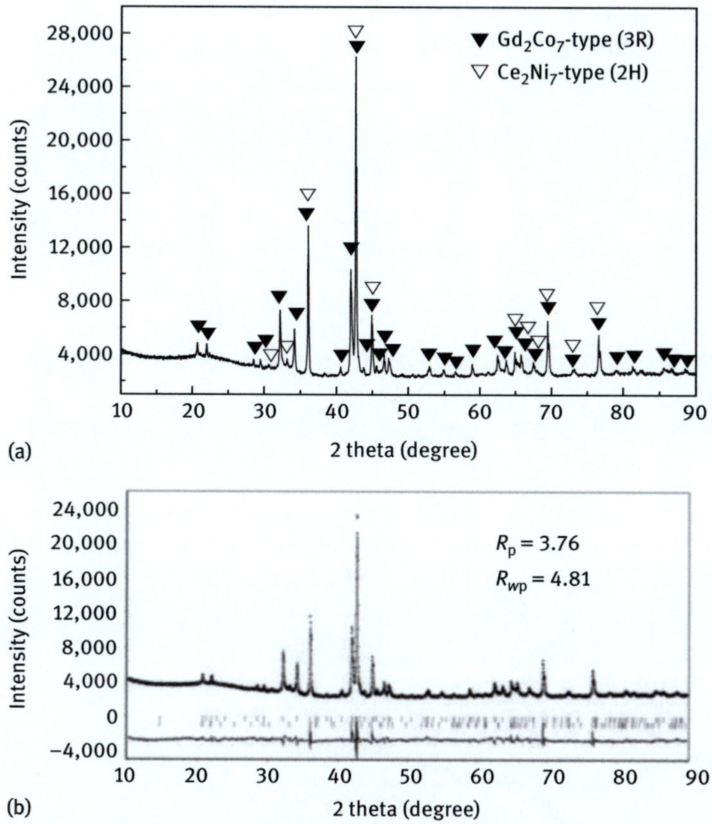

Figure 2.30: (a) X-ray diffraction (XRD) profile of the $Sm_{1.6}Mg_{0.4}Ni_7$ compound in 2θ range of 10–90° and a low angle range of 3–17°. (b) Rietveld refinement of the $Sm_{1.6}Mg_{0.4}Ni_7$ compound. Vertical bars below the pattern show the positions of all possible reflection peaks of the Gd_2Co_7-type (3R) and Ce_2Ni_7-type (2H) phases.

2.4.3 The Gaseous Hydrogen Storage Properties of the Gd_2Co_7-Type Single-Phase Superlattice Sm–Mg–Ni-Based Hydrogen Storage Alloys

2.4.3.1 Hydrogen Storage Capacity of the Gd_2Co_7-Type Single-Phase Superlattice Pr–Mg–Ni-Based Hydrogen Storage Alloys

Figure 2.32 shows the activation curve of the Gd_2Co_7-type $Sm_{1.6}Mg_{0.4}Ni_7$ alloy under 10 MPa H_2 at 298 K. It takes ca. 1.5 h to reach a maximum hydrogen content of 1.57 wt.% for the activation cycle. After activation process, the sample can store the maximum hydrogen content of 1.88 wt.% (1.37 H/M) under the same condition, and it only takes 15 min to store a hydrogen content of 1.58 wt.% (1.15 H/M). It can be concluded that the $Sm_{1.6}Mg_{0.4}Ni_7$ alloy is easy to be activated and stands out in hydrogen storage capacity.

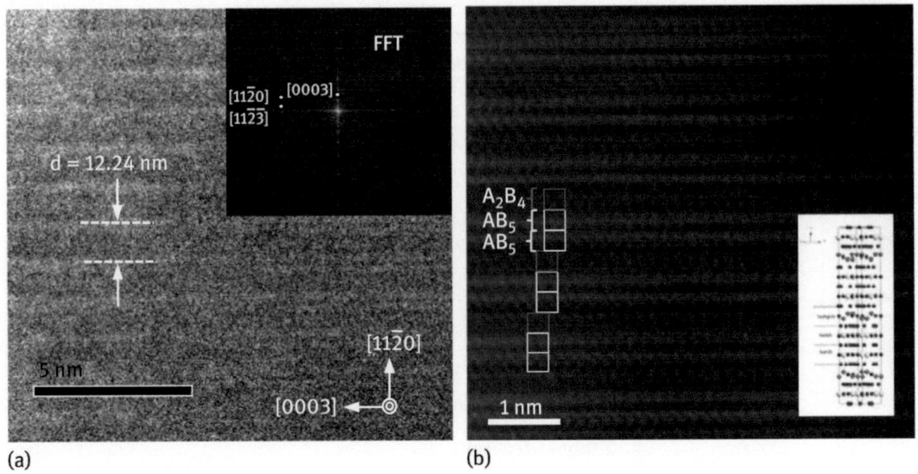

Figure 2.31: High-resolution TEM (HRTEM) image with the corresponding SAED pattern after fast Fourier transformation (FFT) (a) and inverse FFT image (b) for the crystal grain of the Gd_2Co_7-type phase in the $Sm_{1.6}Mg_{0.4}Ni_7$ sample.

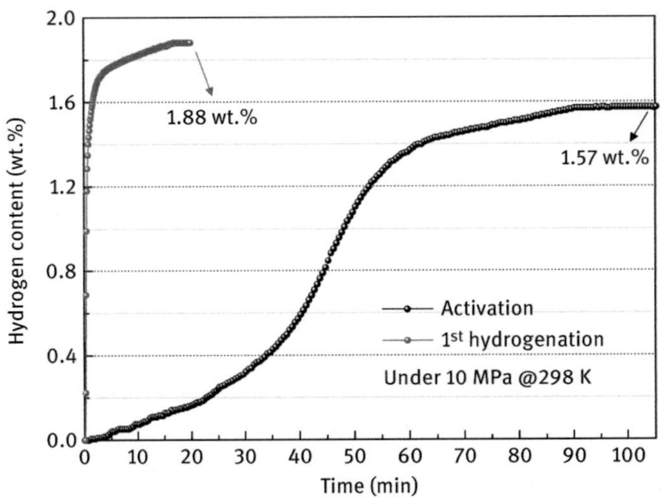

Figure 2.32: Activation curve and hydrogen absorption curve for the first cycle of the $Sm_{1.6}Mg_{0.4}Ni_7$ sample under 10 MPa H_2 at 298 K.

2.4.3.2 Hydrogen Storage Kinetics of the Gd_2Co_7-Type Single-Phase Superlattice Pr–Mg–Ni-Based Hydrogen Storage Alloys

Figure 2.33 shows the hydrogen absorption and desorption curves of the $Sm_{1.6}Mg_{0.4}Ni_7$ sample under 3 MPa H_2 at 298 K for different cycles after activation. The maximum hydrogen content of the alloy is 1.45 wt.% (1.05 H/M) at ca. 35 min, and it takes 9.4 min

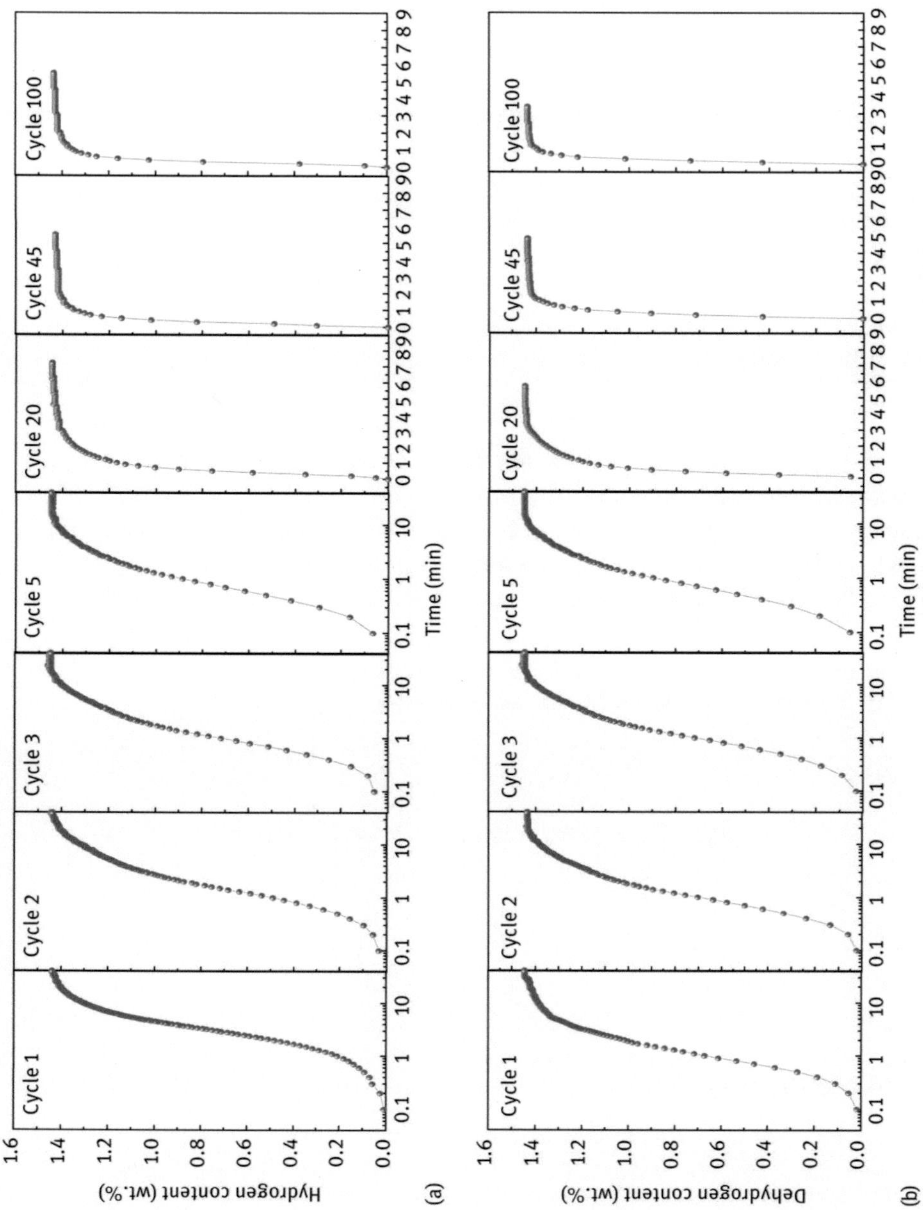

Figure 2.33: Hydrogen absorption (a) and (b) desorption curves of the activated $Sm_{1.6}Mg_{0.4}Ni_7$ alloy under 3 MPa H_2 at 298 K for the 1st, 2nd, 3rd, 5th, 20th, 45th and 100th cycles.

to absorb 90 % of the maximum hydrogen content for the first hydrogen absorption cycle. During hydrogen desorption, the alloy releases 1.44 wt.% (1.04 H/M) at ca. 30 min and takes ca. 5 min to desorb 90 % of the maximum hydrogen content for the first cycle. Moreover, the hydrogen absorption and desorption speeds accelerate with cycling, where the slopes of the curves appear more inclined. It only needs 3.4 min to achieve a maximum hydrogen content of 1.44 wt.% (1.05 H/M) at the 20th cycle and keeps this hydrogen absorption rate until to the 100th cycle for the hydrogenation, and only needs ca. 1.5 min to fully release the hydrogen at the 20th cycle and speeds to 0.7 min until to the 100th cycle for the dehydrogenation.

2.4.3.3 Cyclic Property of the Gd_2Co_7-Type Single-Phase Superlattice Pr–Mg–Ni-Based Hydrogen Storage Alloys

The cycling stability of the activated $Sm_{1.6}Mg_{0.4}Ni_7$ alloy is presented in Figure 2.34. The hydrogen retention rate of the alloy keeps 99.3 % after 100 hydrogenation/dehydrogenation cycles under 3 MPa H_2 at 298 K. The cycling stability of the $Sm_{1.6}Mg_{0.4}Ni_7$ alloy is superior in comparison with the previously reported ternary superlattice alloys of Sm_3MgNi_{14} [35], Pr_3MgNi_{14} [27] and La_4MgNi_{19} [45], which are 95, 98 and 89 % after 30 cycles, respectively.

2.4.3.4 The Hydrogen Absorption–Desorption Thermodynamics of the Gd_2Co_7-Type Single-Phase Superlattice Sm–Mg–Ni-Based Hydrogen Storage Alloys

To study the hydrogen absorption and desorption thermodynamics, the *P–C* isotherms of the activated $Sm_{1.6}Mg_{0.4}Ni_7$ alloy have been measured at temperatures of 298, 323 and 348 K, as shown in Figure 2.35(a), and van't Hoff plots based on the *P–C* isotherms are shown in Figure 2.35(b).

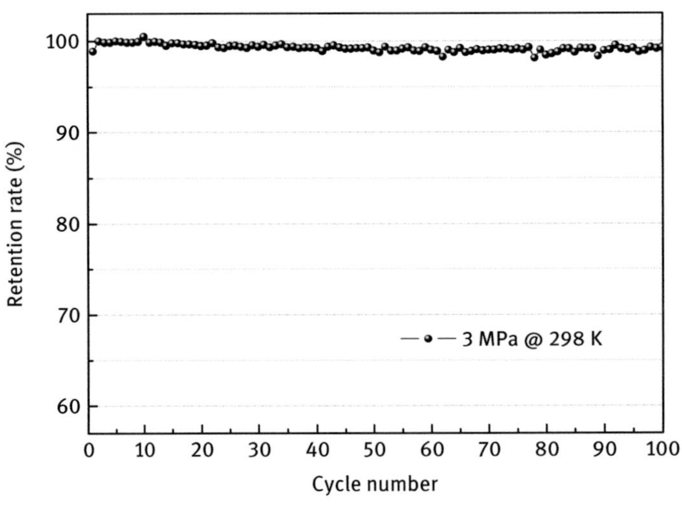

Figure 2.34: Cycling stability of the $Sm_{1.6}Mg_{0.4}Ni_7$ alloy under 3 MPa H_2 at 298 K.

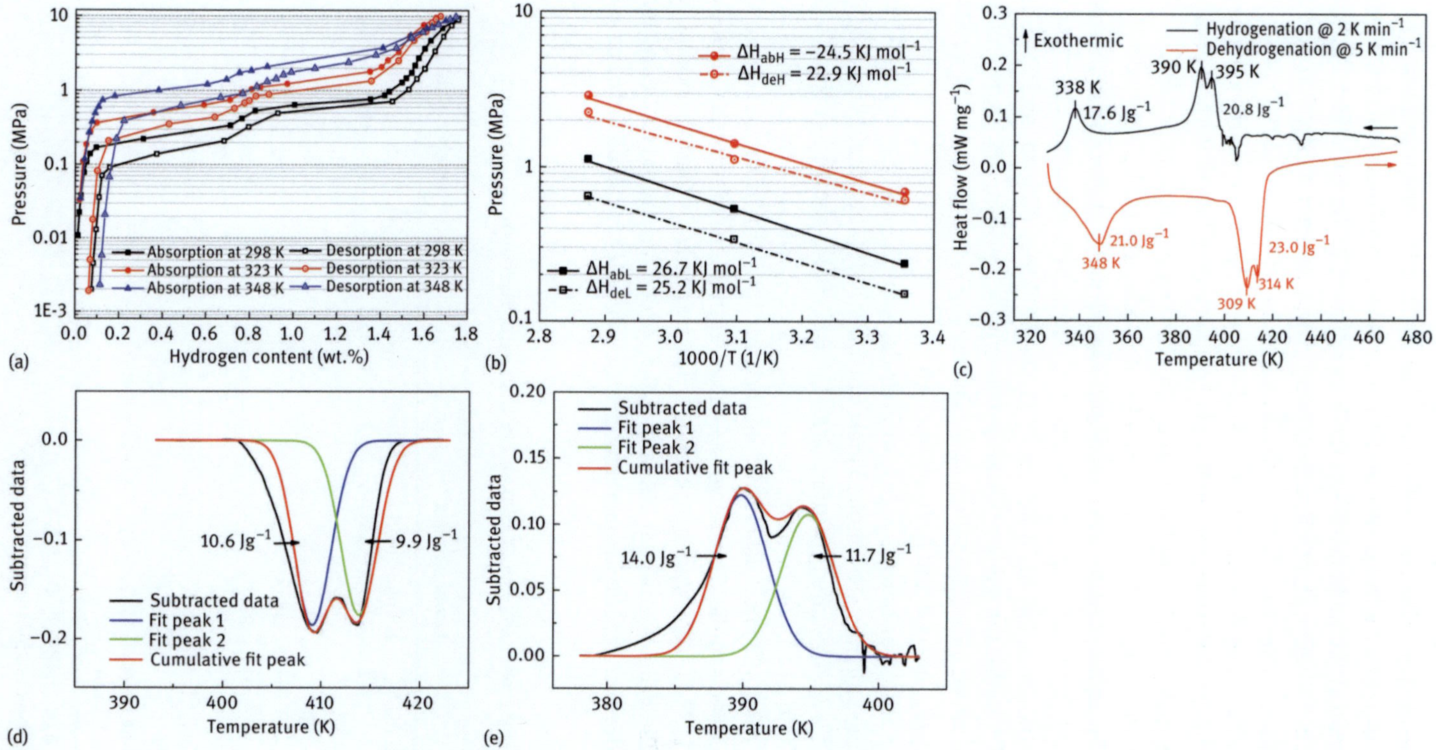

Figure 2.35: $Sm_{1.6}Mg_{0.4}Ni_7$–H_2 system: *P*–*C* isotherms for hydrogen absorption and desorption (a), van't Hoff plots log P_{H2} versus 1,000/T (b). DSC curves of the $Sm_{1.6}Mg_{0.4}Ni_7$ alloy for the hydrogenation and dehydrogenation under 3 MPa hydrogen pressure (c). Fitting results of the overlapping peaks for dehydrogenation (d) and hydrogenation (e).

Two plateaus are visible on each *P–C* curve at each temperature for hydrogen absorption and desorption. The two plateaus can be divided into the regions at a lower concentration H_2 and a higher concentration H_2, respectively. The hydrogen absorption and desorption plateau pressures of the lower plateau at the lower concentration H_2 are 0.23 and 0.15 MPa at 298 K, respectively. The hydrogen absorption and desorption plateau pressures of the higher plateau at the higher concentration H_2 are 0.68 and 0.61 MPa at 298 K, respectively. Based on the midpoint pressures at different temperatures, the enthalpy change (ΔH) and entropy change (ΔS) of the two plateaus are respectively calculated according to the van't Hoff equation (Figure 2.35(b)). The hydrogenation and dehydrogenation enthalpies are determined to be −26.7 and 25.2 kJ mol^{-1} H_2 for the lower plateau, and −24.5 and 22.9 kJ mol^{-1} H_2 for the higher plateau, respectively (Table 2.12), and hydrogenation and dehydrogenation entropies are determined to be −96.6 and 87.7 J K^{-1} mol^{-1} H_2 for the lower plateau, and −97.9 and 88.8 J K^{-1} mol^{-1} H_2 for the higher plateau, respectively. Besides, the higher plateau shows superior reversibility than the lower plateau, where the hysteresis H_f at 298 K, defined as ln P_a/P_d, are 0.11 and 0.45, respectively.

To further verify the changes of ΔH, differential scanning calorimetry (DSC) was performed on the $Sm_{1.6}Mg_{0.4}Ni_7$ alloy to measure the ΔH of hydrogenation and dehydrogenation. ΔH was calculated by integrating the DSC peaks shown in Figure 2.35(c) and listed in Table 2.12. Upon heating, two endothermic events with the onset temperatures near 333 and 403 K are observed, and they are due to the hydrogen desorption of the hydrogenated sample. The dehydrogenation enthalpies ΔH are calculated to be 21.0 and 23.0 J g^{-1} H_2 (27.8 and 30.4 kJ mol^{-1} H_2), respectively. During cooling, two exothermic signals are observed at 400 and 340 K, which are related to the hydrogen absorption of the sample. The hydrogenation enthalpies ΔH are calculated to be −20.8

Table 2.12: Thermodynamic and kinetic data for the $Sm_{1.6}Mg_{0.4}Ni_7$–H system

	Lower plateau	Higher plateau
Transformation of H_2	H-solid solution (α-phase) to hydride (β-phase)	H_2 to hydride (β-phase)
Hydride	$Sm_{1.6}Mg_{0.4}Ni_7H_{4.8}$	$Sm_{1.6}Mg_{0.4}Ni_7H_{9.0}$
P_{Amin} @ 298 K (MPa)	0.23	0.68
P_{Dmin} @ 298 K (MPa)	0.15	0.61
H_f @ 298 K	0.45	0.11
ΔH_A (kJ mol^{-1})[a]	−26.7	−24.5
ΔH_D (kJ mol^{-1})[a]	25.2	22.9
ΔH_A (kJ mol^{-1})[b]	−27.5	−23.3
ΔH_D (kJ mol^{-1})[b]	30.4	27.8
Hydrogenation time reach 90 % H_{max} @ 298 K (min)	11	1.5

[a]Calculated based on the *P–C* isotherms.
[b]Calculated based on the DSC analysis.

and −17.6 J g^{-1} H_2 (27.5 and 23.3 kJ mol^{-1} H_2), respectively. As shown in Figure 2.35(a), the hydrogen desorption pressure of the higher plateau is ca. four times than that of the lower plateau. During the heating process, the dehydrogenation pressure of the sample and the hydrogen pressure of the sample room both increase with the increasing temperature, but the former one increases faster than the latter one. When the atmosphere hydrogen pressure increases to the value below to the higher plateau dehydrogenation pressure of the sample, the hydrogen desorption process begins for the higher plateau (near 333 K). But the hydrogen desorption for the lower plateau appears much later with the rising temperature because of the significant lower dehydrogenation pressure for the lower plateau, which needs more time as the atmosphere hydrogen pressure is further below to the dehydrogenation pressure of the lower plateau. Therefore, the dehydrogenation process of the sample exhibits two endothermic behaviours. The two endothermic peaks belong to the higher and lower plateaus during heating process, respectively. The hydrogen absorption process during cooling process exhibits two exothermic peaks with the same reason and the two peaks relate to the lower and higher plateaus, respectively. Based on the analysis of the DSC curves, it can be seen that the calculated hydrogenation enthalpies for the lower and higher plateaus are close to PCT measurements, but the dehydrogenation enthalpies are a little higher than the values from PCT measurements, which may be relevant to the heating rate. Besides, it can be noticed that the endothermic and exothermic peaks for the lower plateau show overlapped peaks with two peak values (magnified and fitted as Figure 2.35(d–e)), differing from the peaks for the lower plateau. This is closely related to the hydrogenation and dehydrogenation process, which will be discussed later.

2.4.4 The Hydrogen Storage Capacity Degradation of the Gd_2Co_7-Type Single-Phase Superlattice Sm–Mg–Ni-Based Hydrogen Storage Alloys

To clarify the degradation mechanism and the good cycling stability of the $Sm_{1.6}Mg_{0.4}Ni_7$ alloy under present circumstances, the structures of the fully hydrogenated and dehydrogenated samples after 1, 2, 3, 15 and 100 cycles are examined. The Rietveld refinements of the XRD profiles of the samples are shown in Figure 2.36(a–j).

It can be noticed that no new peak is observed for the fully hydrogenated samples, but all peaks of the hydrogenated samples move left and have broadened (Figure 2.36(a, c, e, g, i)), indicating that no new substance forms and the cell volume of the structure becomes larger due to hydrogenation. And after dehydrogenation, the diffraction peaks of the samples keep almost unchanged without appearance of new peak (Figure 2.36(b, d, f, h, j)). Besides, the SAED pattern of the sample after 100 hydrogenation/dehydrogenation cycles shows that the sample still has a complete lattice structure without the appearance of amorphous rings (Figure 2.37). The above results indicate that neither disproportionation nor amorphization occurs after hydrogenation/dehydrogenation. Generally, hydrogen storage capacity

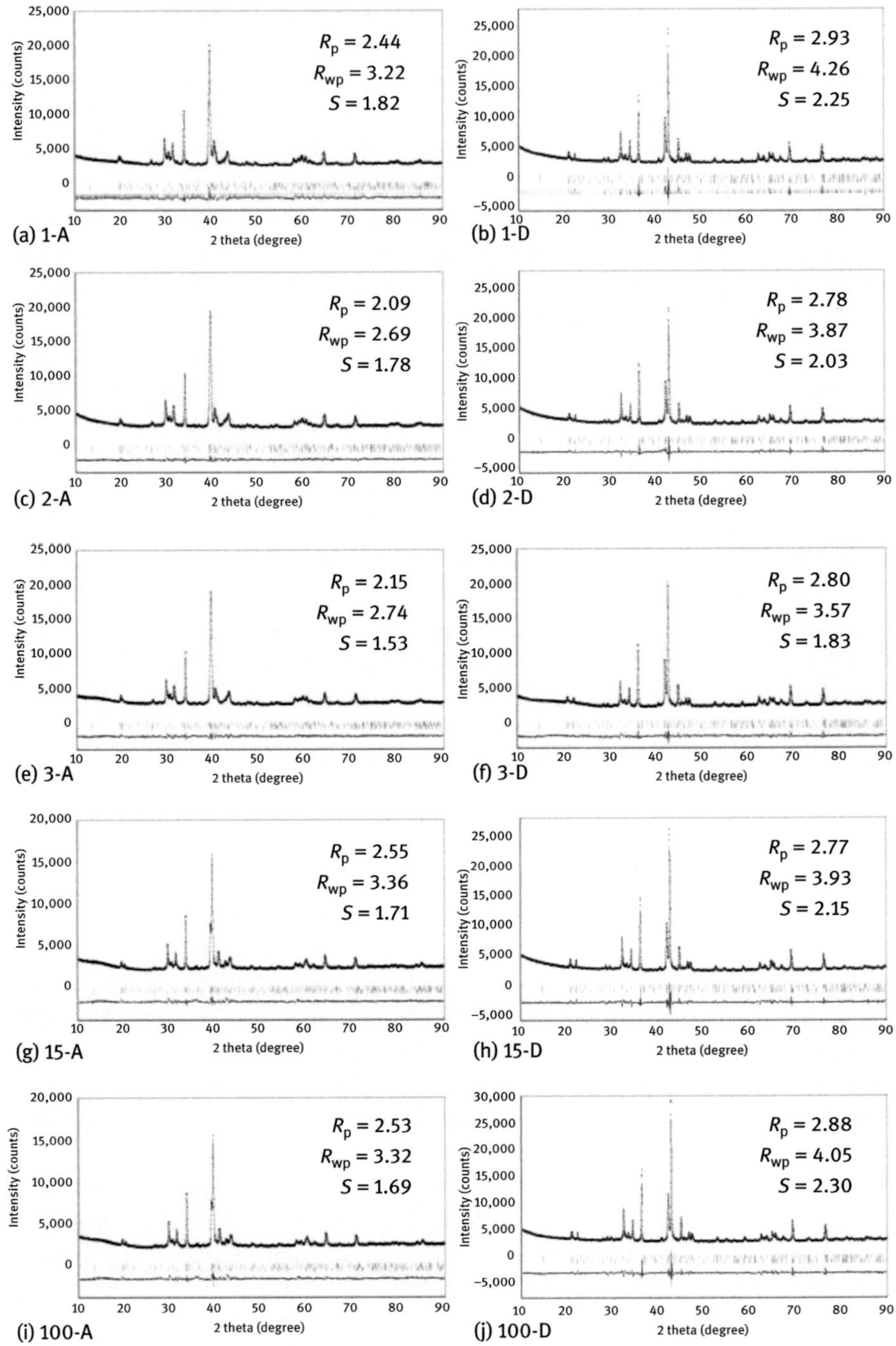

Figure 2.36: Rietveld refinements of the hydrogenated (A) and dehydrogenated (D) $Sm_{1.6}Mg_{0.4}Ni_7$ alloy after 1 cycle (a, b), 2 cycles (c, d), 3 cycles (e, f), 15 cycles (g, h) and 100 cycles (i, j). Vertical bars below the pattern show the positions of all possible reflection peaks of the Ce_2Ni_7-type (2H) and Gd_2Co_7-type (3R) phases for A, and 3R and 2H phases for *D*.

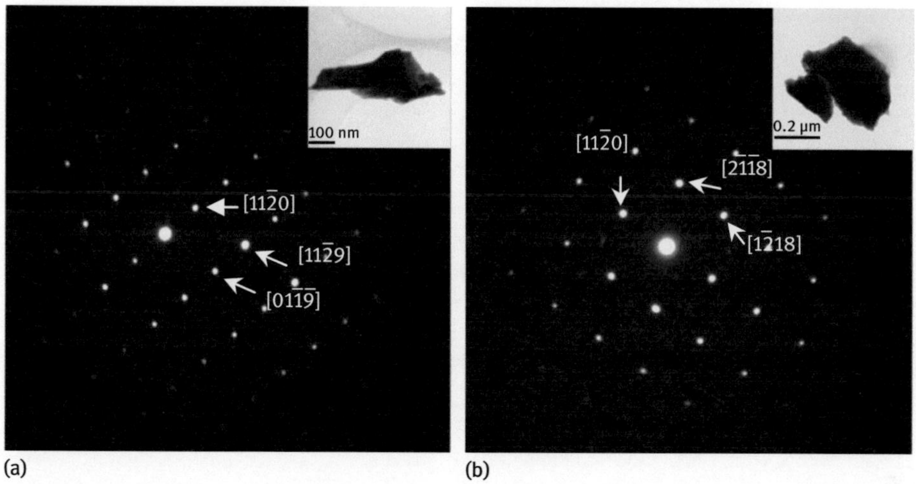

Figure 2.37: SAED patterns of the Gd_2Co_7-type structure for the original (a) and after 100 hydrogenation/dehydrogenation cycles (b) of the $Sm_{1.6}Mg_{0.4}Ni_7$ alloy.

degradation of superlattice RE–Mg–Ni alloys is mainly related to the formation of dislocation or distortion, and the occurrence of partial disproportionation or amorphization during cycles [35, 46]. Hence, based on the above results, the formation of dislocation may be the important cause of the capacity degradation because the accumulation of dislocation can lead to the enlargement of lattice strain [47, 48], which will be discussed below.

Figure 2.38 shows the $[A_2B_4]$ and $[AB_5]$ subunit volumes and their changes after hydrogenation and dehydrogenation at the 1–100 cycles, as well as the lattice strain of the $Sm_{1.6}Mg_{0.4}Ni_7$ alloy, calculated from the refined XRD profile parameters. The original volume of the $[A_2B_4]$ subunit is larger than that of the $[AB_5]$ subunit, which is 87.02 and 83.57 Å^3, respectively. After the first hydrogenation process, the volumes of the $[A_2B_4]$ and $[AB_5]$ subunits expand to 111.6 and 104.9 Å^3, and keep within the range of 104.8–106.0 Å^3 and 110.3–111.6 Å^3 with cycling, respectively (Figure 2.38(a)). After dehydrogenation, the volumes of the $[A_2B_4]$ and $[AB_5]$ subunits shrink to lower values than in the original state, which are within the range of 84.52–85.51 Å^3 and 82.74–83.65 Å^3 with cycling. Figure 2.38(b) shows the volume changes of the $[A_2B_4]$ and $[AB_5]$ subunits based on the last state within the first four cycles. It is clear that the volume expansion and contraction rates of the $[A_2B_4]$ subunit are all larger than that of the $[AB_5]$ subunit, but the rates of two subunits change little with the increase of hydrogenation/dehydrogenation cycles, where the difference of two subunits remaining within is ca. 3.37 and 2.70 % for hydrogenation and dehydrogenation, respectively. The above results indicate that the Gd_2Co_7-type structure of the $Sm_{1.6}Mg_{0.4}Ni_7$ alloy has good restorability and stability. Therefore, disproportionation or amorphization was not observed for the structure. However, accumulation of lattice strain is observed

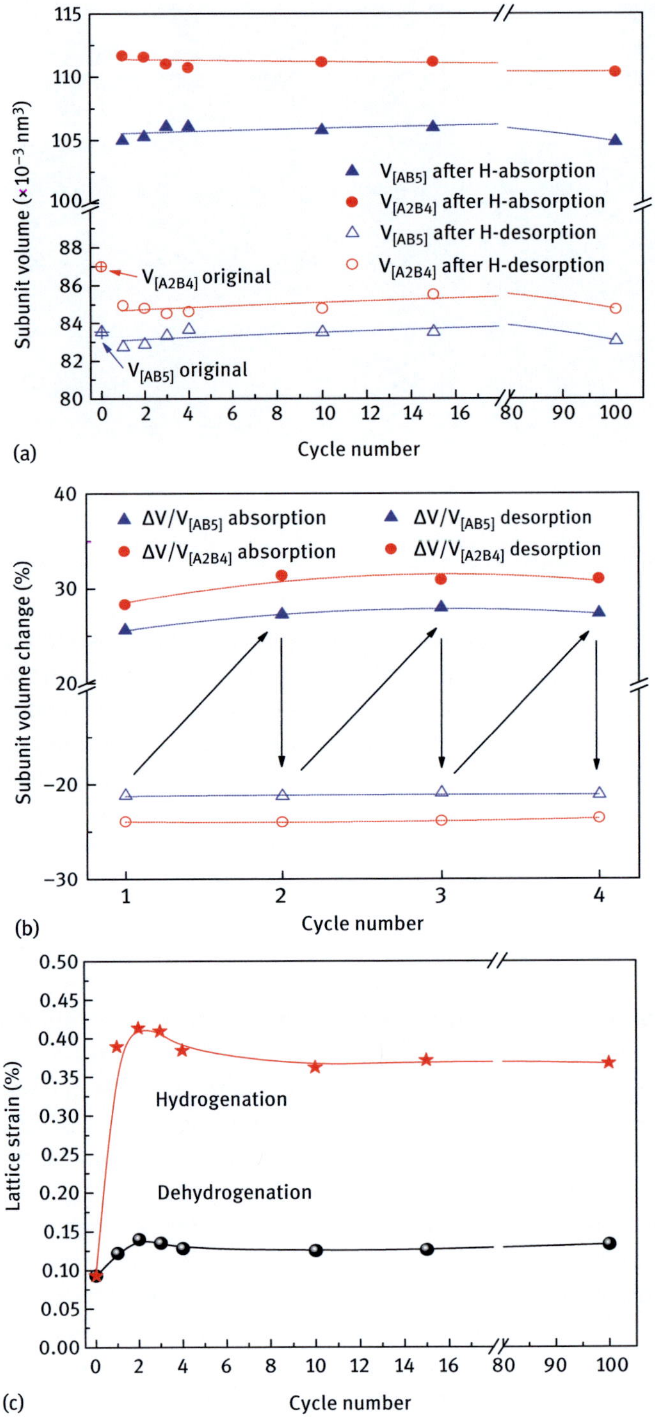

Figure 2.38: Subunit volume (a) and subunit volume change (b) of the $[A_2B_4]$ and $[AB_5]$ subunits of the $Sm_{1.6}Mg_{0.4}Ni_7$ alloy after hydrogenation and dehydrogenation at different cycles. Lattice strain after hydrogenation/dehydrogenation at different cycles of the $Sm_{1.6}Mg_{0.4}Ni_7$ alloy (c).

in Figure 2.38(c). The initial lattice strain of the sample is 0.093 %, while the lattice strain firstly increases to the maximum values of 0.413 and 0.140 % at the second hydrogenation and dehydrogenation cycles, and then decreases to stable values of ca. 0.38 and 0.13 % with further rising cycles for hydrogenation and dehydrogenation, respectively. The accumulation lattice strain after hydrogenation and dehydrogenation process relates to the subunit changes of the crystal structure, namely the formation of dislocation or distortion of the crystal structure. Hence it can be concluded that the formation of the dislocation or distortion of the crystal structure is the dominant factor for degradation in hydrogen storage capacity, similar to the reported $Pr_{0.6}Mg_{1.4}Ni_4$ [49] and V–Ti alloys [50].

2.4.5 The Hydrogen Storage Capacity Degradation of the Gd_2Co_7-Type Single-Phase Superlattice Sm–Mg–Ni-Based Hydrogen Storage Alloys

To determine the reaction during the two plateau courses, the structural changes of the $Sm_{1.6}Mg_{0.4}Ni_7$ alloy during hydrogen absorption and desorption processes are analysed. Although the equilibrium pressures of hydrogen absorption and desorption are quite a bit higher than 0.1 MPa at 25°C for the $Sm_{1.6}Mg_{0.4}Ni_7$ alloy, however, the hydrogen in the alloy bulk did not decompose and release under ambient condition as can be seen in Figure 2.39, in which the peaks of Gd_2Co_7- (3R) and Ce_2Ni_7-type (2H) phases had moved correspondingly to lower and higher angle during hydrogen absorption and desorption, respectively.

As shown in Figure 2.39(a), three processes can be exacted: when the hydrogen content is at 0.06 and 0.10 wt.%, the peaks of the 2H and 3R phases move to lower angle without peak broadening and keep original shapes. When the hydrogen content increases from 0.31 to 0.71 wt.%, except for the original peaks of the 2H and 3R phases, broaden peaks of the 2H and 3R phases are significantly observed left besides the original sharp peaks. However, because the content of 2H phase is much less than that of the 3R phase, broaden peaks of the 2H phase are not so clear as those of the 3R phase until the hydrogen content increases to 0.52 wt.% ($2\theta = 31.6°$). As the hydrogen content further increases to 0.82 wt.%, the Bragg peaks abruptly sharpen again but change in shape at the 2θ range at ca. 40° and 46°, and then further move to lower angle with the new shape until to be fully hydrogenated with the 1.81 wt.%. Hydrogen desorption follows the reverse process, as shown in Figure 2.39(b).

For hydrogen storage alloys, hydrogen absorption is accompanied with the phase transformation from H-dissolved solid solution phase to full hydride phase, and the hydrogen desorption follows the reverse process. And the transformation in the hydrogen of different states leads to appearance of hydrogenation and dehydrogenation plateaus in the *P–C* isotherms. Generally, ternary RE–Mg–Ni alloys with superlattice structures are found to have one plateau, corresponding to the conversation of H-dissolved solid solution phase (α phase) to full hydride phase (β phase), such as

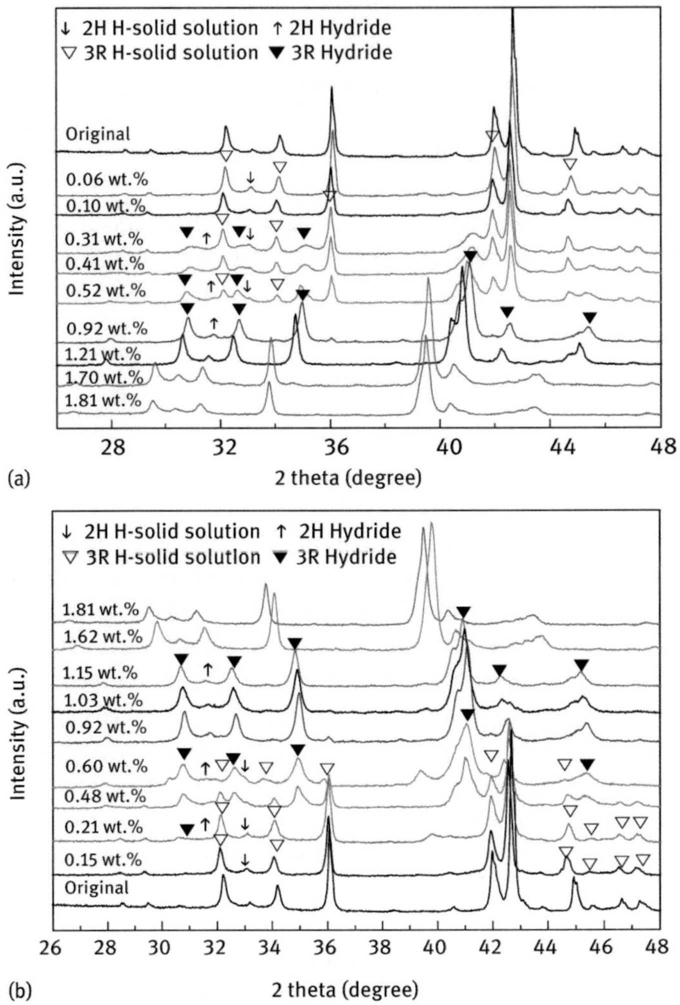

Figure 2.39: XRD patterns of the $Sm_{1.6}Mg_{0.4}Ni_7$ alloy during hydrogen absorption (a) and desorption (b).

La_2MgNi_9 [51], Sm_3MgNi_{14} [35], $Nd_{1.5}Mg_{0.5}Ni_7$ [21] and La_4MgNi_{19} [24]. However, for some binary alloys, such as the RE_2Ni_7 [52] and RE_2Co_7 [53, 54] alloys, two plateaus are shown in the *P–C* curves, which undergoes the $\alpha \leftrightarrow \beta$- and $\beta \leftrightarrow \gamma$-phase transformations during the lower and higher plateau [53, 55]. Cao et al. found that two kinds of hydrides with different structure patterns in the XRD analysis formed at the lower and higher ones respectively ($Sm_2Co_7H_{2.9}$ and $Sm_2Co_7H_{6.4}$) corresponding to the β and γ phases, respectivel [54]. In our study, by relating the *P–C* curves of the alloy in Figure 2.35(a) to XRD evolution with increase of H_2 content in Figure 2.39, it can be determined that no γ-phase appears during the higher plateau regio because the XRD pattern of the hydride phase at the higher plateau region is same with that of the hydride phase at the lower plateau region after transformation of H-dissolved solid solution phase. Therefore, the above three hydrogenation processes present the progress of hydrogen in solid solution phase (α-phase), corresponded to the slope region at a lower hydrogen content, the solid solution and hydride phases (α-phase +β-phase), related to the lower broad plateaus regions, and the hydride phase (β-phase), corresponded to the higher broad plateaus region and slope region at a higher hydrogen content, respectively. The dehydrogenation obeys the reverse processes. The hydrogenation and dehydrogenation processes are different from that reported for the hydrogen storage alloys.

Based on absorbed hydrogen contents obtained from *P–C* runs, the creating β-phase hydrides for the lower and higher plateaus are calculated to be $Sm_{1.6}Mg_{0.4}Ni_7H_{4.8}$ and $Sm_{1.6}Mg_{0.4}Ni_7H_{9.0}$, respectively (Table 2.12). The two hydrides are of the same structure type in view of the same diffraction peaks, but have different thermodynamic stability with regard to the unlike enthalpy changes. Studies found that hydrogen atoms in a crystal lattice can occupy both the tetrahedral and octahedral sites [55], but the two sites exhibited different thermodynamic stabilities, those occupying the octahedral sites being more stable than those at the tetrahedral sites. Hence, the formation of the two kinds of β-phase hydrides ($Sm_{1.6}Mg_{0.4}Ni_7H_{4.8}$ and $Sm_{1.6}Mg_{0.4}Ni_7H_{9.0}$) may be closely related to the different location for the entrance of hydrogen atoms with increase of H_2 concentration. The hydrogen absorption process of the Gd_2Co_7-type $Sm_{1.6}Mg_{0.4}Ni_7$ alloy may be as follows: hydrogen atoms firstly enter into the octahedral sites of the Gd_2Co_7-type structure with the *R*-3*m* space group during the hydrogenation, forming a stable hydride $Sm_{1.6}Mg_{0.4}Ni_7H_{4.8}$ with higher enthalpy changes. After the octahedral sites are fully occupied, the hydrogen atoms further enter into the unstable tetrahedral sites, forming the $Sm_{1.6}Mg_{0.4}Ni_7H_{9.0}$ hydride with less stability after the lower reaction enthalpy change. Besides, the development of the $Sm_{1.6}Mg_{0.4}Ni_7H_{9.0}$ hydride is without the appearance of the H-dissolved solid solution phase as in case of the $Sm_{1.6}Mg_{0.4}Ni_7H_{4.8}$. Therefore, it can be deduced that the H_2 directly reacts to the $Sm_{1.6}Mg_{0.4}Ni_7H_{9.0}$ hydride phase after decomposition to atoms based on the $Sm_{1.6}Mg_{0.4}Ni_7H_{4.8}$ alloy. The DSC analysis in Figure 2.35(c) proves the two-step modes for the hydrogenation/dehydrogenation of the $Sm_{1.6}Mg_{0.4}Ni_7$ alloy. As the above DSC analysis, the single and overlapped endothermic peaks belong to the higher and lower plateaus, respectively. That is, the single and overlapped endothermic peaks are due to the hydrogen desorption of $Sm_{1.6}Mg_{0.4}Ni_7H_{9.0}$ and $Sm_{1.6}Mg_{0.4}Ni_7H_{4.8}$ hydrides (Figure 2.35(c)). During the dehydrogenation of $Sm_{1.6}Mg_{0.4}Ni_7H_{4.8}$, the peak at 309 K is caused by the phase transformation from hydride phase to the H-dissolved solid solution phase, and the peak at 314 K is a result of the hydrogen being released from the H-dissolved solid solution phase. The enthalpy changes ΔH of the two transformation processes of H_2 are calculated to be 10.6 and 9.9 J g^{-1} H_2 (14.0 and 13.1 kJ mol^{-1} H_2), respectively, which are obtained by fitting the peaks after separation of the overlapped peaks (Figure 2.35(d)). However, the dehydrogenation process of $Sm_{1.6}Mg_{0.4}Ni_7H_{9.0}$ shows one endothermic peak, differing from the dehydrogenation process of the $Sm_{1.6}Mg_{0.4}Ni_7H_{4.8}$ hydride, which indicates that H-dissolved solid solution phase does not appear. Besides, the small hysteresis value also confirms the deduction, where large hysteresis is accompanied by the formation of hydride from the H-solid solution [56].

Since a difference in thermodynamic stability exists between the two plateaus, it is also important to determine the difference existing in the kinetic properties of the two plateaus. Figure 2.40(a) and (b) show the hydrogen absorption rate of the lower and higher plateaus at 298 K, respectively. Ninety per cent hydrogen absorbed across the upper plateau is reached in 1.5 min, whereas 11 min is required for the same fraction of hydrogen to absorb across the lower plateau under identical conditions (Table 2.12). And the same result also can be seen for the hydrogen desorption process in Figure 2.40(c) and (d), in which the hydrogen desorption times for the lower and the upper plateaus are 7.6 and 1.2 min. The results indicate that the hydrogen absorption and desorption reaction rates across the upper plateau are ca. 7 and 6 times than those of the lower plateau, respectively.

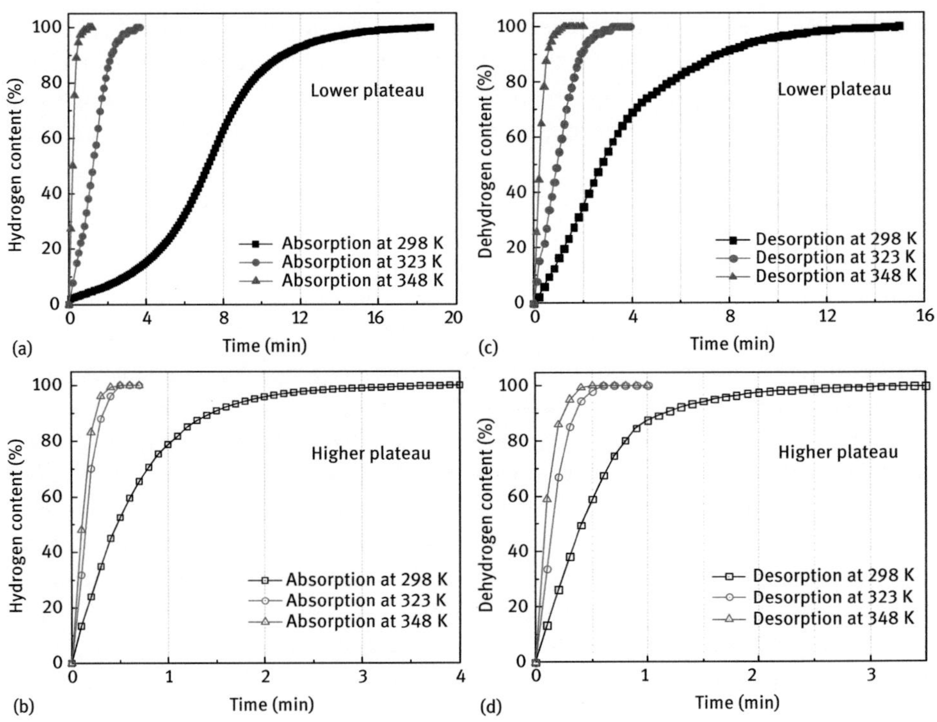

Figure 2.40: Hydrogen absorption and desorption kinetic curves for the $Sm_{1.6}Mg_{0.4}Ni_7$ alloy across the lower and higher plateaus measured in the range of 298–348 K.

2.4.6 The Effect of Hydrogen Absorption and Desorption on Lattice Structure of the Gd_2Co_7-Type Single-Phase Superlattice Sm–Mg–Ni-Based Hydrogen Storage Alloys

To understand the effect of the hydrogen absorption and desorption on lattice structure of the Gd_2Co_7-type $Sm_{1.6}Mg_{0.4}Ni_7$ alloy, the subunit volume changes and lattice strain during the hydrogen absorption and desorption processes are calculated and extracted to Figure 2.41 based on the Rietveld refinement analysis. For hydrogen

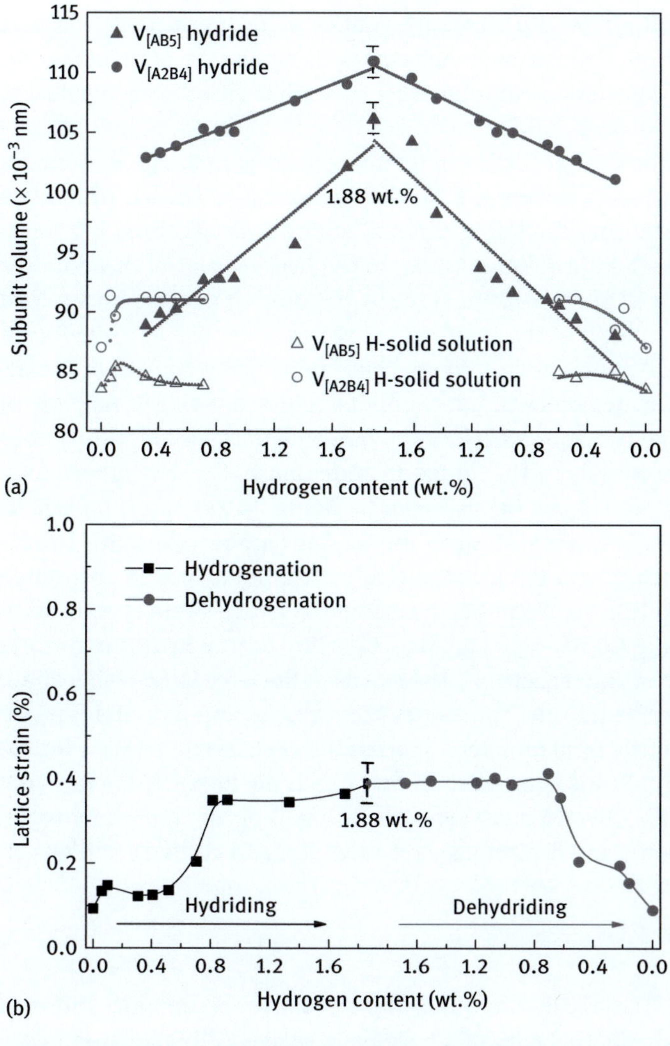

Figure 2.41: Subunit volumes of the $[A_2B_4]$ and $[AB_5]$ subunits during hydrogenation and dehydrogenation of the $Sm_{1.6}Mg_{0.4}Ni_7$ alloy (a). The lattice strain of the structure during hydrogenation and dehydrogenation of the $Sm_{1.6}Mg_{0.4}Ni_7$ alloy (b).

absorption process, as shown in Figure 2.41(a), the volumes of $[A_2B_4]$ and $[AB_5]$ subunits change little in the α-phase with the increasing H_2 content, and the mismatch between the two subunits, which is defined as $V_{[A2B4]} - V_{[AB5]}$, is ca. 5 Å^3. The cause of lattice strain by the entrance of H_2 is no more than 0.15 % when H_2 is only in α-phase state. Meanwhile, as the α-phase begins to transform to β-phase, the $[A_2B_4]$ subunit volume rapidly expands by 11.7 Å^3 and that for the $[AB_5]$ subunit is 4.3 Å^3, and the mismatch between the two subunits in β-phase reaches 1 Å^3 at this moment. Meanwhile, as the α-phase continuously transforms to β-phase, the mismatch between the two subunits in β-phase reduces slightly to 12.7 Å^3, but the lattice strain gradually increases to 0.35 % with the rising content of H_2. Besides, when the α-phase disappears and the H_2 directly forms β-phase with increasing content of H_2, the mismatch value is significantly reduced to 4.86 Å^3 at the maximum content of H_2 and the inner-molecular strain almost remains constant at around 0.35 % during the direct conversion process of H_2.

The above results indicate that the H_2 firstly forms hydride in the $[A_2B_4]$ subunit and then in the $[AB_5]$ subunit, accounting for the decreasing mismatch of the two subunits in hydride phase. And the lattice internal strain is caused from the transformation of H_2 from the H-solid solution phase to the hydride phase, in which the transformation of H induced the expansion of the $[A_2B_4]$ subunit but was constrained by the connected $[AB_5]$ subunit at the basal plane, which gives rise to a significant lattice strain. The lattice strain increases due to continuous rising content of the hydride phase through the above process, but cannot be released with the decrease of the mismatch between the two subunits since the formed deformation for the lattice structure during direct conversion of H_2 to the hydride phase. For dehydrogenation process, the same phenomenon can be also seen in Figure 2.41(a) and (b). Besides, we also found that the lattice internal strain of the Gd_2Co_7-type $Sm_{1.6}Mg_{0.4}Ni_7$ alloy is much smaller in comparison with the La–Mg–Ni alloys during hydrogen absorption and desorption processes [13]. For example, the maximum lattice strain is reported to be 0.67 and 0.86 % for the Ce_2Ni_7-type $La_{1.5}Mg_{0.5}Ni_7$ alloy during hydrogen absorption and desorption processes, respectively. The reason of the large lattice inner strain for La–Mg–Ni alloys may be that the transformation stage between α- and β-phase is much longer. However, the transformation process between α- and β-phase for the G_2Co_7-type $Sm_{1.6}Mg_{0.4}Ni_7$ alloy is shortened, during which the lattice strain does not further increase for the direct conversion between H_2 and β-phase, thus resulting in less lattice strain of the alloy and exhibiting good structural and cycling stabilities.

2.5 Conclusions

Single-phase La_2MgNi_9, La_3MgNi_{14} and La_4MgNi_{19} alloys containing different amounts of $[LaNi_5]$ and $[LaMgNi_4]$ subunits have been successfully prepared by sequential sintering and induction melting. The root cause of capacity degradation in the alloys have been studied. It is found that the discrete expansion/contraction of [$LaNi_5$] and [$LaMgNi_4$] subunits during hydrogen absorption/desorption leads to the volume mismatch between [$LaNi_5$] and [$LaMgNi_4$], which leads to alloy pulverization. Higher [$LaNi_5$]/[$LaMgNi_4$] subunit ratios can improve the accommodation capability for the volume mismatch between [$LaNi_5$] and [$LaMgNi_4$] subunits during charge/discharge; thus, increasing the cycling stability by reducing the pulverization and oxidation.

Single-phase Pr_2MgNi_9, Pr_3MgNi_{14} and Pr_4MgNi_{19} alloys with different proportions between [$PrNi_5$] and [$PrMgNi_4$] subunits have been successfully prepared via strictly controlling the zoning sintering and annealing processes. Phase formation process and capacity degradation of the single-phase Pr–Mg–Ni alloys are elucidated. The AB_3, A_2B_7 and A_5B_{19}-type single-phase alloys are successively formed at the accordingly peritectic temperatures of 900, 925 and 950°C, respectively, starting from the molten $PrMgNi_4$ phase and the solid phase $PrNi_5$. The pulverization of the alloy particle results from the mismatch between [$PrMgNi_4$] and [$PrNi_5$] subunits. It is found that the mismatch degree of the subunits for Pr_4MgNi_{19} alloy is lowest (22.0 %), smaller than those of the Pr_3MgNi_{14} (30.0 %) and Pr_2MgNi_9 (48.6 %) at the 100th cycle. The oxidation degree of the alloy electrode is obtained by analysing the structure of the alloy samples, and it is found that the phase abundance of the hydroxides, $Pr(OH)_3$ and $Mg(OH)_2$, both enhanced with increasing the subunit ratio $PrMgNi_4/PrNi_5$ in Pr_nMgNi_{5n-1} ($n = 2, 3, 4$).

A single-phase $PuNi_3$-type Nd_2MgNi_9 alloy has been obtained by induction melting followed by zoning annealing treatment method. The single-phase alloy electrode exhibits superior overall electrochemical properties. The discharge capacity and *HRD* (HRD_{1440}) of the alloy electrode are 370.6 mAh g^{-1} and 49.6 %, respectively. The cycling life particularly reaches upwards of 92 % at the 100th charge/discharge cycle, which is significantly higher than that of the single-phase $PuNi_3$-type La_2MgNi_9 alloy (S_{100}, 79 %). Besides, the volume deformation rate $\Delta V/V(\%)$ of [AB_5] and [A_2B_4] subunits and the lattice strain after hydrogenation/dehydrogenation is much smaller than those of the La_2MgNi_9 alloy. Hence, the pulverization degree of the alloy particle and oxidization/corrosion of the active material of Nd_2MgNi_9 alloy during the charge/discharge process is evidently weaker than those of the La_2MgNi_9 alloy.

A new Gd_2Co_7-type $Sm_{1.6}Mg_{0.4}Ni_7$ alloy is obtained by powder sintering method. The Gd_2Co_7-type $Sm_{1.6}Mg_{0.4}Ni_7$ alloy is found to be with good hydrogen storage ability and has special hydrogen absorption/desorption process. The compound can be activated at the first cycle and absorb 1.881 wt.% H_2 within 17 min at 298 K under 10 MPa H_2 after activation. And the hydrogen absorption speed increases to 3.4 min after 20 absorption/desorption cycles with a 1.436 wt.% H_2 under 3 MPa H_2. Especially, 99.3 % of the maximum hydrogen storage capacity is held for the Gd_2Co_7-type $Sm_{1.6}Mg_{0.4}Ni_7$ alloy at the 100th cycle. The hydrogen absorption/desorption of the Gd_2Co_7-type $Sm_{1.6}Mg_{0.4}Ni_7$ alloy undergoes two equilibrium stages and exhibiting two plateaus in the *P–C* curves. At the lower concentration H_2 stage, H_2 transforms between H-solid solution phase and hydride phase with a lower rate and higher enthalpy change in lower pressure, and at the higher concentration H_2 stage, H_2 directly converts between atom states and the hydride phase with a higher rate and lower enthalpy change in higher pressure. We found the transformation of H_2 at the lower concentration of H_2 stage results in inner-molecular strain and mismatch in subunit volumes in hydrogenation/dehydrogenation, and the shortened lower concentration of H_2 stage in the two-step mode lowers the values of the alloy, thus leading to good structural stability and excellent cycling stability. These findings are expected to provide an important guidance for developing viable intermetallic materials as high-pressure tank materials for hydrogen storage with nice hydrogen storage properties.

References

[1] Kadir K, Structural investigation and hydrogen capacity of YMg_2Ni_9 and $(Y_{0.5}Ca_{0.5})(MgCa)Ni$: new phases in the ABC system isostructural with $LaMg_2Ni_9$. J Alloys Compd 1999;287: 264–70.

[2] Kadir K, Sakai T, Uehara I. Structural investigation and hydrogen storage capacity of $LaMg_2Ni_9$ and $(La_{0.65}Ca_{0.35})(Mg_{1.32}Ca_{0.68})_2Ni_9$ of the AB_2C_9 type structure. J Alloys Compd 2000;302:112–17.

[3] Reilly JJ, Wiswall RH. Reaction of hydrogen with alloys of magnesium and nickel and the formation of Mg_2NiH_4. Inorg Chem 1968;7:2254–6.

[4] Liao B, Lei YQ, Lu GL, Chen LX, Pan HG, Wang QD. The electrochemical properties of $La_xMg_{3-x}Ni_9$ ($x = 1.0$–2.0) hydrogen storage alloys. J Alloys Compd 2003;356–357:746–9.

[5] Zhang FL, Luo YC, Chen JP, Yan RX, Chen JH. La–Mg–Ni ternary hydrogen storage alloys with Ce_2Ni_7-type and Gd_2Co_7-type structure as negative electrodes for Ni/Mh batteries. J Alloys Compd 2007;430:302–7.

[6] Zhang FL, Luo YC, Chen JP, Yan RX, Kang L, Chen JH. Effect of annealing treatment on structure and electrochemical properties of $La_{0.67}Mg_{0.33}Ni_{2.5}Co_{0.5}$ alloy electrodes. J Power Sources 2005;150:247–54.

[7] Denys RV, Yartys VA. Effect of magnesium on the crystal structure and thermodynamics of the $La_{3-x}Mg_xNi_9$ hydrides. J. Alloys Compd 2011;509:S540–8.

[8] Denys RV, Riabov AB, Yartys VA, Sato M, Delaplane RG. Mg substitution effect on the hydrogenation behaviour, thermodynamic and structural properties of the La_2Ni_7–$H(D)_2$ system. J Solid State Chem 2008;181:812–21.

[9] A. Férey, Cuevas F, Latroche M, Knosp B, Bernard P. Elaboration and characterization of magnesium-substituted La_5Ni_{19} hydride forming alloys as active materials for negative electrode in Ni-MH battery. Electrochim Acta 2009;54:1710–14.

[10] Yamamoto T, Inui H, Yamaguchi M, Sato K, Fujitani S, Yonezu I, Nishio K. Microstructures and hydrogen absorption/desorption properties of LaNi alloys in the composition range of La77.8 ~ 83.2 at.%Ni. Acta Mater 1997;45:5213–21.

[11] Kurtuldu G, Jessner P, Rappaz M. Peritectic reaction on the Al-rich side of Al–Cr system. J Alloys Compd 2015;621:283–6.

[12] Kroupa A, Mishra R, Rajamohan D, Flandorfer H, Watson A, Ipser H. Phase equilibria in the ternary Ni–Sb–Sn system: experiments and calculations. Calphad 2014;45:151–66.

[13] Zhang QA, Fang M, Si TZ, Fang F, Sun DL, Ouyang L, Zhu M. Phase stability, structural transition, and hydrogen absorption–desorption features of the polymorphic La_4MgNi_{19} compound. J Phys Chem C 2010;114::11686–92.

[14] Liu JJ, Han SM, Li Y, Yang SQ, Shen WZ, Zhang L, Zhou Y. An investigation on phase transformation and electrochemical properties of as-cast and annealed $La_{0.75}Mg_{0.25}Ni_x$ ($x = 3.0$, 3.3, 3.5, 3.8) alloys. J Alloys Compd 2013;552:119–26.

[15] Guzik MN, Hauback BC, Yvon K. Hydrogen atom distribution and hydrogen induced site depopulation for the $La_{2-x}Mg_xNi_7$–H system. J Solid State Chem 2012;186:9–16.

[16] Denys RV, Riabov B, Yartys VA, Delaplane RG, Sato M. Hydrogen storage properties and structure of $La_{1-x}Mg_x(Ni_{1-y}Mn_y)_3$ intermetallics and their hydrides. J Alloys Compd 2007;446–447:166–72.

[17] Liu YF, Pan HG, Gao MX, Zhu Y, Lei YQ. Hydrogen storage and electrochemical properties of the $La_{0.7}Mg_{0.3}Ni_{3.825-x}Co_{0.675}Mn_x$ hydrogen storage electrode alloys. J Alloys Compd 2004;365:246–52.

[18] Cheng LF, Wang YX, Wang RB, Pu ZH, Zhang XG, He DN. Microstructure and electrochemical investigations of $La_{0.76-x}Ce_xMg_{0.24}Ni_{3.15}Co_{0.245}Al_{0.105}$ ($x = 0$, 0.05, 0.1, 0.2, 0.3, 0.4) hydrogen storage alloys. Int J Hydrogen Energy 2009;34:8073–8.

[19] Iwakura C, Oura T, Inoue H, Matsuoka M. Effects of substitution with foreign metals on the crystallographic, thermodynamic and electrochemical properties of AB_5-type hydrogen storage alloys. Electrochim Acta 1996;41:117–21.

[20] Lartigue C, Percheron-Guégan A, Achard JC, Tasset F. Thermodynamic and structural properties of $LaNi_{5-x}Mn_x$ compounds and their related hydrides. J Less-Common Met 1980;75:23–9.

[21] Zhang QA, Zhao B, Fang MH, Liu C, Hu QM, Fang F, Sun DL, Ouyang LZ, Zhu M. $(Nd_{1.5}Mg_{0.5})Ni_7$-based compounds: structural and hydrogen storage properties. Inorg Chem 2012;51:2976–83.

[22] Liu YF, Pan HG, Gao MX, Lei YQ, Wang QD. Degradation mechanism of the La–Mg–Ni-based metal hydride electrode $La_{0.7}Mg_{0.3}Ni_{3.4}Mn_{0.1}$. J Electrochem Soc 2005;152:A1089–95.

[23] Liu YF, Pan HG, Yue YJ, Wu XF, Chen N, Lei YQ. Cycling durability and degradation behavior of La–Mg–Ni–Co-type metal hydride electrodes. J Alloys Compd 2005;395:291–9.

[24] Nakamura J, Iwase K, Hayakawa H, Nakamura Y, Akiba E. Structural study of La_4MgNi_{19}hydride by in situ x-ray and neutron powder diffraction. J Phys Chem 2009;C113:5853–9.

[25] Li Y, Han D, Han SM, Zhu XL, Hu L, Zhang Z, Liu Y. Effect of rare earth elements on electrochemical properties of La–Mg–Ni-based hydrogen storage alloys. Int J Hydrogen Energy 2009;34:1399–404.

[26] Lemort L, Latroche M, Knosp B, Bernard P. Elaboration and characterization of new pseudo-binary hydride-forming phases $Pr_{1.5}Mg_{0.5}Ni_7$ and $Pr_{3.75}Mg_{1.25}Ni_{19}$: a comparison to the binary $Pr_2Ni_7$and Pr_5Ni_{19} ones. J Phys Chem 2011;C115:19437–44.

[27] Iwase K, Terashita N, Mori K, Ishigaki T. Structural parameters of Pr_3MgNi_{14} during hydrogen absorption-desorption process. Inorg Chem 2012;51:11805–10.

[28] Iwase K, Terashita N, Mori K, Tsunokake S, Ishigaki T. Crystal structure and cyclic properties of hydrogen absorption–desorption in Pr_2MgNi_9. Int J Hydrogen Energy 2012;37:18095–100.

[29] Iwase K, Terashita N, Mori K, Yokota H, Suzuki T. Crystal structure and cyclic hydrogenation property of Pr_4MgNi_{19}. Inorg Chem 2013;52:14270–4.

[30] Iwase K, Terashita N, Mori K, Tashiro S, Yokota H, Suzuki T. Effects of Mg substitution on crystal structure and hydrogenation properties of $Pr_{1-x}Mg_xNi_3$. Int J Hydrogen Energy 2014;39:12773–7.

[31] Ozaki T, Kanemoto M, Kakeya T, Kitano Y, Kuzuhara M, Watada M, Tanase S, Sakai T. Stacking structures and electrode performances of rare earth–Mg–Ni-based alloys for advanced nickel–metal hydride battery. J Alloys Compd 2007;446–447:620–4.

[32] Denys RV, Yartys VA, Webb CJ. Hydrogen in $La_2MgNi_9D_{13}$: the role of magnesium. Inorg Chem 2012;51:4231–4238.

[33] Zhai TT, Yang T, Yuan ZM, Zhang YH. An investigation on electrochemical and gaseous hydrogen storage performances of as-cast $La_{1-x}Pr_xMgNi_{3.6}Co_{0.4}$ ($x = 0$–0.4) alloys. Int J Hydrogen Energy 2014;39:14282–7.

[34] Zhang YH, Hou ZH, Li BW, Ren HP, Zhang GF, Zhao DL. An investigation on electrochemical hydrogen storage performances of the as-cast and annealed $La_{0.8-x}Sm_xMg_{0.2}Ni_{3.35}Al_{0.1}Si_{0.05}$ ($x = 0–0.4$) alloys. J Alloys Compd 2012;537:175–82.
[35] Zhang QA, Chen ZL, Li YT, Fang F, Sun DL, Ouyang LZ, Zhu M. Comparative Investigations on hydrogen absorption–desorption properties of Sm–Mg–Ni compounds: the effect of $[SmNi_5]/[SmMgNi_4]$ unit ratio. J Phys Chem 2015;C119:4719–27.
[36] Willems JJ, Buschow KH. From permanent magnets to rechargeable hydride electrodes. J Less-Common Met 1987;129:13–30.
[37] Ma S, Gao MX, Li R, Pan HG, Lei YQ. A study on the structural and electrochemical properties of $La_{0\cdot7-x}Nd_xMg_{0.3}Ni_{2.45}Co_{0.75}Mn_{0.1}Al_{0.2}$ ($x = 0.0–3.0$) hydrogen storage alloys. J Alloys Compd 2008;457:457–64.
[38] Zhang FL, Luo YC, Deng AQ, Tang ZH, Kang L, Chen JH. A study on structure and electrochemical properties of $(La,Ce,Pr,Nd)_2MgNi_9$ hydrogen storage electrode alloys, Electrochim Acta 2006;52:24–32.
[39] Zhang YH, Ren HP, Li BW, Guo SH, Wang QC, Wang XL. Structures and electrochemical hydrogen storage behaviours of $La_{0.75-x}Pr_xMg_{0.25}Ni_{3.2}Co_{0.2}Al_{0.1}$ ($x = 0–0.4$) alloys prepared by melt spinning. Int J Hydrogen Energy 2009;34:6335–42.
[40] Zhang L, Du WK, Han SM, Li Y, Yang SQ, Zhao YM, Wu C, Mu HZ. Study on solid solubility of Mg in $Pr_{3-x}Mg_xNi_9$ and electrochemical properties of $PuNi_3$-type single phase RE–Mg–Ni (RE = La, Pr, Nd) hydrogen storage alloys. Electrochim Acta 2015173:200–8.
[41] Liu YF, Pan HG, Gao MX, Miao H, Lei YQ, Wang QD. Function of Al on the cycling behavior of the La–Mg–Ni–Co-type alloy electrodes. Int J Hydrogen Energy 2008;33:124–33.
[42] Young K, Ouchi T, Huang B. Effects of annealing and stoichiometry to $(Nd, Mg) (Ni, Al)_{3.5}$ metal hydride alloys. J Power Sources 2012;215:152–9.
[43] Gao Z, Luo YC, Li RF, Lin Z, Kang L. Phase structures and electrochemical properties of $La_{0.8-x}Gd_{0.2}Mg_xNi_{3.1}Co_{0.3}Al_{0.1}$ hydrogen storage alloys. J Power Sources 2013;241:509–16.
[44] Chen ZL, Si TZ, Zhang QA. Hydrogen absorption–desorption cycle durability of $SmMgNi_4$. J Alloys Compd 2015;621:42–6.
[45] Liu ZY, Yan XL, Wang N, Chai YJ, Hou DL. Cyclic stability and high rate discharge performance of $(La,Mg)_5Ni_{19}$ multiphase alloy. Int J Hydrogen Energy 2011;36:4370–4.
[46] Li Y, Ren HP, Zhang YH, Liu ZC, Zhang HW. Hydrogen induced amorphization behaviors of multiphase $La_{0.8}Mg_{0.2}Ni_{3.5}$ alloy. Int J Hydrogen Energy 2015;40:7093–102.
[47] Bowman R, Lindensmith C, Luo S, Flanagan TB, Vogt T. Degradation behavior of $LaNi_{5-x}$ Sn_xHz ($x = 0.20–0.25$) at elevated temperatures. J Alloys Compd 2002;330:271–5.
[48] Wang BP, Zhao LM, Cai CJ, Wang SX. Effects of surface coating with polyaniline on electrochemical properties of La–Mg–Ni-based electrode alloys. Int J Hydrogen Energy 2014;39:10374–9.
[49] Sakaki K, Terashita N, Kim H, Majzoub E, Machida A, Watanuki T, Tsunokake S, Nakamura Y, Akiba E. Degradation mechanism against hydrogenation cycles in $Mg_{2-x}Pr_xNi_4$ ($x = 0.6$ and 1.0). J Phys Chem C 2014;118:6697–705.
[50] Kim H, Sakaki K, Ogawa H, Nakamura Y, Nakamura J, Akiba E, Machida A, Watanuki T, Proffen T. Origin of degradation in the reversible hydrogen storage capacity of $V_{1-x}Ti_x$ alloys from the atomic pair distribution function analysis. J Phys Chem C 2013;117:26543–50.
[51] Denys RV, Yartys VA. Effect of magnesium on the crystal structure and thermodynamics of the $La_{3-x}Mg_xNi_9$ hydrides. J Alloys Compd 2011;509(2):S540–8.
[52] Iwase K, Mori K, Hoshikawa A, Ishigaki T. Hydrogenation and structural properties of Gd_2Ni_7 with superlattice structure. Int J Hydrogen Energy 2012;37:5122–7.

[53] Goudy A, Wallace WE, Craig RS, Takeshita T. Thermodynamics and kinetics of hydrogen absorption in rare earth-cobalt ($R_2Co_7$and RCo_3) and rare earth-Iron (RFe_3) compounds. Trans Met Hydrides. 1978167:312–26.

[54] Cao ZJ, Ouyang LZ, Wang H, Liu JW, Sun DL, Zhang QA, Zhu M. Structural characteristics and hydrogen storage properties of Sm_2Co_7. J Alloys Compd 2014;608:14–18.

[55] Ming L, Goudy AJ. Hydriding and dehydriding kinetics of Dy_2Co_7 hydride. J Alloys Compd 1999;283:146–50.

[56] Kuijpers FA, van Mal HH. Sorption hysteresis in the $LaNi_5$-H and $SmCo_5$-H systems. J Less-Common Met 1971;23:395–8.

3 Effect of Multiphase Structures on Electrochemical Properties of the Superlattice RE–Mg–Ni-Based Hydrogen Storage Alloys

This chapter mainly introduces the interaction between multiphase structure and its effect on electrochemical properties of the superlattice RE–Mg–Ni-based hydrogen storage alloys. The interaction between multiphase structures is one of the main factors influencing the electrochemical properties of the superlattice RE–Mg–Ni-based hydrogen storage alloys. As is known, there are several types of phase structures in the superlattice RE–Mg–Ni-based hydrogen storage alloys due to different ratios of [$LaNi_5$] and [$LaMgNi_4$] subunits. Different structures would be with varying B/A proportion. A-site elements include rare earth metals and are able to absorb/desorb hydrogen. B-site elements mainly refer to transitional metals and can adjust the conditions for hydrogen absorption/desorption. The alloy phases with higher ratio of A-site elements generally help increase the hydrogen storage capacity of the alloys, while those with higher B-site content contribute to higher hydrogen absorption/desorption plateau pressure and better high-rate dischargeability [1]. The characteristics in electrochemical properties of the single-phase structure alloys have been present in Chapter 2.

The prepared RE–Mg–Ni-based hydrogen storage alloys often have a multiphase structure and the electrochemical properties of the RE–Mg–Ni-based hydrogen storage alloys would change with the changing of phase component. Not only the phases themselves but also the interactions between different phases contribute to the expression of the electrochemical properties. It is meaningful to clarify the interaction between multiphase structures for the improvement of electrochemical properties of the superlattice RE–Mg–Ni-based hydrogen storage alloys. Optimizing the compositions is proved to be an effective way to improve the overall electrochemical properties of RE–Mg–Ni-based alloys. Finding out the effect of the interaction between multiphase structures on electrochemical properties of the superlattice RE–Mg–Ni-based hydrogen storage alloys would provide theoretical basis for composition optimization (namely adjusting the phase structure and abundances in RE–Mg–Ni-based system to obtain alloys possessing preferable electrochemical performance). It will make the composition optimization more initiative. This chapter will introduce the interaction between multiphase structures from the following aspects: the preparation of the multiphase structure alloys, the interaction between AB_3-type phase structure and other superlattice structures, the interaction between A_2B_7-type phase structure and other superlattice structures and the interaction between A_5B_{19}-type phase structure and other superlattice structures.

DOI 10.1515/9783110501483-003

3.1 The Preparation Method of the Multiphase Structure RE–Mg–Ni-Based Alloys

This section mainly talks about the preparation method of the multiphase structure RE–Mg–Ni-based alloys. The methods for multiphase structure alloys preparation are same with the normal preparation method of the RE–Mg–Ni-based hydrogen storage alloys, including stepwise powder sintering, inductive melting, high-energy ball milling, etc. However, as the preparation of RE–Mg–Ni-based hydrogen storage alloys is becoming more purposive and initiative, the methods also become selective to obtain RE–Mg–Ni-based alloys with specific composition.

The stepwise powder sintering is the most frequently used method. It is convenient to control the phase composition by adjusting the proportion of the sintering precursors. La–Mg–Ni-based alloys with $(La,Mg)Ni_3$ as main phase and $(La,Mg)_2Ni_7$ or $LaNi_5$ as minor phase [2], or $(La,Mg)_2Ni_7$ as main phase and $(La,Mg)_5Ni_{19}$ or $LaNi_5$ as minor phase [3, 4], or with $(La,Mg)_5Ni_{19}$ as main phase and $(La,Mg)_2Ni_7$ or $LaNi_5$ as minor phase [5] have been successfully prepared using this method by adjusting precursor ratios. Besides, La–Mg–Ni-based alloys with three phases including $(La,Mg)_5Ni_{19}$, $(La,Mg)_2Ni_7$ and $LaNi_5$ in varying ratios have also been made and exhibited good electrochemical properties [6].

Stepwise powder sintering is an effective method to prepare La–Mg–Ni-based alloys with single-phase structure and specific multiphase compositions. For example, under certain sintering conditions, Pr_5Co_{19}-type single-phase $La_{0.84}Mg_{0.16}Ni_{3.80}$ alloy would be prepared with precursors, $LaNi_5$ and $LaMgNi_4$, at the molar ratio $(LaNi_5/LaMgNi_4)$ of 1.2 [5]. We could also design and adjust the molar ratio of $LaNi_5$ and $LaMgNi_4$ to obtain La–Mg–Ni-based alloys with Pr_5Co_{19}-type $(La,Mg)_5Ni_{19}$ phase as the main phase and other phases as minor phase (Table 3.1). The alloy samples were prepared by the following steps: the precursors were crushed into powders under 300 meshes and mixed in various molar ratio $(LaNi_5/LaMgNi_4)$ denoted as x, in which x = 0.75, 0.95, 1.2 and 1.5, respectively. To compensate for Mg volatility losses during the sintering process, $LaMgNi_4$ was excessively added. Then the mixtures with

Table 3.1: Phase abundances and Mg occupation ratios of the alloys.

Alloys	Phase	Type	Mass fraction (wt.%)	Mg occupation ratio (%)
x = 0.75	$(La,Mg)_5Ni_{19}$	Pr_5Co_{19}	59.4	46.5
	$(La,Mg)_2Ni_7$	Ce_2Ni_7	40.6	51.9
x = 0.95	$(La,Mg)_5Ni_{19}$	Pr_5Co_{19}	80.2	48.4
	$(La,Mg)_2Ni_7$	Ce_2Ni_7	19.8	48.3
x = 1.2	$(La,Mg)_5Ni_{19}$	Pr_5Co_{19}	100	53.6
x = 1.5	$(La,Mg)_5Ni_{19}$	Pr_5Co_{19}	79.3	50.4
	$LaNi_5$	$CaCu_5$	20.7	–

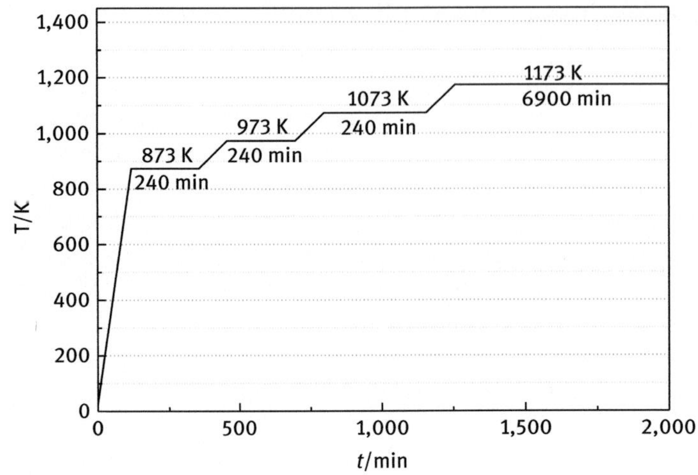

Figure 3.1: Plot of temperature-controlling programme.

a total weight of 2.5 g were pressed into pellets of 10 mm in diameter, 8 mm in thickness under 10 MPa. Each sample pellet was wrapped in Ni foil separately and sintered at 1,173 K for 6,900 min in a quartz tube under 0.04 MPa Ar pressure. There are three heat preservation stages between 873 and 1,073 K in the process of rising temperature (Figure 3.1). Afterwards, the pellets were quenched to room temperature.

Induction melting with different annealing conditions is also a common method to obtained RE–Mg–Ni-based multiphase structure alloys. It is noticeable that the phases near the boundary of the binary phase diagram tend to be consumed by annealing treatment to form the phases in the centre zone of the La–Ni phase diagram. This is reasonable because the phases near La-side in the diagram usually have lower melting points than those near the Ni-side. When the temperature increases, the phases close to the La-side melt to liquid phase, and the solid phases which are on the Ni-side of the diagram react with the liquid phase and form the phases between them. Different type phases can be stable at different annealing temperature, so the phase structure would change with the change of annealing temperature. And the decrease and increase of annealing time also has an important effect on the phase composition of the RE–Mg–Ni-based alloys. For example, Liu et al. [7] firstly prepared a $La_{0.75}Mg_{0.25}Ni_{3.5}$ alloy via inductive melting. In order to obtain $La_{0.75}Mg_{0.25}Ni_{3.5}$ alloy with A_2B_7- and A_5B_{19}-type superlattice structure, four stepwise annealing processes were designed (Table 3.2). Table 3.3 shows the phase abundances of the stepwise annealed $La_{0.75}Mg_{0.25}Ni_{3.5}$ alloys.

Stepwise powder sintering and inductive melting followed by annealing are widely applied to practical and experimental applications due to their mild and easily controllable conditions and good quality of the products.

Table 3.2: Temperature programme of stepwise annealing.

Steps	P_1	P_2	P_3	P_4
1	From 298 to 873 K, 5 K min^{-1}			
2	Holding at 873 K for 4 h			
3	From 873 to 1,073 K, 1 K min^{-1}			
4	Holding at 1,073 K for 4 h			
5	1,073–1,223 K, 1 K min^{-1}	1,073–1,173 K, 1 K min^{-1}		
6	1,223 K, 15 h	1,173 K, 15 h	1,173 K, 20 h	1,173 K, 25 h

Table 3.3: Phase abundances of the stepwise annealed $La_{0.75}Mg_{0.25}Ni_{3.5}$ alloys.

Alloys	Phases	Phase abundance (wt.%)
#1	$(La,Mg)_2Ni_7$ (2H)	31.5
	$(La,Mg)_2Ni_7$ (3R)	16.8
	$(La,Mg)_5Ni_{19}$	51.7
#2	$(La,Mg)_2Ni_7$ (2H)	32.7
	$(La,Mg)_2Ni_7$ (3R)	27.7
	$(La,Mg)_5Ni_{19}$	39.6
#3	$(La,Mg)_2Ni_7$ (2H)	32.5
	$(La,Mg)_2Ni_7$ (3R)	23.0
	$(La,Mg)_5Ni_{19}$	37.6
	$LaNi_5$	6.9
#4	$(La,Mg)_2Ni_7$ (2H)	22.3
	$(La,Mg)_2Ni_7$ (3R)	20.2
	$(La,Mg)_5Ni_{19}$	42.4
	$LaNi_5$	15.1

3.2 The Interaction between AB_3-Type Phase Structure and Other Superlattice Structures

This section mainly talks about the interaction between AB_3-type phase structure and other superlattice structures. The powder sintering is an effective technology for controlling preparation of the super-stacking phases in La–Mg–Ni-based hydrogen storage alloys [8–10]. According to the La–Ni binary phase diagram [11], it can be known that super-stacking phases (e. g. $LaNi_3$) can be formed during the process of heating by the peritectic reactions.

3.2.1 The Interaction between A_2B_7-Type Phase Structure and A_5B_{19}-Type Phase Structures

Our group [12] has revealed that $LaNi_5$ and $(La,Mg)Ni_3$ phases can transform into $(La,Mg)_2Ni_7$ phase during annealing treatment. Based on the above theory and researches, ternary La–Mg–Ni-based alloys with different $PuNi_3$- and Ce_2Ni_7-type phase abundance are prepared by a stepwise sintering method. The effects of different $PuNi_3$- and Ce_2Ni_7-type phase abundance on electrochemical behaviours of La–Mg–Ni-based alloy electrodes were studied in detail. Zhang et al. [13] prepared a series of La–Mg–Ni-based hydrogen storage alloys consisted of $PuNi_3$-type (La,Mg)Ni_3 and Ce_2Ni_7-type $(La,Mg)_2Ni_7$ phase. The alloys were prepared by a stepwise sintering technique using precursors: $LaNi_5$ and $LaMgNi_4$ intermetallic compounds. Four alloys were prepared in four kinds of $LaNi_5/LaMgNi_4$ molar ratio(x) (x = 0.56, 0.67, 0.74, 0.89). The two alloys were induction-melted from the individual metals, La, Mg and Ni (purity ≥ 99.5 %), under a protective atmosphere of purified argon gas. The $LaNi_5$ and $LaMgNi_4$ powders were then pressed into a pellet suitable weight ratios, wrapped in Ni foil and sintered at 973–1,123 K for 7 h followed by annealing treatment at 1,023 K for 4 days under an argon atmosphere with subsequent slow cooling to room temperature in the tube furnace.

3.2.1.1 Crystal Structure

Figure 3.2 shows the XRD patterns of the studied samples. It can be seen that the alloy has a single-phase microstructure when x is 0.56, namely $PuNi_3$-type (La,Mg)Ni_3 phase, all the other alloys consist of double phases which can be identified as $PuNi_3$-type (La,Mg)Ni_3 phase and Ce_2Ni_7-type $(La,Mg)_2Ni_7$ phase. Because there is a small difference in peaks between the Ce_2Ni_7-type $(La,Mg)_2Ni_7$ phase and $PuNi_3$-type (La,Mg)Ni_3 phase, many peaks are explained as the overlap of the both phases. The four samples were refined with a double-phase model including Ce_2Ni_7-type $(La,Mg)_2Ni_7$ and $PuNi_3$-type (La,Mg)Ni_3. Figure 3.2 shows the results of alloys with

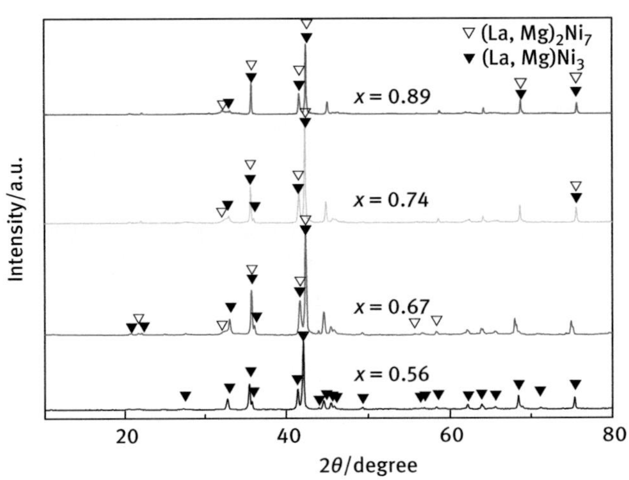

Figure 3.2: XRD patterns of alloys sintered with different $LaNi_5/LaMgNi_4$ molar ratios (x).

x = 0.56 and 0.67, and the Lattice parameters and phase abundances of each phase are listed in Table 3.4. As can be seen, it varies due to the various molar ratios of $LaNi_5$ to $LaMgNi_4$, the $(La,Mg)Ni_3$ phase abundance decreases from 100 % to 28 % with increasing x from 0.56 to 0.89. On the contrary, the $(La,Mg)_2Ni_7$ phase increases to 72 %.

From the La–Ni binary phase diagram, we can know that super-stacking phases can be formed during the process of heating by the peritectic reactions. It is deduced that the $PuNi_3$-type $(La,Mg)Ni_3$ single phase can be formed by strictly controlling the molar ratio of two precursors ($LaNi_5$ and $LaMgNi_4$) and preparation conditions. According to our previous research results [4], peritectic reaction between the $LaNi_5$ and $(La,Mg)Ni_3$ phases can occur during the process of sintering, thus generating the $(La,Mg)_2Ni_7$ phase. The reason is that the $(La,Mg)Ni_3$ phase whose melting point is far below the $LaNi_5$ phase becomes liquid firstly, while the $LaNi_5$ phase still remains solid. So if x is slightly higher compared to that of the single-phase alloy, the redundant $LaNi_5$ solid phase will react with the generated $(La,Mg)Ni_3$ liquid phase to become $(La,Mg)_2Ni_7$ phase. As a result, $(La,Mg)Ni_3$ phase will coexist with $(La,Mg)_2Ni_7$ phase in the alloys, which account for the above experimental results. The $(La,Mg)_2Ni_7$ phase abundance for the alloys is closely related to the molar ratio of $LaNi_5$ to $LaMgNi_4$. What's more, there exist no $LaMgNi_4$ and $LaNi_5$ phases in the XRD patterns, indicating that the peritectic reaction is completed.

It is worth noting that both a and c parameters of the $(La,Mg)Ni_3$ phase increase with the increase of x, leading to a subsequent expansion of cell volume from 0.5293 to 0.5343 nm^3, as shown in Table 3.4. However, for the $(La,Mg)_2Ni_7$ phase, the parameter of a increases, conversely, the parameter of c decreases, and the cell volume expansion from 0.5486 to 0.5508 nm^3 is also observed noticeably. Besides, the c/a ratio for both phases remains almost constant. The changes of lattice parameters may be attributed to the decrease for the content of Mg element in the alloys.

Table 3.4: Lattice parameters and phase abundances of alloys sintered with different $LaNi_5/LaMgNi_4$ molar ratios (x).

Alloys	Phases	Structure type	Lattice constants				Phase abundance (wt.%)
			a (nm)	c (nm)	$V(nm^3)$	c/a	
x = 0.56	$(La,Mg)Ni_3$	$PuNi_3$	0.5019	2.4242	0.5293	4.83	100
x = 0.67	$(La,Mg)Ni_3$	$PuNi_3$	0.5026	2.4243	0.5304	4.82	66
	$(La,Mg)_2Ni_7$	Ce_2Ni_7	0.5042	2.4917	0.5486	4.94	34
x = 0.74	$(La,Mg)Ni_3$	$PuNi_3$	0.5034	2.4258	0.5324	4.82	58
	$(La,Mg)_2Ni_7$	Ce_2Ni_7	0.5048	2.4898	0.5495	4.93	42
x = 0.89	$(La,Mg)Ni_3$	$PuNi_3$	0.5036	2.4274	0.5343	4.82	28
	$(La,Mg)_2Ni_7$	Ce_2Ni_7	0.5049	2.4893	0.5508	4.93	72

3.2.1.2 Electrochemical *P–C* Isotherms

In order to study the effect of different phase compositions on the discharge plateau pressure, the *P–C* isotherms of the alloys were measured and depicted in Figure 3.3 For the $x = 0.56$ alloy with a $PuNi_3$-type single phase structure, a wide plateau region is observed between 0.15 and 0.63 H/M, the plateau pressure is approximately 0.025 atm and the maximum hydrogen capacity is 0.8 H/M. What's more, all the other alloys also show one plateau, which indicates that there exists one hydrogen desorption phase. However, as mentioned above, all alloys, except for the $x = 0.56$ alloy, have double phases, and the equilibrium pressures of $PuNi_3$-type (La,Mg)Ni_3 phase and Ce_2Ni_7-type $(La,Mg)_2Ni_7$ phase are close to each other due to their similar stacking structures [14], so it is deduced that the plateau is the combination of both (La,Mg)Ni_3 and $(La,Mg)_2Ni_7$ phase plateaus. In addition, it can be seen from Figure 3.3 that the plateau pressure increases from 0.025 to 0.108 atm as x increases from 0.56 to 0.89. This means that hydrides of (La,Mg)Ni_3 phase are more stable than that of $(La,Mg)_2Ni_7$ phase, and the (La,Mg)Ni_3 phase is unfavourable for hydrogen desorbed. On the other hand, it is noticeable that the desorption plateaus become flatter and wider with x increasing from 0.56 to 0.74; accordingly, the hydrogen content (H/M) increases from 0.80 to 0.93, just as shown in Table 3.4, which are related to the increase of unit cell volumes for the both (La,Mg)Ni_3 and $(La,Mg)_2Ni_7$ phases, making the alloys more sites for hydrogen storage. Whereas the desorption plateau becomes more sloping and more narrow, and the H/M decreases to 0.89 when further increasing x to 0.89 in spite of the increased unit cell volume. This may result from interactions between two phases, which need further investigation.

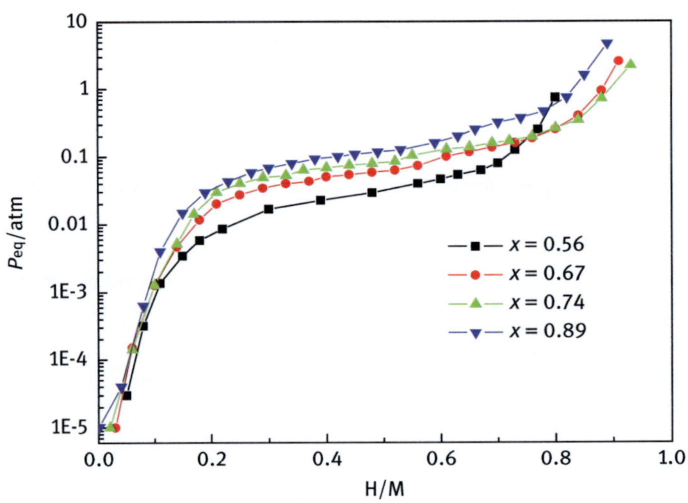

Figure 3.3: Electrochemical desorption *P–C* isotherms for alloys sintered with different $LaNi_5/LaMgNi_4$ molar ratios (x) at 298 K.

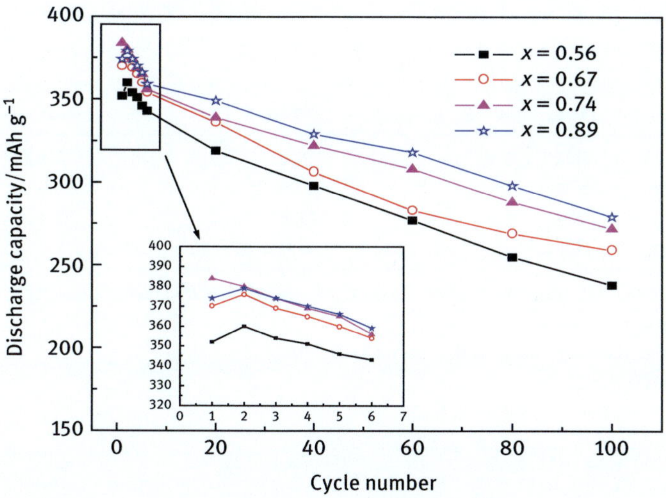

Figure 3.4: Evolution of the discharge capacity with cycle number of alloy electrodes with different $LaNi_5$/$LaMgNi_4$ molar ratios (x).

3.2.1.3 Charge and Discharge Performances

Figure 3.4 illustrates the relationship between discharge capacity and cycle number of the alloy electrodes at a discharge density of 72 mA g^{-1}, and the electrochemical characteristics of different alloy electrodes are summarized in Table 3.5. It is evident that the phase abundance has a great effect on the electrochemical properties of alloy electrodes. It can be seen from the inset of Figure 3.4 and Table 3.5 that all the alloy electrodes possess excellent activation characteristic and can be fully activated within one or two cycles to reach the maximum discharge capacity. Among the studied electrodes, the x = 0.56 electrode exhibits relatively lower discharge capacity of 360 mAh g^{-1}, while the maximum discharge capacity goes up to 384 mAh g^{-1} (x = 0.74) and then decreases to 379 mAh g^{-1} (x = 0.89), which is well in agreement with the above-reported hydrogen desorption capacity (H/M). In a word, the electrode with the phase abundance ratio of $(La,Mg)_2Ni_7$ to $(La,Mg)Ni_3$ close to 1:1 exhibits the highest discharge capacity.

The capacity retention rate can be calculated by the following formula:

$$S_n\,(\%) = C_n/C_{max} \times 100\ \% \tag{3.1}$$

where S_n and C_n are the capacity retention and discharge capacity at the nth cycle, respectively, and C_{max} is the maximum discharge capacity. It can be derived from Table 3.5 that the capacity retention rate of $x = 0.56$ alloy electrode at the 100th charge/discharge cycle (S_{100}) is only 66.1 %, whereas the capacity retention rate of $x = 0.89$ alloy electrode increases to 73.6 %, which is not only attributed to the increase of $(La,Mg)_2Ni_7$ phase abundance whose cycling stability is better than that of $(La,Mg)Ni_3$ phase [15], but also results from the influence of unit cell volume of two

Table 3.5: The electrochemical properties of alloys sintered with different $LaNi_5/LaMgNi_4$ molar ratios (*x*).

Alloys	N^a	$C_{max}{}^b$ (mAh g^{-1})	$HRD_{1440}{}^c$ (%)	$S_{100}{}^d$ (%)
$x = 0.56$	2	360	36.6	66.1
$x = 0.67$	2	376	45.5	70.0
$x = 0.74$	1	384	47.5	70.8
$x = 0.89$	2	379	45.0	73.6

[a]The cycling numbers for the different alloy electrodes to reach their maximum discharge capacity.
[b]The maximum discharge capacity of alloy electrodes.
[c]The high discharge ability at a discharge current density of 1,440 mA g^{-1}.
[d]The capacity retention at the 100th cycle.

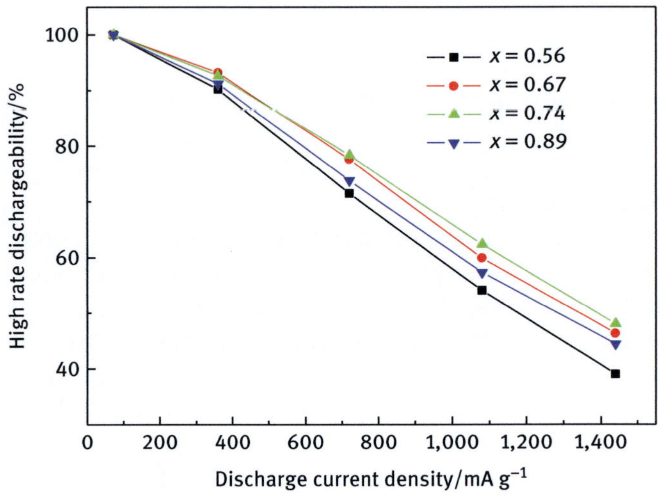

Figure 3.5: *HRD* curves of alloy electrodes with different $LaNi_5/LaMgNi_4$ molar ratios (*x*).

phases. The unit cell volume enlarges gradually as evidenced from Rietveld refinement data, hence, the expansion ratio of unit cell in the charge/discharge process will be smaller than that of $x = 0.56$ alloy electrode, thereby less pulverization [16].

Figure 3.5 shows the *HRD* of the La–Mg–Ni-based alloys. The *HRD* at a discharge current density of 1,440 mA g^{-1} are listed in Table 3.5. The *HRD* can be calculated by the following expression:

$$HRD(\%) = \frac{C_d}{C_{72}} \times 100\% \tag{3.2}$$

where in C_d is the discharge capacity at the discharge current density I_d and C_{72} is the discharge capacity at 72 mA g^{-1}. From Figure 3.5, we can see that there is almost no difference in *HRD* with the increase of discharge current density. The *HRD* at 1,440 mA g^{-1}

firstly increases from 36.6 % (x = 0.56) to 47.5 % (x = 0.74), and then decreases to 45.0 % (x = 0.89). The *HRD* at 1,440 mA g^{-1} of the electrode with the ratio of (La,Mg)Ni_3 to $(La,Mg)_2Ni_7$ phase abundance close to 1:1 is 10.9 % higher than the electrode with a $PuNi_3$-type single-phase structure. It is well known that the high discharge ability of a metal-hydride electrode is determined by two factors, that is, charge–transfer reaction at the interface of metal/electrolyte and hydrogen diffusion rate in alloy bulk [17]. The two factors are closely bound up with exchange current density, hydrogen diffusion coefficient.

Figure 3.6(a) presents the linear polarization curves of alloy electrodes at 50 % DOD. The polarization resistance (R_p) can be estimated from the slope of the obtained curves, and the exchange current density (I_0) which can be used to judge the speed of charge–transfer on the surface of alloy electrodes can be calculated using the following formula:

$$I_0 = \frac{RTI_d}{F\eta} \tag{3.3}$$

where R is the gas constant (J mol^{-1} K^{-1}), T the absolute temperature (K), I_d the applied current density (mA g^{-1}), F the Faraday constant (C mol^{-1}) and η the total over potential. A high value of I_0 corresponds to good kinetics for hydrating/dehydrating. The I_0 of the x = 0.56 electrode is 325.9 mA g^{-1}, whereas that of the x = 0.89 electrode mounts up to 570.5 mA g^{-1}. This indicates that increasing the content of $(La,Mg)_2Ni_7$ phase can improve the electroactivity at surface of the alloy electrodes. The $PuNi_3$-type phase which has more [$LaMgNi_4$] units is more likely to be oxidized and pulverized than the Ce_2Ni_7-type phase, so the hydroxides (such as $Mg(OH)_2$ and $La(OH)_3$) are easier to be formed by mixing Mg^{2+}, La^{3+} ions and OH^- ion in alkaline solution,

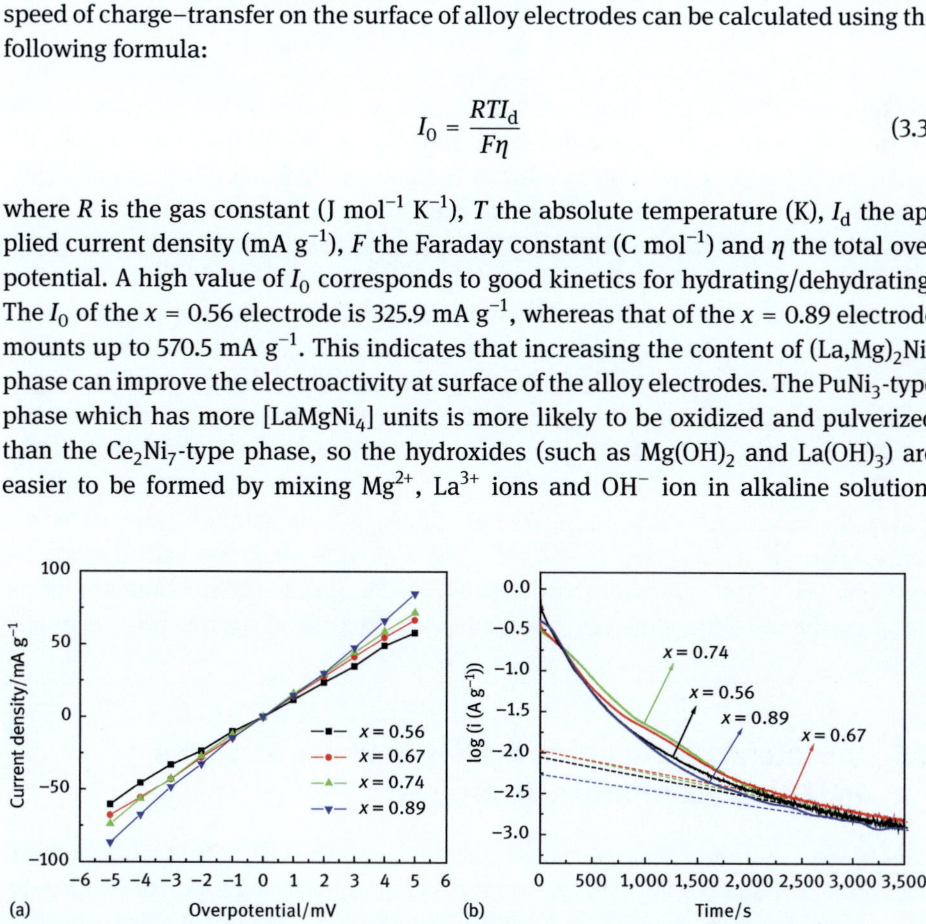

Figure 3.6: Linear polarization curves of alloy electrodes with different $LaNi_5/LaMgNi_4$ molar ratios (x) at 50 % DOD and 298 K (a); correspondence of anodic current density versus discharge time of alloy electrodes with different $LaNi_5/LaMgNi_4$ molar ratios (x) after +500 mV potential steps at 298 K (b).

respectively, and continuously cover the surface of alloy particles, thus an oxidation-passivation layer appears. The existence of layers reduces the electronic conductibility on the surface of alloy electrodes. So the decrease in content of (La,Mg)Ni_3 phase and the increase in content of $(La,Mg)_2Ni_7$ phase contribute to the increase of the speed of charge–transfer on the surface of alloy electrodes.

The hydrogen diffusion rate in alloy bulk can be characterized by means of potential-step method. Figure 3.6(b) illustrates the correspondence of anodic current density to discharge time of the alloy electrodes at fully charged state and 298 K. The hydrogen diffusion coefficient D in the bulk of the alloys can be calculated by the following formula:

$$\log i = \log\left(\frac{6FD}{da^2}(C_0 - C_s)\right) - \frac{\pi^2 Dt}{2.303a^2} \tag{3.4}$$

wherein i, D, d, a, C_0, C_s and t are anodic current density (mA g^{-1}), hydrogen diffusion coefficient ($cm^2\ s^{-1}$), the density of the alloy, radius of the alloy particle, the initial hydrogen concentration in alloy bulk (mol cm^{-3}), the surface hydrogen concentration of alloys (mol cm^{-3}) and the discharge time (s), respectively. Assuming that a is 13 μm, D can be calculated according to the slope of log(i) versus t. It can be seen that the D of alloys increases from $1.36 \times 10^{-11}\ cm^2\ s^{-1}$ (x = 0.56) to $1.50 \times 10^{-11}\ cm^2\ s^{-1}$ (x = 0.74), and then decreases to $1.07 \times 10^{-11}\ cm^2\ s^{-1}$ (x = 0.89), whose order is consistent with that of HRD_{1440} values. The result indicates that the hydrogen diffusion in bulk of alloys is a more important factor than the charge–transfer on surface of alloy electrodes. At the same time, the alloy electrode with the ratio of $(La,Mg)_2Ni_7$ to (La,Mg)Ni_3 phase abundance close to 1:1 exhibits the best kinetic characteristics. Based on the above alloy structure analysis, the phase boundary maybe the major factor accounting for the change of hydrogen diffusion coefficient, just as Dong et al. reported before [16]. The x = 0.74 electrode with $(La,Mg)_2Ni_7$ to (La,Mg)Ni_3 phase abundance ratio tending to 1:1, whose phase boundary maybe reach maximum value, thus can provide more tunnels for the diffusion of hydrogen atoms.

3.3 The Interaction between A_2B_7-Type Phase Structure and Other Superlattice Structures

This section mainly discusses the interaction between A_2B_7-type phase structure and other superlattice structures. Among the phases in superlattice RE–Mg–Ni-based hydrogen storage alloys, A_2B_7-type phase has superior overall electrochemical properties. As we know, alloys with $(La,Mg)_2Ni_7$ main phase have superior overall electrochemical properties. According to the La–Ni phase diagrams, La_2Ni_7 phase has two allotropes with hexagonal 2H-type (*P63*/*mmc*) and rhombohedral 3R-type (*R3*-*m*) structures [18]. Their unit cells consist of two and three blocks stacking along the *c* axis, respectively. In both structures, each block is composed of one layer of [La_2Ni_4] unit and two layers of [$LaNi_5$] units. The structural distinction between the two variants exists only in the stacking mode of subunit layers. However, the *HRD* and cycling stability of the alloy electrodes still need to be further improved before practical applications. For this purpose, it would be beneficial to reveal the functions of alloy phases and thus optimize the phase compositions in the alloys.

3.3.1 The Interaction between A_2B_7-Type Phase Structure and A_5B_{19}-Type Phase Structures

The $(La,Mg)_5Ni_{19}$ phase, a neighbouring phase of the $(La,Mg)_2Ni_7$ phase, not only possesses super-stacking structure that is similar to the $(La,Mg)_2Ni_7$ phase, but also has high Ni content that acts as catalyst for hydrogen absorption/desorption. Moreover, the $(La,Mg)_5Ni_{19}$ phase itself is believed to have superior *HRD* and cycling stability. It is believed that the $(La,Mg)_5Ni_{19}$ phase exists with A_2B_7-type phase in RE–Mg–Ni-based hydrogen storage alloys; the electrochemical properties of the alloys would show some improvement. But the prime solution is to find out the optimal content of the $(La,Mg)_5Ni_{19}$ phase in A_2B_7–type alloys.

Liu et al. [3] obtained a group of A_2B_7-type La–Mg–Ni-based alloys with single $(La,Mg)_2Ni_7$ phase and different amounts of $(La,Mg)_5Ni_{19}$ minor phase. The alloy samples including the single $(La,Mg)_2Ni_7$ phase alloy, the alloys with different amounts of $(La,Mg)_2Ni_7$ and $(La,Mg)_5Ni_{19}$ phases and the single $(La,Mg)_5Ni_{19}$ phase alloy were obtained by stepwise sintering using $La_{0.75}Mg_{0.25}Ni_{3.5}$ and $La_{0.80}Mg_{0.20}Ni_{3.8}$ as-cast alloys as precursors. The detailed preparation methods are as follows: (1) for the single $(La,Mg)_2Ni_7$ phase alloy, $La_{0.75}Mg_{0.25}Ni_{3.5}$ as-cast alloy was firstly crushed and ground to −300 mesh and the alloy powder was ball milled with supplementary Mg powder for 1 h. Then the mixture was squashed into a pellet and sintered in a tube furnace at the temperature ranging from 600 to 980°C, and subsequently annealed for 4 days at 900 °C. (2) For the alloys with various amounts of $(La,Mg)_2Ni_7$ and $(La,Mg)_5Ni_{19}$ phases, the $La_{0.75}Mg_{0.25}Ni_{3.5}$ alloy powder was ball milled individually with 5, 10, 20 and 40 wt.% of $La_{0.80}Mg_{0.20}Ni_{3.8}$ alloy powder as well as extra Mg powder for 1 h, and then the mixtures were cold squashed into pellets followed by sintering and annealing. The parameters such as extra Mg amount, pressing pressure and the heating treatment scheme were the same as those for the preparation of single $(La,Mg)_2Ni_7$ phase alloy. (3) For the single $(La,Mg)_5Ni_{19}$ phase alloy, the as-cast $La_{0.80}Mg_{0.20}Ni_{3.8}$ alloy was grounded into powder and ball milled with extra Mg powder. Then the mixture was cold squashed into a pellet and sintered using the same heating scheme in (1) and (2). For convenience, the alloys mixed with 0, 5, 10, 20 and 40 wt.% of $La_{0.80}Mg_{0.20}Ni_{3.8}$ alloy powder during preparation process were labelled as #1, #2, #3, #4 and #5. The single $(La,Mg)_5Ni_{19}$ phase alloy was notified as #6.

3.3.1.1 Microstructures

The XRD patterns of the #1–#6 alloys are shown in Figure 3.7(a–b). As shown in the figure, the XRD patterns exhibit sharp peaks, indicating long-range crystallographic orders and good crystallinity of the alloys [19]. Both (La,Mg)$_2$Ni$_7$ and (La,Mg)$_5$Ni$_{19}$ phases contained in the alloys exist with a mixture of hexagonal (2H) and rhombohedra (3R) polymorphic structures. This result is in agreement with our previous work related to the A$_2$B$_7$-type La–Mg–Ni-based alloy system [6], and such coexistence is believed to be a general trend for those pseudo binary R–Mg–Ni-based (R = La [20], Nd [21], Pr [22]) compounds. It is also noted that no decomposition of the (La,Mg)$_5$Ni$_{19}$ phase occurs in the alloys because that the final annealing temperature in our study is 900°C which is within the stable temperature range of (La,Mg)$_5$Ni$_{19}$ phase (840–960°C).

For Rietveld refinements of the XRD patterns, the starting structure models were taken from the reported data for 2H- and 3R-type (La$_{1.5}$Mg$_{0.5}$)Ni$_7$ phase [15] and 2H- and

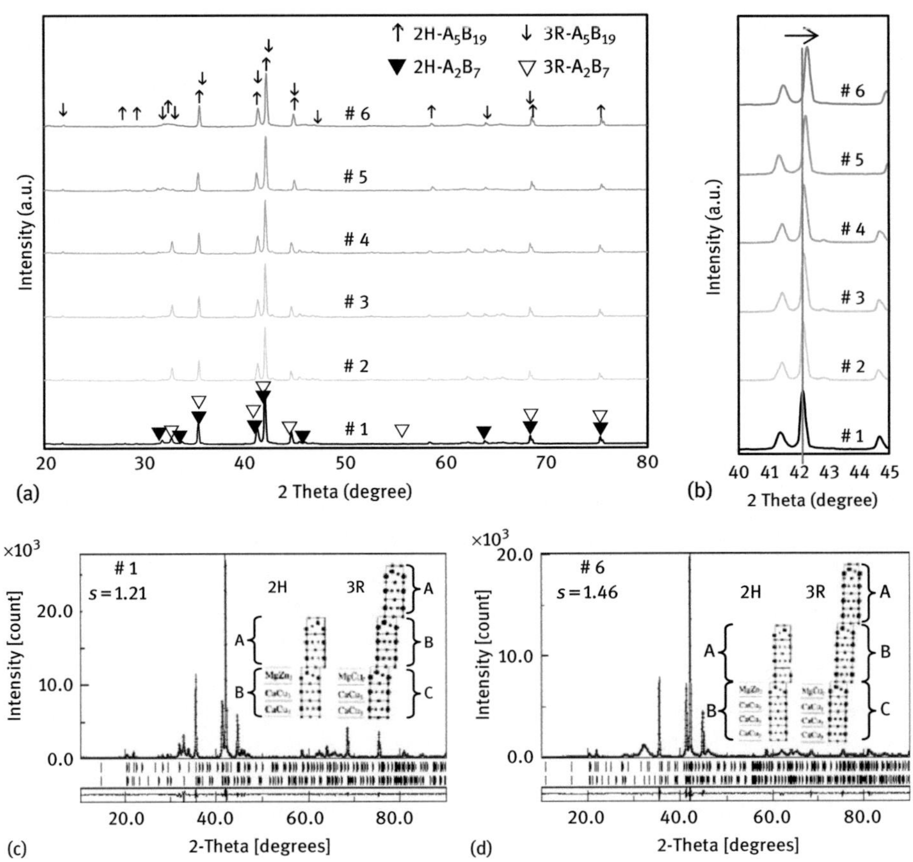

Figure 3.7: XRD patterns of the #1–#6 alloys in the 2θ range of 20–80° (a), zoom of (a) in the 2θ range of 40–46° (b) and Rietveld refinement of #1 (c) and #6 (d).

Table 3.6: Phase abundance, lattice parameters and strain of #1–#6 alloys.

	Phase	Phase abundance (%)	*a* (nm)	*c* (nm)	*V* (nm³)	Strain (%)
#1	A_2B_7-2H	55.1	5.0437	24.3462	536.365	0.123
	A_2B_7-3R	44.9	5.0437	36.5175	804.508	
#2	A_2B_7-2H	41.6	5.0432	24.3364	536.043	0.130
	A_2B_7-3R	53.1	5.0430	36.5022	803.947	
	A_5B_{19}-2H	2.89	5.0435	32.4350	714.510	
	A_5B_{19}-3R	2.40	5.0432	48.6480	1071.539	
#3	A_2B_7-2H	40.4	5.0437	24.3107	535.583	0.139
	A_2B_7-3R	49.1	5.0436	36.4653	803.326	
	A_5B_{19}-2H	7.3	5.0436	32.4016	713.803	
	A_5B_{19}-3R	3.2	5.0432	48.6048	1070.588	
#4	A_2B_7-2H	51.3	5.0434	24.3086	535.473	0.149
	A_2B_7-3R	29.4	5.0433	36.4599	803.111	
	A_5B_{19}-2H	7.04	5.0432	32.3896	713.425	
	A_5B_{19}-3R	12.3	5.0432	48.5832	1070.112	
#5	A_2B_7-2H	18.4	5.0430	24.2237	533.518	0.157
	A_2B_7-3R	44.2	5.0433	36.3357	800.375	
	A_5B_{19}-2H	12.5	5.0430	32.2648	710.620	
	A_5B_{19}-3R	24.9	5.0431	48.4075	1066.199	
#6	A_5B_{19}-2H	59.5	5.0335	32.2288	707.155	0.124
	A_5B_{19}-3R	40.4	5.0335	48.3392	1060.645	

3R-type $(La,Mg)_5Ni_{19}$ phase [23], respectively. The results from Rietveld analysis are summarized in Table 3.6 and the Rietveld refinements of #1 and #6 alloys are presented in Figure 3.7(c–d) as typical examples. #1 alloy only contains $(La,Mg)_2Ni_7$ phase and #6 alloy only contains $(La,Mg)_5Ni_{19}$ phase. However, #2–#5 alloys contain different ratios of $(La,Mg)_2Ni_7$ and $(La,Mg)_5Ni_{19}$ phases with the $(La,Mg)_5Ni_{19}$ phase content of 5.3, 10.5, 19.3 and 37.4, respectively. These numbers are close to the amount of A_5B_{19}-type alloy added in the alloy preparation process. Moreover, it is noticeable from both Table 3.6 and Figure 3.7 (b) with the peak shifts marked by the arrow that the unit cells of both $(La,Mg)_2Ni_7$ and $(La,Mg)_5Ni_{19}$ phases shrinks with increasing $(La,Mg)_5Ni_{19}$ phase abundance. The regular change in both $(La,Mg)_5Ni_{19}$ phase abundance and unit cell volume indicates that the $(La,Mg)_5Ni_{19}$ phase was successfully incorporated to the alloys.

As mentioned above, the unit cells' size of both $(La,Mg)_2Ni_7$ and $(La,Mg)_5Ni_{19}$ phases decreases with the addition of the $(La,Mg)_5Ni_{19}$ phase. Two reasons may lead to this change in the size of the cell volume: (1) Mg content in the phases of the alloys changes with increasing amount of $(La,Mg)_5Ni_{19}$ phase; (2) the constituent subunit lattice parameters and volume size are different, causing the changes in cell volumes of $(La,Mg)_2Ni_7$ and $(La,Mg)_5Ni_{19}$ phases with increasing amount of $(La,Mg)_5Ni_{19}$ phase. Below is the detailed discussion about the factors.

As we know, $(La,Mg)_2Ni_7$ and $(La,Mg)_5Ni_{19}$ phases are composed of $[LaNi_5]$ and $[(La,Mg)_2Ni_4]$ subunits stacking along *c* axis. Mg exclusively shares *4f* site (2H-type phases) or *6c* site (3R-type phases) of the $[(La,Mg)_2Ni_4]$ subunit with La. Because the atomic radius of Mg (0.160 nm) is much smaller than that of La (0.187 nm), the larger ratio of Mg to La at the *4f* or *6c* site is, the smaller the phase cell volume will be. The occupation ratios of La to Mg (La/Mg) at the *4f* site for 2H-type $(La,Mg)_2Ni_7$ and $(La,Mg)_5Ni_{19}$ phases and those at the *6c* site for 3R-type $(La,Mg)_2Ni_7$ and $(La,Mg)_5Ni_{19}$ phases are listed in Table 3.7. It can be found that the La/Mg ratios at the corresponding sites of both $(La,Mg)_2Ni_7$ and $(La,Mg)_5Ni_{19}$ phases of different alloy samples are very similar, being approximately 1:1, indicating that the compositions of $[(La,Mg)_2Ni_4]$ subunits for all the alloy samples are the same, being $[LaMgNi_4]$. Thus, it is believed that the solubility of Mg in the $(La,Mg)_2Ni_7$ and $(La,Mg)_5Ni_{19}$ phases is not the reason causing the decrease of the phase cell volumes with more $(La,Mg)_5Ni_{19}$ phase amount in the alloys.

To further explore the reason for the cell volume contraction of the $(La,Mg)_2Ni_7$ and $(La,Mg)_5Ni_{19}$ phases of the alloys with increasing $(La,Mg)_5Ni_{19}$ phase abundance, below we analysed the average lattice parameters as well as volume size of the constituent $[LaNi_5]$ and $[La,MgNi_4]$ subunits of the alloy phases, and investigated the internal strain in the alloys through which the lattice parameters of the constituent subunits are supposed to affect each other. From a crystallographic point of view, the unit cells of 2H and 3R structures are consisted of two and three blocks, respectively, and each block is composed of one $[LaMgNi_4]$ subunit and two $[LaNi_5]$ subunits for $(La,Mg)_2Ni_7$ phase and one $[LaMgNi_4]$ subunit and three $[LaNi_5]$ subunits for $(La,Mg)_5Ni_{19}$ phase, as illustrated in the insets of Figure 3.7 (c) and (d). The average volume sizes of the constituent subunits of the $(La,Mg)_2Ni_7$ and $(La,Mg)_5Ni_{19}$ phases for the #2–#5 alloys were calculated, and the evolution trends of the average subunit volumes of $(La,Mg)_2Ni_7$ phase and $(La,Mg)_5Ni_{19}$ phase with increasing $(La,Mg)_5Ni_{19}$ phase abundance are shown in Figure 3.8(a). It can be seen that the average subunit

Table 3.7: La and Mg occupation ratio at the *4f* site (2H-type phases) or *6c* site (3R-type phases) of the alloy phases calculated by Rietveld method.

Alloy	Phase type	La/Mg ratio at *4f* site	Phase type	La/Mg ratio at *6c* site
#1	A_2B_7-2H	0.497/0.503	A_2B_7-3R	0.495/0.505
#2	A_2B_7-2H	0.495/0.505	A_2B_7-3R	0.492/0.508
	A_5B_{19}-2H	0.495/0.505	A_5B_{19}-3R	0.487/0.513
#3	A_2B_7-2H	0.497/0.503	A_2B_7-3R	0.501/0.499
	A_5B_{19}-2H	0.500/0.500	A_5B_{19}-3R	0.502/0.498
#4	A_2B_7-2H	0.502/0.498	A_2B_7-3R	0.506/0.494
	A_5B_{19}-2H	0.497/0.503	A_5B_{19}-3R	0.499/0.501
#5	A_2B_7-2H	0.503/0.497	A_2B_7-3R	0.510/0.490
	A_5B_{19}-2H	0.498/0.502	A_5B_{19}-3R	0.501/0.499
#6	A_2B_7-2H	0.498/0.502	A_2B_7-3R	0.493/0.507

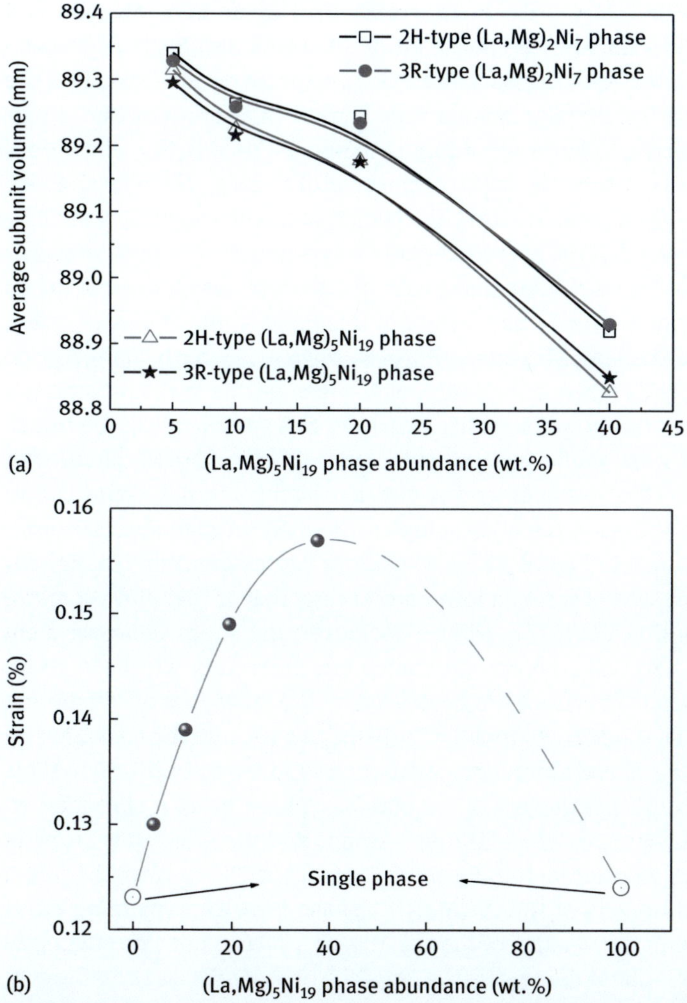

Figure 3.8: Comparison of the average volumes of the constituent subunits in the $(La,Mg)_2Ni_7$ and $(La,Mg)_5Ni_{19}$ phases (a); the evolution trend of the stains of the alloy samples with increasing $(La,Mg)_5Ni_{19}$ phase abundance (b).

volumes of both $(La,Mg)_2Ni_7$ and $(La,Mg)_5Ni_{19}$ phases decrease as the $(La,Mg)_5Ni_{19}$ phase abundance increases, which is in agreement with the changing trend of the phase cell volumes. Moreover, it is seen that for each alloy sample, the average subunit volumes of 2H- and 3R-type structures of the same phase are very similar. Particularly, it is noteworthy that for every alloy the curves of the average subunit volumes of 2H- and 3R-type $(La,Mg)_5Ni_{19}$ phase lie below those of 2H- and 3R-type $(La,Mg)_2Ni_7$ phase, indicating that for each sample, the average subunit volume of $(La,Mg)_5Ni_{19}$ phase is smaller than that of $(La,Mg)_2Ni_7$ phase. As reported, the $[A_2B_4]$ subunit volumes ($V_{[A2B4]}$) of the $La_{1.63}Mg_{0.37}Ni_7$ [24], $La_{1.6}Mg_{0.4}Ni_7$ [25], and $La_{1.5}Mg_{0.5}Ni_7$ [10] alloys with single 2H-type $(La,Mg)_2Ni_7$ phase are 91.05, 90.69 and 89.01 Å, respectively, while the $V_{[AB5]}$ values are 88.6, 88.3 and 87.9 Å, respectively. It is seen that the volume of $[AB_5]$ subunit is generally smaller than that of $[A_2B_4]$ subunit for a particular phase. Consequently, the average subunit volume of $(La,Mg)_5Ni_{19}$ phase with higher $[AB_5]/[A_2B_4]$ ratio tends to be smaller than that of $(La,Mg)_2Ni_7$ phase. As we know, constituent subunit volumes are directly dependent on the constituent subunit parameters. It is calculated that the average subunit *c* axis length of both $(La,Mg)_2Ni_7$ and $(La,Mg)_5Ni_{19}$ phases has a similar trend with the average subunit cell volumes of the phases. The average subunit *c* axis length of $(La,Mg)_5Ni_{19}$ phase is also generally smaller than that of $(La,Mg)_2Ni_7$ phase. However, the *a* axis length changes little with increasing $(La,Mg)_5Ni_{19}$ phase abundance, which may be due to the constriction of the basin plane of the alloy super-stacking structure. In a specific alloy, the bridge that $(La,Mg)_5Ni_{19}$ phase with smaller subunit parameters and $(La,Mg)_2Ni_7$ phase with larger subunit parameters are interacted is considered to be the internal strains within the alloys. The internal strains of the alloy samples obtained through spectra correlated to XRD data are shown in Figure 3.8(b). The alloys containing only $(La,Mg)_2Ni_7$ phase or $(La,Mg)_5Ni_{19}$ phase show much lower strain than that of the alloys containing both $(La,Mg)_2Ni_7$ and $(La,Mg)_5Ni_{19}$ phases. Moreover, the strain increases from 0.123 to 0.151 as the $(La,Mg)_5Ni_{19}$ phase abundance increases from 5.1 to 37.4 wt.%. The increase of the strain is considered to be caused by the mismatch of constituent subunit parameters of the $(La,Mg)_2Ni_7$ and $(La,Mg)_5Ni_{19}$ phases, and the mismatch increases with $(La,Mg)_5Ni_{19}$ phase abundance getting close to 50 wt.%. In each alloy, smaller constituent subunit parameters of $(La,Mg)_5Ni_{19}$ phase bring a shrinking effect on its coexisting $(La,Mg)_2Ni_7$ phase through strain, and the effect gets stronger with $(La,Mg)_5Ni_{19}$ phase amount increasing from 5.1 to 37.4 wt.%. In turn, the larger constituent subunit parameters of the $(La,Mg)_2Ni_7$ phase have an expanding effect on the constituent subunit parameters of the $(La,Mg)_5Ni_{19}$ phase; but this effect gets weaker when $(La,Mg)_2Ni_7$ phase decreases ($(La,Mg)_5Ni_{19}$ phase increases). So the constituent subunit parameters of both $(La,Mg)_2Ni_7$ and $(La,Mg)_5Ni_{19}$ phases in the alloys decrease with increasing amount of $(La,Mg)_5Ni_{19}$ phase, leading to the decrease of the constituent subunit volumes and cell volumes of both $(La,Mg)_2Ni_7$ and $(La,Mg)_5Ni_{19}$ phases.

The backscattered SEM images of the #1, #5 and #6 alloys are presented in Figure 3.9(a–c). It can be seen that #1 and #6 exhibit one uniform area while #5 exhibits two different areas. The EDS result indicates that areas A and B are $(La,Mg)_2Ni_7$ phase with compositions of $La_{0.746}Mg_{0.254}Ni_{3.52}$ and $La_{0.754}Mg_{0.246}Ni_{3.54}$, respectively. Areas C and D are $(La,Mg)_5Ni_{19}$ phase with compositions of $La_{0.807}Mg_{0.193}Ni_{3.77}$ and $La_{0.804}Mg_{0.196}Ni_{3.79}$, respectively. The compositions of the same phase for different alloy samples are in good agreement with the above XRD analysis in terms of La/Mg ratios. Furthermore, from the inset of Figure 3.9(b) it can be seen that the $(La,Mg)_5Ni_{19}$ phase distributes among the $(La,Mg)_2Ni_7$ main phase with a continuous net shape,

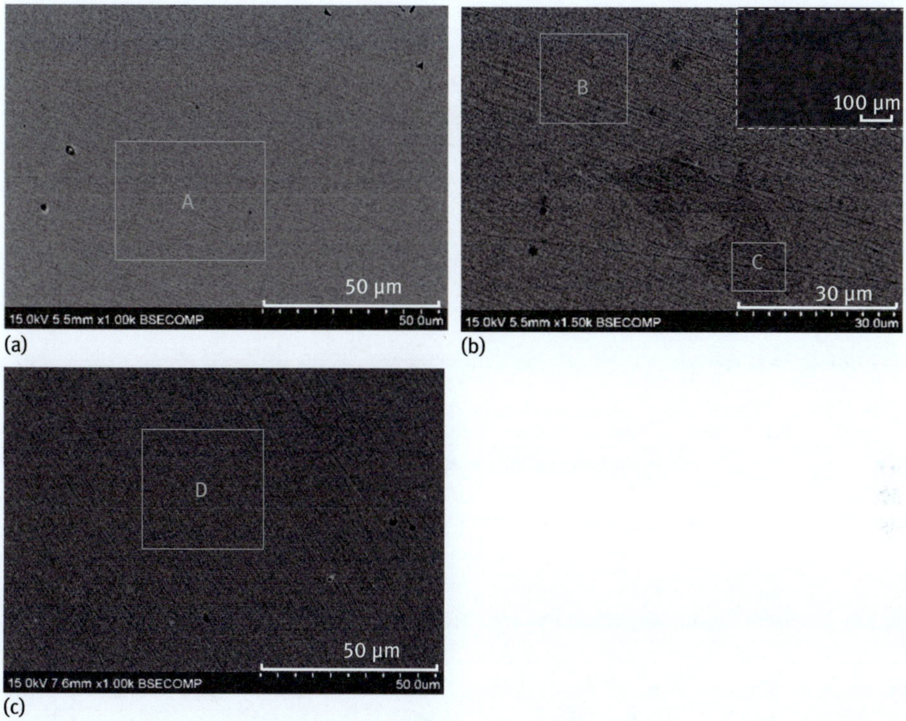

Figure 3.9: SEM backscattering electron images of #1 (a); #5 (b); #6 (c).

which is responsible to its function in the electrochemical performance. Besides, although both $(La,Mg)_2Ni_7$ and $(La,Mg)_5Ni_{19}$ phases contain 2H- and 3R-type structures, the two types cannot be identified in the SEM photograph because they have almost the same chemical compositions.

3.3.1.2 Electrochemical Pressure–Composition Isotherms

For applications in Ni/MH rechargeable batteries, both high hydrogen capacity and moderate hydride stability are expected for hydrogen storage alloys. Information about these properties for Ni/MH batteries can be described using pressure–composition (*P–C*) isotherms [26]. The value of hydrogen absorption/desorption equilibrium plateau pressure is used to predict the electrochemical potential of hydrogen storage alloys; and the length of plateau is used to estimate the theoretical electrochemical capacity of hydrogen storage alloys in Ni/MH batteries.

The resulting electrochemical *P–C* isotherms of the alloy samples at room temperature are shown in Figure 3.10, and the information is summarized in Table 3.8. For the alloy with single $(La,Mg)_2Ni_7$ phase, a wide plateau region is observed and the plateau pressure of hydrogen desorption is about 0.012 MPa. For the alloy with single $(La,Mg)_5Ni_{19}$ phase, the curve is with much higher slope and the plateau

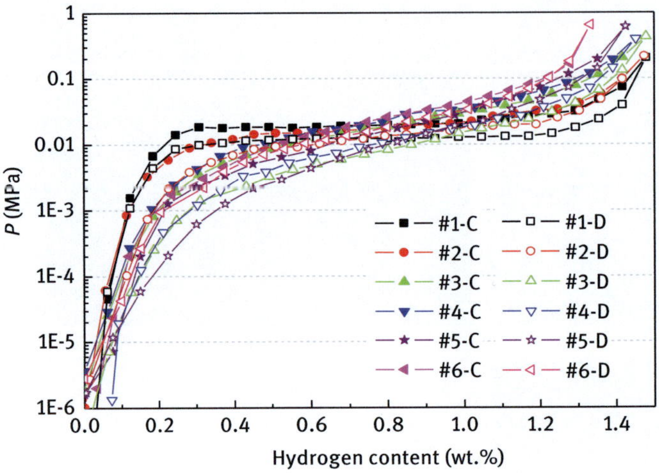

Figure 3.10: Electrochemical P–C isotherms of the alloy electrodes during charge/discharge process.

Table 3.8: Hydrogen absorption/desorption and electrochemical properties of #1–#6 alloys.

Alloy	C_{abs} (wt. %)	C_{rev} (wt. %)	P_{abs} (MPa)	P_{des} (MPa)	H_f	C_{max} (mAh g^{-1})	S_{100} (%)	HRD_{900} (%)	HRD_{1200} (%)	HRD_{1500} (%)
#1	1.47	1.46	0.0196	0.0123	0.466	384	81.8	77.6	63.6	46.2
#2	1.47	1.46	0.0222	0.0135	0.497	389	82.5	79.5	66.2	47.5
#3	1.48	1.47	0.0263	0.0186	0.346	393	83.2	81.4	67.1	51.0
#4	1.45	1.45	0.0310	0.0233	0.286	384	84.1	82.2	71.6	54.4
#5	1.43	1.42	0.0438	0.0343	0.305	372	85.3	85	74.9	58.9
#6	1.33	1.33	0.0590	0.0459	0.251	353	86.7	89.1	79.7	63.9

pressure of the hydrogen desorption reaches 0.046 MPa. For the alloys containing both $(La,Mg)_2Ni_7$ and $(La,Mg)_5Ni_{19}$ phases, the plateau pressure is between those of single $(La,Mg)_2Ni_7$ phase and $(La,Mg)_5Ni_{19}$ alloys, and the plateau pressure of those alloys increases with more amount of $(La,Mg)_5Ni_{19}$ phase. This observation is in accordance with the trend of decreasing of phase cell parameters with increasing $(La,Mg)_5Ni_{19}$ phase. Moreover, the isotherm curves of alloys with both $(La,Mg)_2Ni_7$ and $(La,Mg)_5Ni_{19}$ phases also exhibit one hydrogen absorption/desorption plateau, indicating that the $(La,Mg)_2Ni_7$ and $(La,Mg)_5Ni_{19}$ phases have close equilibrium pressures due to their similar super-stacking structures. This combined plateau with both $(La,Mg)_2Ni_7$ and $(La,Mg)_5Ni_{19}$ phases is different from that of the alloys containing both $(La,Mg)_2Ni_7$ stacking phase and $LaNi_5$ non-stacking phase whose P–C isotherms exhibit two discharge plateaus due to the significant difference of the plateau pressures in $(La,Mg)_2Ni_7$ and $LaNi_5$ phases.

From Table 3.8 and Figure 3.10, it can be seen that the hydrogen storage capacity is initially increased and then decreased with increasing amount of $(La,Mg)_5Ni_{19}$ phase abundance. The optimal hydrogen storage capacity (1.479 wt.%) is obtained when the $(La,Mg)_5Ni_{19}$ phase abundance is 10.5 wt.%. The following factors contribute to the increase of the hydrogen storage capacity: firstly, $(La,Mg)_5Ni_{19}$ phase with higher Ni content has a catalytic effect on the hydrogen absorption/desorption of $(La,Mg)_2Ni_7$ phase. Besides, when the $(La,Mg)_5Ni_{19}$ phase appears as a secondary phase, the amount of the crystal boundaries increases. The high density of the phase boundaries promotes the formations of coherent and catalytic interfaces between $(La,Mg)_2Ni_7$ and $(La,Mg)_5Ni_{19}$ phases, thus improving the hydrogen storage performance of the alloys. However, as we know, the theoretical hydrogen storage capacity of $(La,Mg)_5Ni_{19}$ phase is lower than that of the $(La,Mg)_2Ni_7$ phase, so the hydrogen absorption/desorption capacity decreases when $(La,Mg)_5Ni_{19}$ phase amount further increases.

Reversibility between hydrogen absorption and desorption is an important criterion in the practical application of hydrogen storage alloys. Hysteresis factor (H_f) is usually employed to evaluate the reversibility between hydrogen absorption and desorption, and it can be calculated by the following equation [19]:

$$H_f = \ln\left(P_{abs}/P_{des}\right) \tag{3.5}$$

where P_{abs} is the equilibrium pressure of hydrogen absorption andP_{des} is the equilibrium pressure of hydrogen desorption. The calculated H_f values of the alloys at room temperature are presented in Table 3.8. It is shown that H_f values of the tested alloys decreased with increasing $(La,Mg)_5Ni_{19}$ phase abundance, indicating that the reversibility of the alloys is enhanced by the addition of $(La,Mg)_5Ni_{19}$ phase.

3.3.1.3 Discharge Capacity and Cycle Stability

The discharge capacities of the alloy electrodes within 100 cycles are shown in Figure 3.11, and the summary information of the curves is presented in Table 3.8. When the negative electrodes of Ni/MH batteries are charged for the first time, they absorb relatively small amount of hydrogen, leading to the expansion of cell volume, thus some of the large particles of the original alloys break into smaller ones and fresh particle surfaces are created. Thanks to the expansion of the cell volume and the increase of the specific surface area, more hydrogen atoms are absorbed in the subsequent cycles until the maximum hydrogen storage capacity is obtained. This process is called 'activation'. The fewer cycles needed for the batteries to reach the maximum discharge capacity, the better the activation of the batteries is. It can be seen from the inset of Figure 3.11 that the alloy electrodes show good activation performance and they can be fully activated within two cycles. As expected, the evolution of electrode discharge capacity essentially follows the same trend as the hydrogen storage capacity that is demonstrated by the *P–C* isotherms. The maximum discharge capacity (C_{max}) of the alloy electrodes initially increases and then decreases with increasing $(La,Mg)_5Ni_{19}$

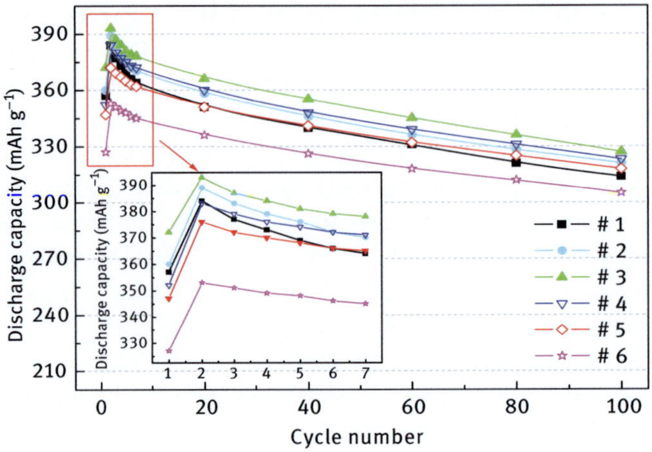

Figure 3.11: Electrochemical discharge capacity within 100 cycles of the alloy electrodes.

phase abundance, and the maximum value is obtained when $(La,Mg)_5Ni_{19}$ phase abundance is 10.5 wt.%. Furthermore, it is noted that the maximum discharge capacity drops fast when $(La,Mg)_5Ni_{19}$ phase abundance exceeds 10 wt.%, thereby confirming that the amount of $(La,Mg)_5Ni_{19}$ phase should not be too high in the A_2B_7-type La–Mg–Ni-based alloys to guarantee a satisfying discharge capacity.

The cycling stability of the alloy electrodes within 100 cycles is also shown in Figure 3.11 by the slopes of the curves. Smaller slope means better cycling stability. It is seen that the slopes of the curves become gentle as $(La,Mg)_5Ni_{19}$ phase abundance increases. The cycling stability of the alloy electrodes at the 100th cycle (S_{100}) increases from 81.8 to 85.3 % when $(La,Mg)_5Ni_{19}$ phase abundance increases from 0 to 37.4 wt.%, as shown in Table 3.8. To explore the mechanism of cycle life enhancement of the A_2B_7-type alloys by introducing $(La,Mg)_5Ni_{19}$ phase, the microstructure of the #1 and #5 alloy powders after cycling are compared and the possible reasons are discussed below.

The lattice stress due to the expansion/contraction of the phase cells is inevitable when hydrogen atoms enter and leave the interstitial sites of the lattice, and it is the real driving force that leads to the pulverization of La–Mg–Ni-based alloys [27]. The XRD patterns of the #1 and #5 alloy powders after 100 electrochemical charge/discharge cycles are presented in Figure 3.12(a), and the zoom of the 2θ range between 34–44° is shown in the inset the figure. It can be seen that after 100 cycles the XRD pattern of the #5 alloy keeps almost unchanged, while the peaks of the #1 alloy are apparently widened and the peak intensity is much smaller than that of the #5 alloy. Moreover, the strain of the cycled #1 alloy labelled in the figure is much higher than that of the #5 alloy after 100 cycling. These phenomena indicate that the #1 alloy suffers more lattice distortion and deformation during hydrogen atoms drifting in and out of the cell lattice. As for the comparison of FWHM, because the

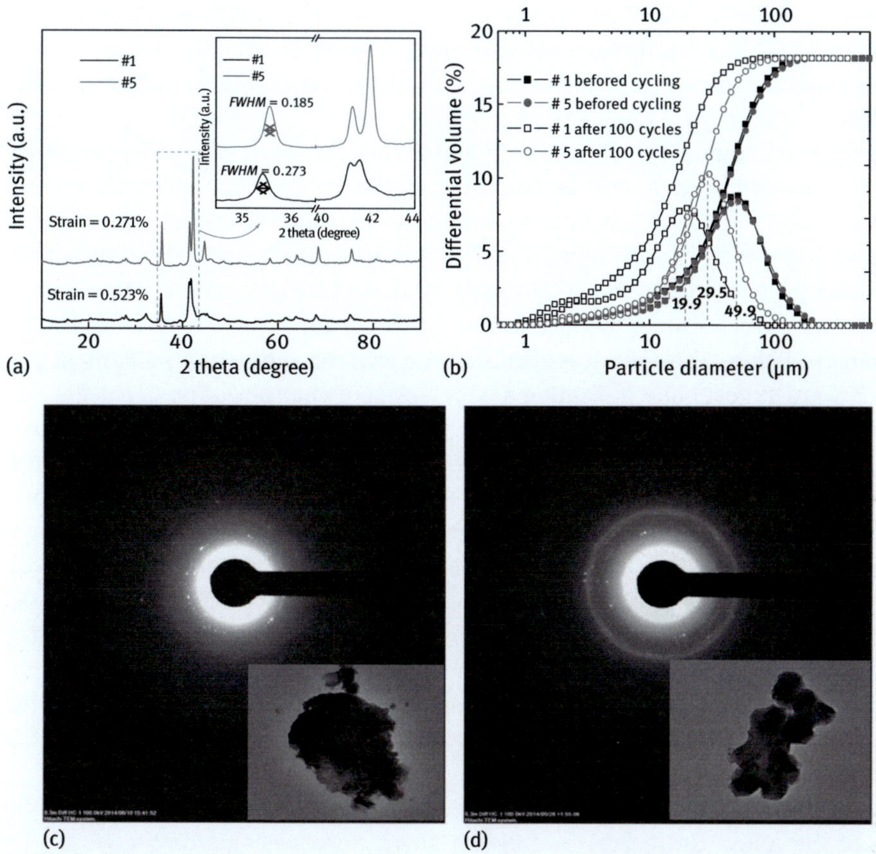

Figure 3.12: XRD patterns of the #1 and #5 alloys after 100 charge/discharge cycles (a); particle size distributions of the #1 and #5 alloys before and after 100 charge/discharge cycles (b); and TEM electron diffraction patterns of the #1 and #5 alloys after 200 hydrogen absorption/desorption cycles (c–d).

first main peak of the #1 alloy is not clearly separated with its neighbouring peak, the second main peak at around 35° was used. It can be seen that the second main peak of the #1 alloy is apparently wider than that of the #5 alloy, and the calculated FWHM value of the #1 alloy is higher, indicating finer grains and larger amorphous tendency of the #1 alloy after 100 charge/discharge cycles. Correspondingly, the particle size distribution of the #1 and #5 alloys before and after 100 charge/discharge cycles are presented in Figure 3.12(b). It can be seen that the mean particle diameter of the alloys before cycling is similar; while after 100 charge/discharge cycles, the mean particle diameter of the #1 alloy powder is 19.9 μm while that of the #5 alloy powder is 29.5 μm, demonstrating less pulverization of the #5 alloy during charge/discharge cycling. Alloys with less pulverization are usually less sensitive to electrolyte corrosion during charge/discharge due to their bigger particles. The oxygen contents for the #1 and #5 alloy powders after 100 electrochemical cycles in 6 M KOH alkaline solution are 3.71 and 2.82 mass%, respectively. The lower oxygen amount of the cycled #5 alloy electrode indicates less consumption of the active materials, resulting in better cycle life.

In addition, our previous research found that the alloy showed amorphization accompanied with phase decomposition after 20 gaseous hydrogen absorption/desorption cycles. This is also considered to be a significant factor leading to the capacity decrease of A_2B_7-type La–Mg–Ni-based alloys. To show the status of the alloys after gaseous cycles, the TEM images and electron diffraction patterns of #1 and #5 alloy particles after 200 gaseous hydrogen absorption/desorption cycles are presented in Figure 3.12(c–d). It is seen that the electron diffraction patterns in Figure 3.12(c) are almost halo, indicating a large extent of amorphization of #1 after 200 cycles. While clear rings can be seen from Figure 3.12(d), which demonstrates that most part of the #5 alloy still keeps in the form of crystallization after 200 cycles. It was previously reported that the $[A_2B_4]$ subunit was the main hydrogen accommodating part in the super-stacking structure and this subunit is unstable when expanding and contracting during hydrogen absorption/desorption, resulting in collapse of the stacking structure [28]. The $(La,Mg)_5Ni_{19}$ phase with higher $[AB_5]/[A_2B_4]$ ratio than $(La,Mg)_2Ni_7$ phase possesses stronger structural stability during repeated hydrogen absorption/desorption, and the network of the $(La,Mg)_5Ni_{19}$ phase distributing among $(La,Mg)_2Ni_7$ main phase (demonstrated in the SEM part) can stabilize the structure of the alloy during hydrogen absorption/desorption process, and lower the amorphization.

In general, the $(La,Mg)_5Ni_{19}$ phase network distributing in $(La,Mg)_2Ni_7$ main phase of the alloys decreases the internal stress during charge/discharge, alleviates the pulverization and oxidation of the alloy electrodes during battery cycling, as well as stabilizes the stacking structure against amorphization during hydrogen absorption/desorption. Therefore, the charge/discharge cycling stability of the alloys can be improved by the addition of $(La,Mg)_5Ni_{19}$ phase.

3.3.1.4 High-Rate Dischargeability

HRD reflects the discharge capability of the batteries at high discharge current densities, and it can be calculated by the following equation:

$$HRD_n = C_n / (C_n + C_{60}) \times 100\% \quad (3.6)$$

HRD_n represents the *HRD* at a discharge current density of n mA g^{-1}, C_n is the discharge capacity at a discharge current density of n mA g^{-1} and C_{60} is the residue discharge capacity at a discharge current density of 60 mA g^{-1}. The *HRD* values of the alloy electrodes at the discharge current density of 900, 1,200 and 1,500 mA g^{-1} are shown in Table 3.8. It can be seen that the addition of $(La,Mg)_5Ni_{19}$ phase renders a tremendous impact on *HRD*. The *HRD* at various discharge current densities increases with more

$(La,Mg)_5Ni_{19}$ phase content. The HRD_{1500} of the alloy electrode containing 37.4 wt.% of $(La,Mg)_5Ni_{19}$ phase reaches 58.9 % compared to 46.2 % for the alloy electrode with only $(La,Mg)_2Ni_7$ phase.

To find out the reason for the significant improvement in the *HRD* of the alloy electrodes with increasing $(La,Mg)_5Ni_{19}$ phase abundance, the changes in the microstructure of the alloy phases during discharge process are studied. As we know, the discharge process of the negative electrodes of Ni/HM batteries is actually a dehydrogenation process which is accompanied with the phase transformation from full hydride to H-dissolved solid solution. When the alloy is fully hydrogenated, the lattice expansion reaches a peak value, and the XRD peaks shift to the lowest angle correspondingly. In the discharge process, the cell volumes become smaller and the XRD peaks shift to higher angle direction. The XRD patterns of the #5 alloy powder at the discharge depths of 0, 25, 50, 75 and 100 % are shown in Figure 3.13. It can be seen that the main peaks of $(La,Mg)_2Ni_7$ and $(La,Mg)_5Ni_{19}$ phases are almost overlapped before discharge. But the main peaks of the two phases apparently separated during discharge process, and finally, overlapped again after the complete discharge process. However, because the position for the main peaks of $(La,Mg)_2Ni_7$ and $(La,Mg)_5Ni_{19}$ phases are too close to each other, the XRD patterns within the range of 20–38° where the peaks of the two phases can be clearly identified are presented in Figure 3.13(b). It can be seen that the peaks of $(La,Mg)_5Ni_{19}$ phase always move to the high angle (right hand) earlier than that of $(La,Mg)_2Ni_7$ phase in the whole discharging process. The above analysis indicates that the discharge process of the $(La,Mg)_2Ni_7$ and $(La,Mg)_5Ni_{19}$ phases in the alloy is not synchronous. The $(La,Mg)_5Ni_{19}$ phase first starts to desorb hydrogen. The initial discharge of $(La,Mg)_5Ni_{19}$ phase decreases the hydrogen concentration and establishes a lower hydrogen pressure around $(La,Mg)_2Ni_7$ phase, thus leading to the hydrogen desorption of the $(La,Mg)_2Ni_7$ phase. Similar behaviour was also observed for high-Ni $LaNi_5$ phase during the discharge process of $(La,Mg)Ni_3$ phase [29]. It is seen that $(La,Mg)_5Ni_{19}$ phase not only has a faster hydrogen desorption rate, but also generates catalytic effect on the discharge process of $(La,Mg)_2Ni_7$ phase, contributing to the improvement in the *HRD* of the A_2B_7-type alloy electrodes.

3.3.2 The Interaction between A_2B_7-Type Phase Structure and AB_5-Type Phase Structures

As mentioned above, in the A_2B_7-type La–Mg–Ni-based alloys, the $(La,Mg)_2Ni_7$ phase often exists by the accompany of some minor phases among which $LaNi_5$ phase is the most commonly seen due to its relatively high melting point and large solubility range. Moreover, $LaNi_5$ phase is considered to be responsible for catalysing hydrogen absorption/desorption behaviour as well as fast charge/discharge processes of LaeMgeNi-based alloys. In this section, the interaction between A_2B_7-type phase structure and AB_5-type phase structures is discussed by the comparative study in terms

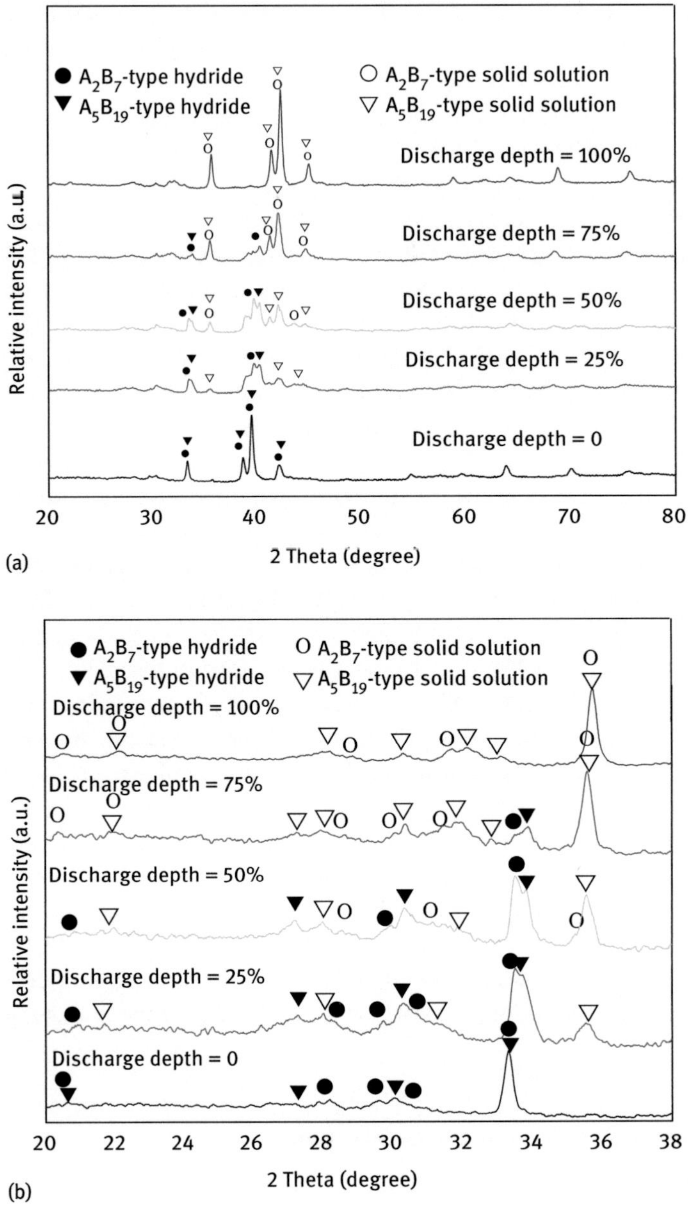

Figure 3.13: XRD patterns of the #5 alloy at various discharge depths: (a) XRD patterns between 20° and 80° for overall observation; (b) XRD patterns between 20° and 38° for detailed observation.

of microstructure and electrochemical characteristics of La–Mg–Ni-based alloys with only $(La,Mg)_2Ni_7$ phase and those with various amounts of $(La,Mg)_2Ni_7$ and $LaNi_5$ phases.

The $La_{0.75}Mg_{0.25}Ni_{3.5}$ compounds with different amounts of 2H- and 3R-type allotropes of $(La,Mg)_2Ni_7$ phase were prepared by the following steps: the as-cast $La_{0.75}Mg_{0.25}Ni_{3.5}$ alloy was ball milled for 1 h during which a small amount of $MgNi_2$ was added to compensate Mg loss in the next sintering process. Afterwards, the powders were cold pressed into pellets which were then stepwise sintered in the temperature range of 873–1,253 K. After the sintering process, the pellets were annealed at 1,073, 1,123, 1,173 and 1,223 K for 24 h, respectively (notified as #A_0, #A_1, #A_2 and #A_3) to obtain various polymorphic structures (2H- and 3R-types). Several attempts had been made before the correct weight ratio and the optimum preparative conditions were determined. The purposes of ball milling were mixing the powers and evening the treating conditions with the $LaNi_5$ doped alloys to make convincing comparisons. The alloys with various amounts of $LaNi_5$ phase were prepared by the following steps: firstly, 5, 10, 20 and 30 wt.% of $LaNi_5$ alloy powders were added into the as-cast $La_{0.75}Mg_{0.25}Ni_{3.5}$ alloy powders, respectively (notified as #B, #C, #D and #E, respectively). Besides, $MgNi_2$ powders were added in the same amount with that in the preparation of $La_{0.75}Mg_{0.25}Ni_{3.5}$ alloy to compensate for Mg loss. Then the mixtures were ball milled for 1 h followed by stepwise sintering in the range 873–1,253 K. The purpose of the ball milling process was mixing the $LaNi_5$, $MgNi_2$ and $La_{0.75}Mg_{0.35}Ni_{3.5}$ compounds. Finally, the alloys were annealed at 1,073 K for 24 h for homogeneity. All the handles were operated under the Ar atmosphere.

3.3.2.1 Microstructure

The XRD patterns of the alloy samples are shown in Figure 3.14. We can see that the alloy samples #A_0–#A_3 which were ball milled without $LaNi_5$ alloy contain only $(La,Mg)_2Ni_7$ phase in both 2H- and 3R-type polymorphic structures within the annealing temperature range of 1,073–1,123 K. It is found from Rietveld analysis results of #A_0–#A_3 that when the annealing temperature are 1,073, 1,123, 1,173 and 1,223 K, the relative abundances of the 2H-type $(La,Mg)_2Ni_7$ phase are 52.10. 50.81, 48.76, and 38.43 wt.%, and those of 3R-type $(La,Mg)_2Ni_7$ phase are 47.90, 49.19, 51.24 and 61.57 wt.%, respectively. It is notable from the results that the change in the 2H- and 3R-type phase abundances is very slightly within the temperature range of 1,073–1,173 K, while the 2H-type abundance decreases and 3R-type abundance increases dramatically once the temperature rises to 1,223 K. This indicates that the crystal transformation from 2H-type $(La,Mg)_2Ni_7$ phase to 3R-type $(La,Mg)_2Ni_7$ phase occurs at 1,223 K while little crystal transformation occurs at lower temperature. In the La–Ni binary diagram, the temperature point of phase transformation between 2H- and 3R-type allotropes in La_2Ni_7 phase is 1,253 K. This distinction in crystal transformation temperature indicates that the addition of the Mg element lowers the transformation temperature of 2H- and 3R-type allotropes of A_2B_7 phase.

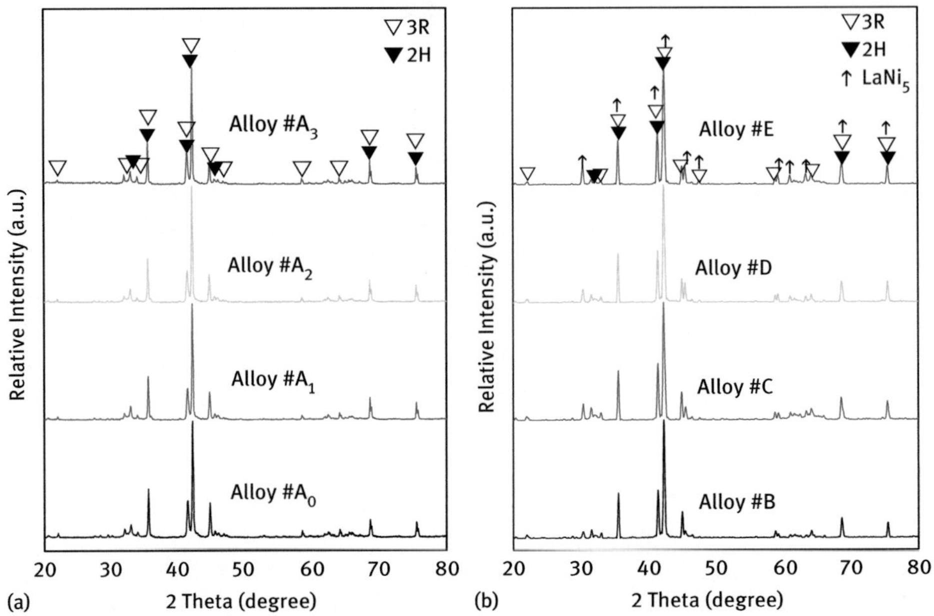

Figure 3.14: XRD patterns of the alloy samples: (a) single compounds annealed at various temperatures: alloy #A_0–# A_3; (b) compounds with $LaNi_5$ addition: alloy #B–#E.

The backscattered SEM image of alloy #A_0 is shown in Figure 3.15(a). It can be seen that the morphology of the alloy is homogenous except for some small pores which might be left during cooling. As we know, multiple-phase compositions were usually found in the La–Mg–Ni-based alloys prepared by a conventional melting method, even annealing is unlikely to obtain alloys with single composition. This is largely because the distance between areas with big compositional difference is relatively far for atomic migration in a bulk alloy. However, by grounding the as-cast bulk alloy into powders, the bulks can be split and thoroughly mixed so that the distance of areas with different compositions is shortened and the compositions can be more homogenous after heat treatment. After thorough mixture, the heat treatment is more effective to eliminate minor phases and homogenize the alloy composition. Figure 3.15(b) is the EDS for elemental composition of area A from Figure 3.15(a). The composition result from the EDS corresponds well with the XRD and ICP results. Figure 3.15(c–e) presents the elemental distribution patterns for La, Mg and Ni elements, respectively. It can be seen that the elements are homogeneously distributed in the alloy. The even composition of the alloy is due to the depletion of minor phases by peritectic reaction during sintering process and the thorough atomic diffusion during long-time annealing process.

Figure 3.14(b) shows the XRD patterns of the alloy with the addition of various amounts of $LaNi_5$ compound. It can be seen that adding $LaNi_5$ alloy leads to the

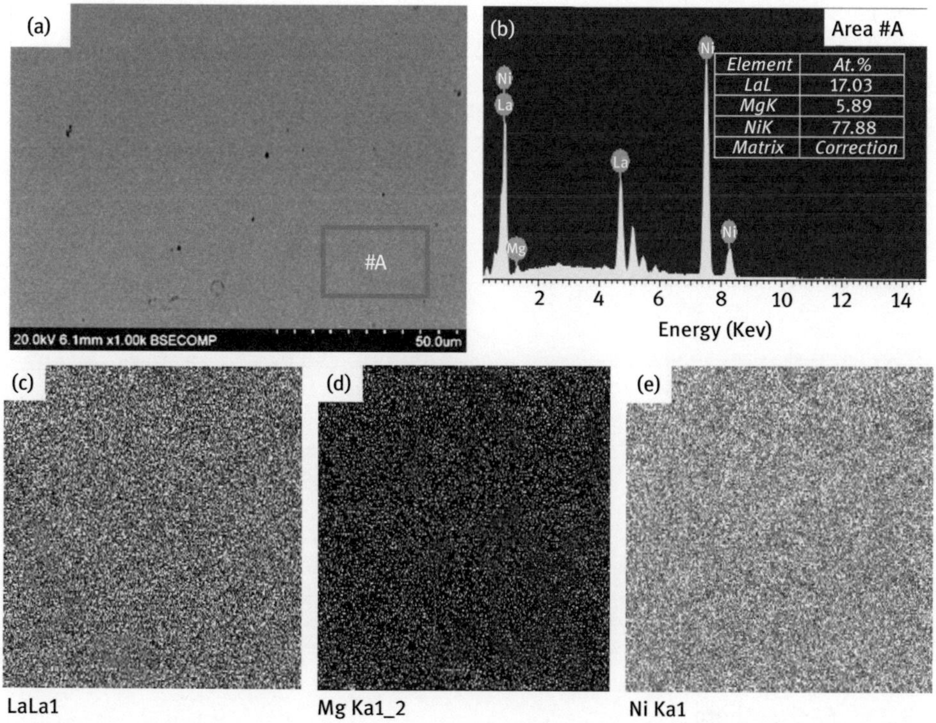

Figure 3.15: (a) SEM backscattering electron image; (b) EDS pattern; (c) EDS elemental mappings of La; (d) Mg; (e) Ni for #A_0.

emergence of $LaNi_5$ phase. But no $(La,Mg)_5Ni_{19}$ phase appears although the chemical composition of the alloys shifts towards Ni direction in the La–Ni diagram after $LaNi_5$ compound is introduced. This is because of the low annealing temperature. Zhang et al. [20] reported that $(La,Mg)_5Ni_{19}$ phase decomposed into $LaNi_5$ and $(La,Mg)_2Ni_7$ phases when the annealing temperature is below 840°C. The phase abundances and lattice parameters of alloy #A_0, #B, #C, #D and #E all of which were annealed at 1,073 K are listed in Table 3.9. The $LaNi_5$ phase abundance increases with the increase of $LaNi_5$ alloy added during ball milling. Moreover, the lattice parameters and cell volumes of the 2H- and 3R-type $(La,Mg)_2Ni_7$ phases decrease with the increase of the $LaNi_5$ phase abundance. Noticeably, the ratio of 2H- to 3R-type $(La,Mg)_2Ni_7$ phase abundance also decreases with the increase of $LaNi_5$ phase abundance. Buschow et al. [30] and Zhang et al. [21] reported that the crystal structure of the R_2Ni_7 compounds is size-dependent. For larger R-atomic radii, the hexagonal form (Ce_2Ni_7-type) is preferred while for the smaller R-atom, the rhombohedra form (Gd_2Co_7-type) is preferred. Therefore, we consider that the change in the abundance ratio of 2H- to 3R-type $(La,Mg)_2Ni_7$ phases in our study is related to the shrinkage in cell parameters of 2H- and 3R-type $(La,Mg)_2Ni_7$ phases.

Table 3.9: Phase abundances and lattice parameters of alloy samples.

Alloys	Phase	Phase abundance (%)	*a* (Å)	*c* (Å)	V(Å^3)	2*H*/3*R*
#A_0	2H	52.10	5.0406	24.2823	534.300	1.08
	3R	47.90	5.0418	36.3039	799.199	
#B	2H	43.74	5.0400	24.2559	533.592	0.89
	3R	48.96	5.0412	36.2723	798.314	
	$LaNi_5$	6.19	5.0357	3.9913	87.653	
	LaNi	1.11	–	–	–	
#C	2H	29.01	5.0399	24.2480	533.397	0.53
	3R	54.14	5.0395	36.2709	797.745	
	$LaNi_5$	15.75	5.0346	3.9861	87.500	
	LaNi	1.10	–	–	–	
#D	2H	9.99	5.0390	24.2138	532.454	0.16
	3R	60.49	5.0362	36.2504	796.250	
	$LaNi_5$	27.65	5.0329	3.9842	87.399	
	LaNi	1.87	–	–	–	
#E	2H	5.80	5.0383	24.2138	532.306	0.10
	3R	57.19	5.0364	36.2432	796.155	
	$LaNi_5$	36.19	5.0323	3.9830	87.352	
	LaNi	0.82	–	–	–	

3.3.2.2 *P–C* Isotherms

A high hydrogen storage capacity and moderate hydride stability are necessary for hydrogen storage alloys to be used as electrodes for nickel metal hydride batteries. These properties can be evaluated from *P–C* isotherms whose plateau pressure is used to predict the electrochemical potential of the hydrogen storage electrodes, and the length of the plateau is used to estimate the theoretical electrochemical capacity of the hydrogen storage alloys in the Ni/MH batteries [31]. The electrochemical discharge *P–C* isotherms of the alloys are shown in Figure 3.16. It is observed from Figure 3.16(a) that alloys #A_0–#A3 present just one plateau in the *P–C* isotherms while from Figure 3.16(b) it can be seen that the alloys containing $LaNi_5$ phase have an additional plateau at higher hydrogen pressure. Moreover, the lower plateau becomes shorter yet the higher plateau becomes longer as the $LaNi_5$ phase abundance increases. After a careful examination of the *P–C* isotherms of Figure 3.16(b), it is found that the second plateau pressure of the alloys initially increases and then decreases as $LaNi_5$ amount increases (0.074, 0.047, 0.033 and 0.049 MPa for 5, 10, 20 and 30 wt.% $LaNi_5$-doped alloys, respectively). It can be seen that the plateau pressure for #D is the lowest and the plateau is also the flattest. The catalytic effect of $LaNi_5$ phase and the crystallization of the alloys may contribute to this phenomenon.

The hydrogen storage capacities of the alloy electrodes calculated from the *P–C* isotherms are listed in Table 3.10. It can be seen that the discharge capacity of the $La_{0.75}Mg_{0.25}Ni_{3.5}$ alloys (#A_0–#A_3) are similar while the hydrogen storage capacity decreases monotonously as the $LaNi_5$ phase abundance increases from 0 to 36.19

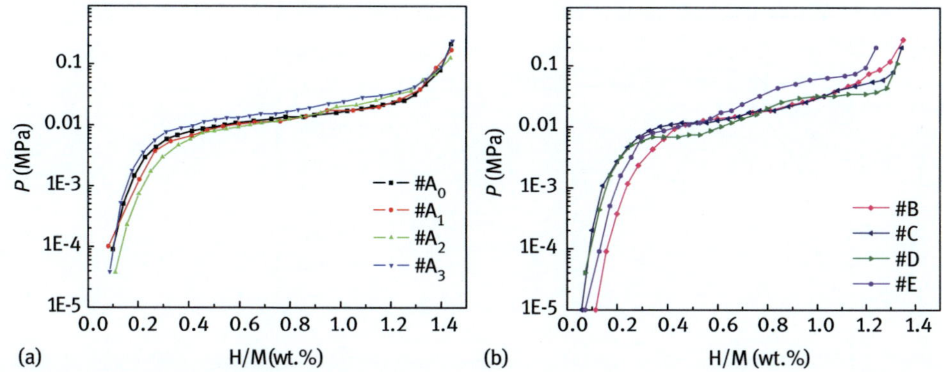

Figure 3.16: *P–C* isotherms of the alloys.

Table 3.10: Electrochemical characteristics of the alloy electrodes.

Samples	H/M[a] (wt.%)	N_a[b]	C_{max}[c] (mAh g^{-1})	HRD_{1500}[d] (%)	S_{100}[e] (%)
$\#A_0$	1.44	3	384	45.3	82.6
$\#A_1$	1.44	3	383	45.7	82.2
$\#A_2$	1.44	3	384	44.8	82.8
$\#A_3$	1.45	3	386	48.2	83.2
B	1.35	3	377	53.0	81.9
C	1.34	3	369	50.6	75.6
D	1.33	3	336	42.2	70.3
E	1.24	3	324	40.4	63.3

[a]Hydrogen storage capacity from *P–C* isotherms.
[b]The cycling numbers for the alloy electrodes to reach their maximum discharge capacity.
[c]The maximum discharge capacity of the alloy electrodes.
[d]The high discharge ability at a discharge current density of 1,500 mA g^{-1}.
[e]the Capacity retention at the 100th cycle.

wt.%, which is due to the fact that the hydrogen storage capacity of $LaNi_5$ phase is much lower than that of $(La,Mg)_2Ni_7$ phase.

3.3.2.3 Electrochemical Properties

The activation capability is characterized by the number of charge–discharge cycles required for attaining the maximum discharge capacity of the batteries. The activation curves are shown in Figure 3.17. It can be seen that all the alloy electrodes display good activation performance and can be fully activated within three cycles. Neither the annealing temperature nor the addition of $LaNi_5$ alloy renders obvious effect on the activating cycle of the alloy electrodes.

The maximum discharge capacity of the alloy electrodes is summarized in Table 3.10. The $La_{0.75}Mg_{0.25}Ni_{3.5}$ alloys $\#A_0$–$\#A_3$ with only $(La,Mg)_2Ni_7$ phase show

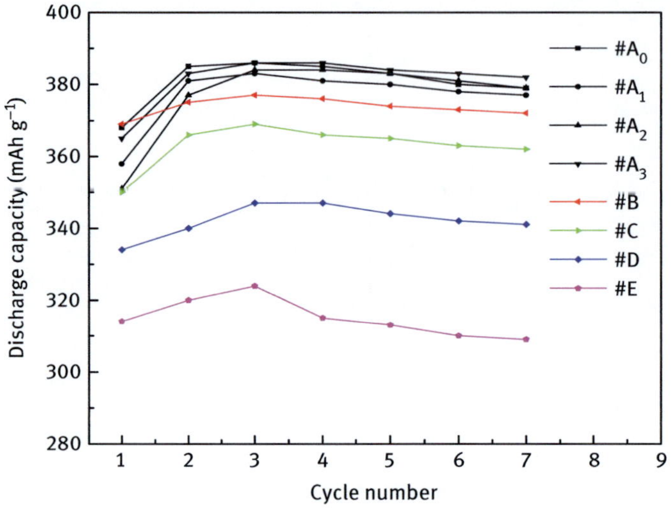

Figure 3.17: Activation curves of the alloy electrodes.

very similar maximum discharge capacity ranging within 5 mAh g^{-1} in spite of varying annealing temperatures and different amounts of 2H- and 3R-type allotropes. In a structure point of view, the two allotropes have very similar stacking structures and have exactly the same elemental compositions. Moreover, Zhang et al. reported that the charge–discharge behaviour of the 2H- and 3R-types is so similar that the difference could not be observed from *P–C–T* curves. The *P–C* isotherms of the alloys with only $(La,Mg)_2Ni_7$ phase in our study also show one plateau, indicating similar discharge behaviour of the two types.

As for the alloys containing various amounts of $LaNi_5$ phase, it is evident from Figure 3.17 that the maximum discharge capacity decreases with the increase of the $LaNi_5$ phase abundance. This changing trend is in agreement with that of the discharge capacity from the *P–C* isotherms.

Figure 3.18 shows the *HRD* curves of the alloy electrodes which reflect the discharge capability of the alloy electrodes at high current densities. *HRD* can be calculated by the following formula:

$$HRD_n = C_n/(C_n + C_{60}) \times 100\% \tag{3.7}$$

where HRD_n represents the *HRD* at a discharge current density of n mA g^{-1}, C_n represents the discharge capacity at a discharge current density of n mA g^{-1} and C_{60} is the residue discharge capacity at a discharge current density of 60 mA g^{-1}. The HRD_{1500} is listed in Table 3.10. From the nearly overlapped curves of #A_0–#A_3 electrodes in Figure 3.18, it can be known that when the annealing temperature is below 1,173 K, annealing can hardly affect the *HRD* of the alloys with 2H- and 3R-type allotropes of $(La,Mg)_2Ni_7$ phase. While when the temperature increases to 1,223 K where 2H-type

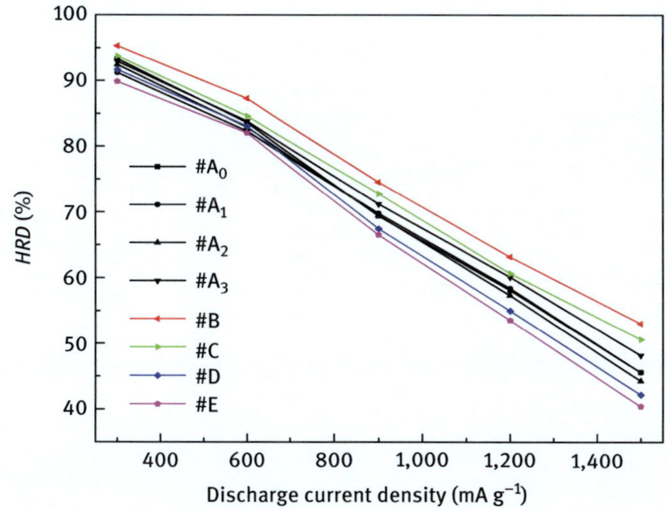

Figure 3.18: *HRD* as a function of discharge current density of the alloy electrodes.

$(La,Mg)_2Ni_7$ phase obviously transforms into 3R-type $(La,Mg)_2Ni_7$ phase, the *HRD* is improved. Zhang et al. [20] have proved with XRD patterns that the 3R-type hydride desorbed hydrogen ahead of 2H-type hydride so that the $La_{0.75}Mg_{0.25}Ni_{3.5}$ alloy with more 3R-type phase tends to discharge faster, and this effect is more apparent at high discharge current densities.

$LaNi_5$ minor phase exerts tremendous influence on the *HRD* property. As $LaNi_5$ phase abundance increases from 0 to 6.19 wt.%, the HRD_{1500} increases from 45.6 to 53.0 %, but further increasing $LaNi_5$ phase engenders an adverse effect. The improvement can be attributed to the following reasons: on one hand, the $LaNi_5$ phase induces higher ratio of 3R-type to 2H-type phase abundance. On the other, the $LaNi_5$ phase has a catalytic effect on the hydrogen storage kinetics [32]. Besides, the appearance of $LaNi_5$ phase offers pathway for hydrogen diffusion during discharging process. However, as the $LaNi_5$ phase amount further increases, the $(La,Mg)_2Ni_7$ phase with superior *HRD* property [33] is considerably reduced, which renders a negative effect on the discharge capability at large current densities.

The cycling stability is an essential factor that needs to be improved for La–Mg–Ni-based alloys to be applied as Ni/MH battery cathode materials. The capacity retention curves at every 20th cycle within 100 cycles are shown in Figure 3.19. The slopes reflect the degradation rate of the discharge capacity during charge–discharge cycling. Smaller slope means better cycling stability. It can be seen that the $La_{0.75}Mg_{0.25}Ni_{3.5}$ alloys with only $(La,Mg)_2Ni_7$ phase in different amounts of 2H- and 3R-type allotropes have very similar degradation behaviour which can be found from the nearly overlap degradation curves in Figure 3.19. The S_{100} of the alloy electrodes is listed in Table 3.10. The S_{100} values of the $La_{0.75}Mg_{0.25}Ni_{3.5}$ alloys $\#A_0$–$\#A_3$ are also very close.

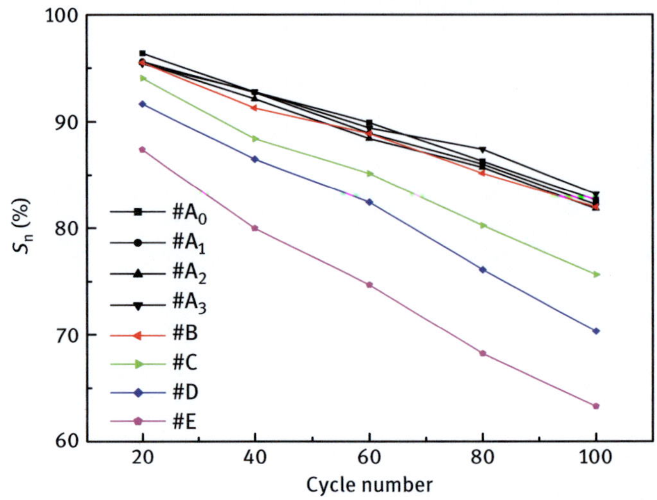

Figure 3.19: Discharge retention of the alloy electrodes within 100 cycles.

Capacity decay is due to the loss of active materials caused by both pulverization and oxidation of the alloys, and can be expressed by the percentage of capacity loss after n cycles ($1–S_n$). It can be seen that the capacity decay becomes more serious as the $LaNi_5$ phase abundance increases. Actually, the lattice internal stress is the real driving force that leads to the pulverization of the alloy. When $LaNi_5$ phase appears in the $La_{0.75}Mg_{0.25}Ni_{3.5}$ alloy, the difference in the expansion rate between the $LaNi_5$ phase and the $(La,Mg)_2Ni_7$ phase during hydrogen absorption/desorption happens, which aggravates the pulverization of the alloys. To prove the explanation above, Figure 3.20 shows the Rietveld analysis of the XRD pattern for hydrogenated alloy #C. The cell parameters and lattice expansion rates after hydrogenation are labelled in the figure. It can be seen that 2H- and 3R-type allotropes of $(La,Mg)_2Ni_7$ phase have close expansion rate in both a and c axis, while the expansion ratio of the lattices after hydrogenation is widely divergent between the $(La,Mg)_2Ni_7$ and $LaNi_5$ phases, especially for the c axis, which makes discordance in expansion during cycling.

Figure 3.21 shows the frequency distributions of particle size and the cumulative size of alloy samples #A_0 and #C after 40 electrochemical cycles. The particle size of alloys #A_0 and #C after cycling is spread in the range of 1.318–101.46 and 1.318–133.10 μm, respectively. It can be seen that the alloy particles of electrode #A_0 with only $(La,Mg)_2Ni_7$ phase spread in narrower range than that of #C, which further proves homogeneous compositions of #A_0. The volume mean particle diameters were measured to be 34.225 and 26.111 μm for #A_0 and #C, respectively. From the volume mean particle diameters it can be known that the general particle size of #A_0 is bigger than that of #C, indicating better pulverization resistance of the single-composition alloy. The pulverization of the alloys further accelerates the oxidation of the active

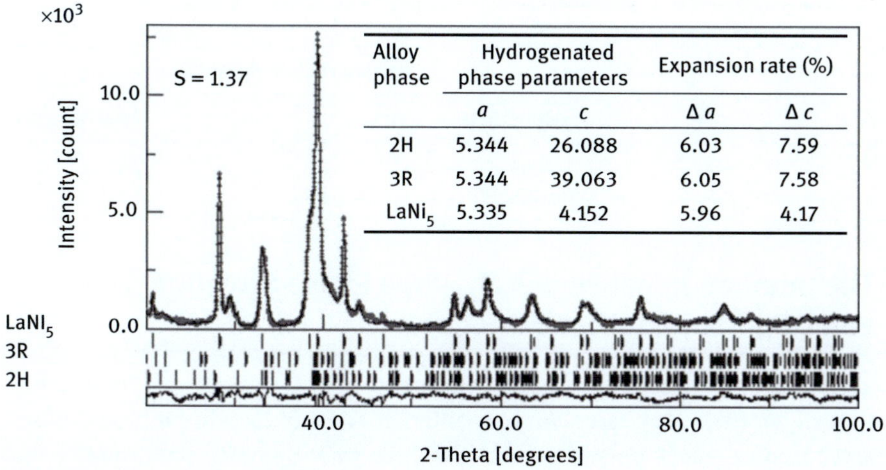

Figure 3.20: Rietveld analysis of alloy #C after hydrogenation.

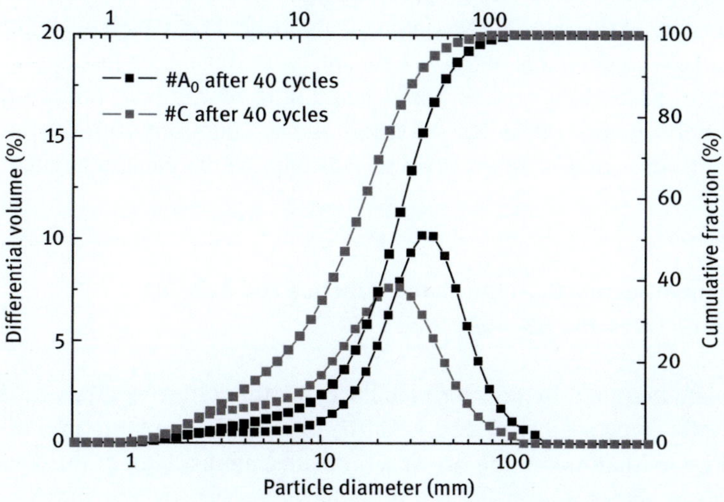

Figure 3.21: Particle size distribution of $\#A_0$ and #C after 40 charge/discharge cycles.

materials. The oxygen content of $\#A_0$ and #C before and after 40 electrochemical cycling in the alkaline electrolyte is shown Table 3.11. It can be seen that the oxygen content of the alloy containing $LaNi_5$ phase is apparently higher than that of the single-composition alloy, indicating that the $LaNi_5$ containing alloy electrode decays faster during cycling in the alkaline electrolyte than the single-composition alloy electrode.

Table 3.11: Oxygen content of the alloy electrodes before cycle and after 40 cycles.

Alloy	Oxygen content (mass%)	
	Before cycle	After 40 cycles
#A_0	0.0019	1.69
#C	0.0016	3.07

3.4 The Interaction between A_5B_{19}-Type Phase Structure and Other Superlattice Structures

This section mainly discusses the interaction between A_5B_{19}-type phase structure and other superlattice structures. Among different types of La–Mg–Ni-based alloys, Pr_5Co_{19}-type phase alloys possess remarkable high-rate discharge performance and satisfying cycling stability. Liu et al. [34] reported that $(La,Mg)_5Ni_{19}$ alloy was able to achieve the discharge capacity of 140 mAh g^{-1} even at the discharge current density as high as 3,600 mA g^{-1}. The alloy also exhibited good cycling stability in both repeated hydriding/dehydriding and charging/discharging processes. However, the hydrogen storage capacity of the $(La,Mg)_5Ni_{19}$ phase is relatively low, which deteriorates the alloy's overall electrochemical performance. The preceding examples suggest that the appearance of secondary phases has a facilitating or catalytic effect on the electrochemical performance of La–Mg–Ni-based alloys. Therefore, introducing a secondary phase into the alloy is an effective way to improve its electrochemical properties.

3.4.1 The Interaction between A_5B_{19}-Type Phase Structure and A_2B_7-Type Superlattice Structures and AB_5-Type Phase

The alloy samples were prepared by powder metallurgy by the following steps: two precursors, $LaNi_5$ ($CaCu_5$-type structure) and $LaMgNi_4$ ($MgCu_4Zn$-type structure) alloys, were obtained by induction melting under a protective atmosphere of purified argon gas. All the elements were with the purity higher than 99.9 %. The precursors were crushed into powders under 300 meshes and mixed in various molar ratio ($LaNi_5/LaMgNi_4$) denoted as x, in which $x = 0.75, 0.95, 1.2$ and 1.5, respectively. To compensate for Mg volatility losses during the sintering process, $LaMgNi_4$ was excessively added. Then the mixtures with a total weight of 2.5 g were pressed into pellets of 10 mm in diameter, 8 mm in thickness under 10 MPa. Each sample pellet was wrapped in Ni foil separately and sintered at 1,173 K for 6,900 min in a quartz tube under 0.04 MPa Ar pressure. There are three heat preservation stages between 873 and 1,073 K in the process of rising temperature. Afterwards, the pellets were quenched to room temperature.

3.4.1.1 Crystal Structure

The XRD patterns of alloy samples with $x = 0.75$, 0.95, 1.2 and 1.5 are shown in Figure 3.22. The Rietveld refinement patterns are presented in Figure 3.23 and the results are summarized in Table 3.12. It can be seen that there is only Pr_5Co_{19}-type phase structure existing in the alloy with $x = 1.2$. The other three alloys all consist of two phases with Pr_5Co_{19}-type phase as the main phase. When x increases from 1.2 to 1.5, the characteristic peaks of $LaNi_5$ phase appear. The abundance of Pr_5Co_{19}-type phase is 79.3 wt.% and that of the $LaNi_5$ phase is 20.7 wt.%. When x decreases to 0.95, the secondary phase turns into Ce_2Ni_7-type phase with 19.8 wt.% content. As x further descends to 0.75, the abundance of Pr_5Co_{19}-type phase decreases to 59.4 wt.% and the Ce_2Ni_7-type phase abundance increases to 40.6 wt.%. In addition, the refinement in the 2θ range 30–35° in Figure 3.23(c) and (d) is relatively poor. It indicates that stacking faults of the sub-units occurs along the c axis of the 5:19 phase owing to the lower Mg content [35–38].

The lattice parameters and unit cell volumes listed in Table 3.12 show that for the Pr_5Co_{19}-type single-phase structure, $a = 0.503$ nm and $c = 3.222$ nm and a volume per formula unit is 0.707 nm^3. For the two-phase alloys, the corresponding cell parameters of the Pr_5Co_{19}-type phase are larger than those of the single Pr_5Co_{19}-type phase. The expansion of the cell parameters is due to the reduction of Mg occupation ratio in Pr_5Co_{19}-type phase. The Mg occupation ratios in corresponding La sites are listed in Table 3.12. As seen, the Mg occupation ratios of Pr_5Co_{19}-type phase in two-phase alloy are less than that of the single Pr_5Co_{19}-type phase. It is known that Mg atoms occupy La1 ($4f_1$) sites in the $[A_2B_4]$ slab for Ce_2Ni_7-type phase and La3 ($4f_2$) sites for Pr_5Co_{19}-type phase. Neither other La sites nor Ni sites are filled with Mg atoms. The atomic radius of Mg (0.1602 nm) is smaller than that of La (0.1897 nm). When the Mg

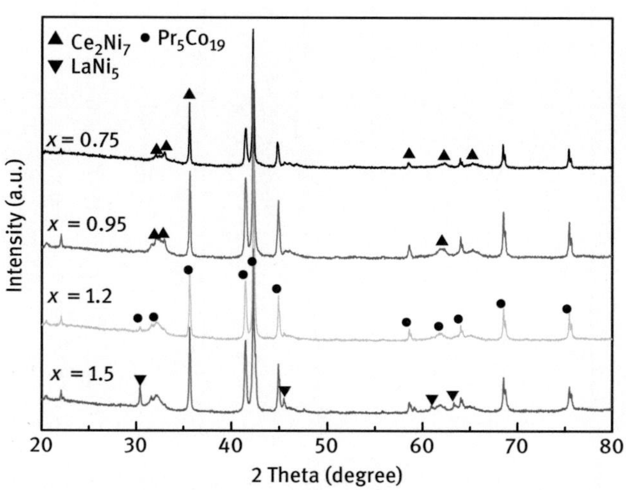

Figure 3.22: XRD patterns of the alloys.

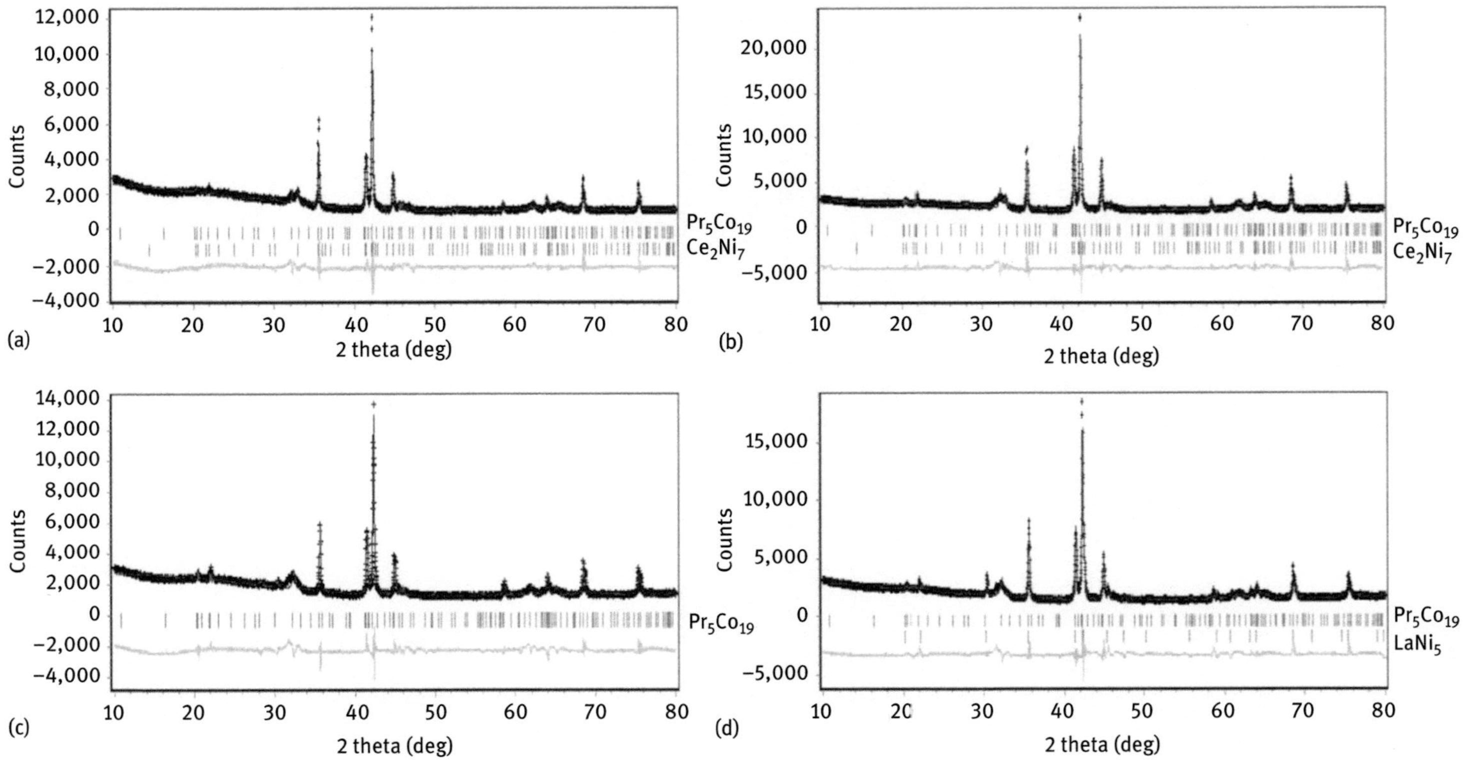

Figure 3.23: Rietveld refinements of the XRD patterns for the alloys with (a) $x = 0.75$ ($R_p = 4.7$, $R_{wp} = 6.5$), (b) $x = 0.95$ ($R_p = 5.5$, $R_{wp} = 7.8$), (c) $x = 1.2$ ($R_p = 5.8$, $R_{wp} = 8.0$) and (d) $x = 1.5$ ($R_p = 5.4$, $R_{wp} = 7.8$).

Table 3.12: Lattice constants and phase abundances of the alloys.

Alloys	Phase	Type	*a* (Å)	*c* (Å)	*v* (Å^3)	*c*/*a*	Mass fraction (wt.%)	Mg occupation ratio (%)
$x = 0.75$	$(La,Mg)_5Ni_{19}$	Pr_5Co_{19}	5.0340	32.302	708.61	6.42	59.4	46.5
	$(La,Mg)_2Ni_7$	Ce_2Ni_7	5.0259	24.179	528.94	4.81	40.6	51.9
$x = 0.95$	$(La,Mg)_5Ni_{19}$	Pr_5Co_{19}	5.0328	32.243	707.03	6.41	80.2	48.4
	$(La,Mg)_2Ni_7$	Ce_2Ni_7	5.0285	24.219	530.34	4.82	19.8	48.3
$x = 1.2$	$(La,Mg)_5Ni_{19}$	Pr_5Co_{19}	5.0317	32.224	706.55	6.40	100	53.6
$x = 1.5$	$(La,Mg)_5Ni_{19}$	Pr_5Co_{19}	5.0321	32.228	706.67	6.40	79.3	50.4
	$LaNi_5$	$CaCu_5$	5.0307	3.9824	87.290	0.79	20.7	–

occupation ratio of Pr_5Co_{19}-type phase decreases, the corresponding cell parameters get larger.

Figure 3.24 shows the SEM images of the alloys and Table 3.13 exhibits the element compositions of the phases. It can be seen that the alloy with $x = 1.2$ has only single phase and the apparent atomic ratio of La, Mg and Ni in mapping E is close to

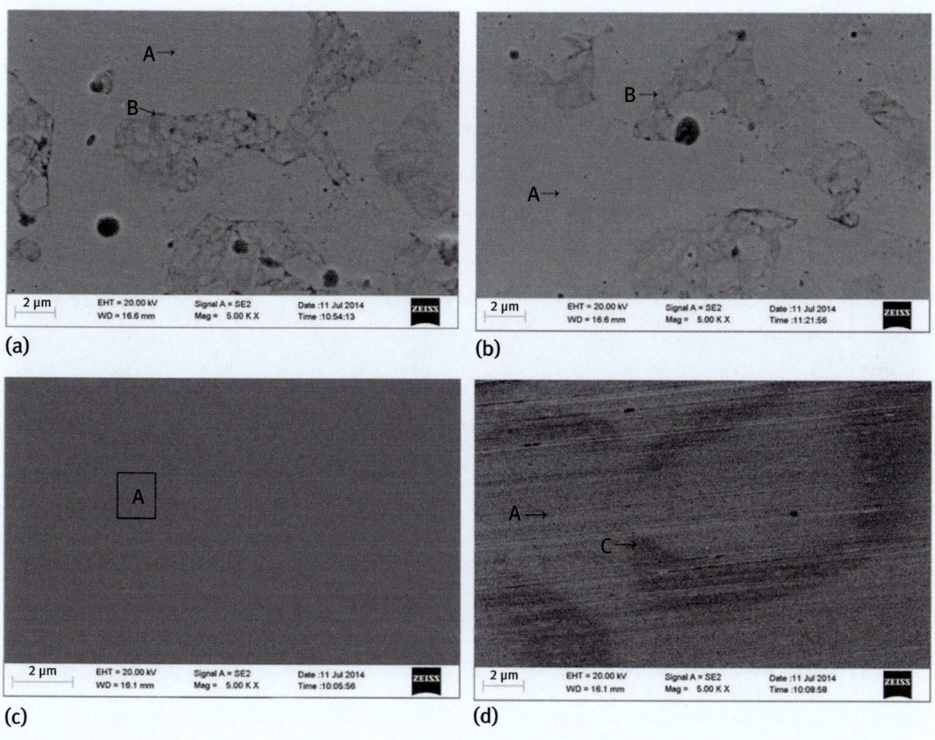

Figure 3.24: SEM images of the alloys with (a) $x = 0.75$, (b) $x = 0.95$, (c) $x = 1.2$ and (d) $x = 1.5$.

Table 3.13: Element compositions of the phases for the alloys.

	Mapping	La (wt.%)	Mg (wt.%)	Ni (wt.%)	La:Mg:Ni (atomic ratio)
$x = 0.75$	A	34.58	0.95	64.47	0.86:0.14:3.81
	B	33.09	1.82	65.08	0.76:0.24:3.52
$x = 0.95$	C	34.27	1.02	64.71	0.85:0.15:3.81
	D	33.52	1.77	64.70	0.77:0.23:3.51
$x = 1.20$	E	33.98	1.16	64.85	0.84:0.16:3.80
$x = 1.50$	F	34.38	0.97	64.65	0.86:0.14:3.83
	G	32.66	0.12	67.22	1:0:4.9

4.2:0.8:19 which coincides with the ICP chemical analysis. The other three alloys are both two-phase structure. For $x = 0.75$ and 0.95 alloys, $(La,Mg)_2Ni_7$ phase appears in $(La,Mg)_5Ni_{19}$ phase and there are distinct phase boundaries observed. The Mg content in $(La,Mg)_2Ni_7$ phase is higher than that in $(La,Mg)_5Ni_{19}$ phase while there are little differences in the Mg amounts of the same phase. As the alloy with $x = 1.5$, some $LaNi_5$ phase distribute on $(La,Mg)_5Ni_{19}$ phase with relatively clear phase boundaries. The phase composition of the alloys shows a good agreement with the XRD analysis.

3.4.1.2 *P–C* Isotherms

Figure 3.25 shows the electrochemical desorption *P–C* isotherms of the alloys. For the Pr_5Co_{19}-type single-phase $La_{0.84}Mg_{0.16}Ni_{3.80}$ alloy, there is only one plateau region between 0.42 and 0.62 H/M in the curve and the plateau pressure is about 0.018 MPa. But the two-phase alloys with $x = 1.5$ and 0.75 exhibit two plateau regions, indicating two hydrogen desorption processes. For the alloy with $x = 1.5$, there are two distinct

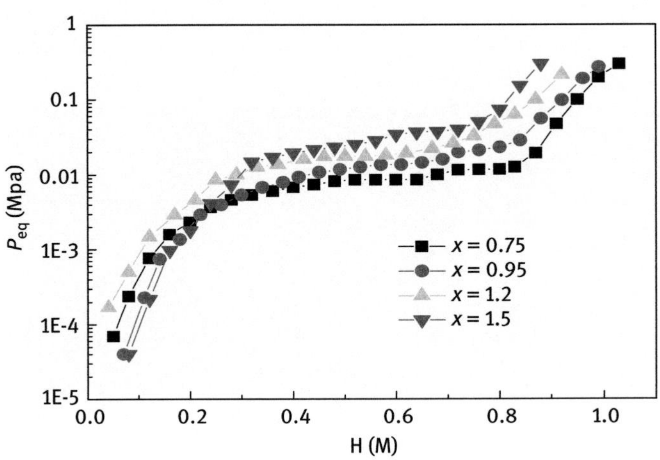

Figure 3.25: Electrochemical desorption *P–C* isotherms of the alloys.

plateaus in the curve: one between 0.4 and 0.52 H/M and the other one between 0.6 and 0.72 H/M. the plateau pressures are around 0.02 MPa and 0.035 MPa, respectively. Gao et al. reported that the $LaNi_5$ phase possesses a higher plateau pressure than the Pr_5Co_{19}-type phase, so the higher plateau region is supposed to belong to the $LaNi_5$ phase and the lower one belongs to the Pr_5Co_{19}-type phase. With regard to $x = 0.95$ alloy, there is a slight of sloping plateau noticed. This is because the plateau pressures of Pr_5Co_{19}-type and Ce_2Ni_7-type phases are close to each other due to their similar superstacking structures. But for the alloys with $x = 0.75$, the plateau region is divided into two parts, one between 0.49 and 0.69 H/M at 0.13 MPa plateau pressures and the other one between 0.72 and 0.8 H/M at 0.21 MPa. The difference between alloys with $x = 0.75$ and 0.95 may be caused by the changed phase abundance that the amount of the Ce_2Ni_7-type secondary phase increases from 19.8 to 40.6 wt.%.

3.4.1.3 Maximum Discharge Capacity and Electrochemical Cycling Stability

The discharge curves (the second cycle) of the alloys at discharge density of 60 mA g^{-1} and 298 K are presented in Figure 3.26. As seen, there are wide discharge potential plateaus in all the discharge curves. However, the discharge potential plateau of the $x = 0.75$ alloy is wider than that of the others and it possesses the highest discharge capacity.

Figure 3.27 shows the activation curves of the alloy electrodes, and the electrochemical properties are summarized in Table 3.14. It can be seen that all the alloy electrodes show good activation property and can reach the maximum discharge capacity within two or three cycles. The maximum discharge capacity of the Pr_5Co_{19}-type single-phase $La_{0.84}Mg_{0.16}Ni_{3.80}$ alloy electrode is 338 mAh g^{-1}, which is close to that reported by Férey [23]. The introduction of Ce_2Ni_7-type phase results in a remarkable

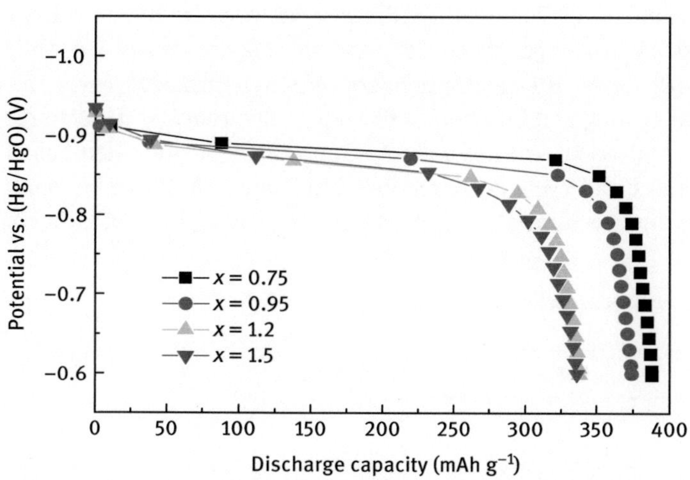

Figure 3.26: Discharge curves of the alloys at discharge density of 60 mA g^{-1} and 298 K.

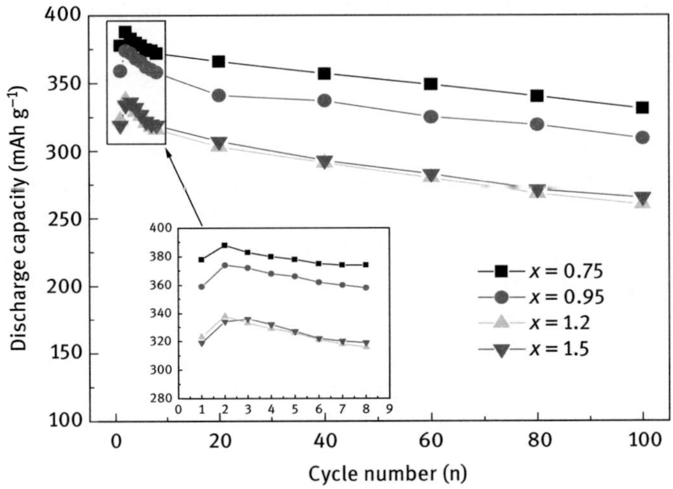

Figure 3.27: Discharge capacity versus cycle number of the alloy electrodes.

increase in the maximum discharge capacity. When the Ce_2Ni_7-type phase abundance increases to 40.6 %, the maximum discharge capacity reaches up to 388 mAh g^{-1}. It can be attributed to the fact that Ce_2Ni_7-type phase possesses a preferable discharge capacity because of the higher scale of the A side elements and more Laves-type [$LaMgNi_4$] units in its super-stacking structures compared with Pr_5Co_{19}-type phase as reported [7]. As seen in the Table 3.12, the Mg content of Ce_2Ni_7-type phase is higher than that of Pr_5Co_{19}-type phase. And the total Mg content in [$LaMgNi_4$] units of the alloy with 40.6 % Ce_2Ni_7-type phase is the highest. This contributes to the highest discharge capacity of 388 mAh g^{-1}. Conversely, the presence of $LaNi_5$ phase slightly reduces the discharge capacity of the alloy ascribing to its lower hydrogen storage capacity compared with Pr_5Co_{19}-type phase [39]. And the Mg content of the alloy containing 20.7 wt.% $LaNi_5$ phase is lower than that of the single-phase alloy.

The electrochemical cycling stability of different alloy electrodes at a discharge current density of 60 mA g^{-1} is shown in Figure 3.27, and the capacity retentions of the alloy electrodes at the 100th cycle (S_{100}) are listed in Table 3.14. It can be found that the S_{100} of all the two-phase alloy electrodes are enhanced compared with that of

Table 3.14: Electrochemical performances of the alloy electrodes.

Sample	N_a	C_{max} (mA g^{-1})	S_{100} (%)
x = 0.75	2	388	85.3
x = 0.95	2	374	82.6
x = 1.2	2	338	77.0
x = 1.5	3	336	79.0

the single-phase alloy electrode (77 %). Especially for the alloy with 40.6 wt.% Ce_2Ni_7-type phase, the S_{100} increases to 85.3 %. This owes to the increase of phase boundaries which can release the stress formed in the charge/discharge process, thus improving the cycling stability. Also for the alloy with 20.7 wt.% $LaNi_5$ phase, the S_{100} slightly augments to 79 %. The results suggest that the presence of secondary phases leads to a better electrochemical cycling stability. This can be explained as follows.

Generally, the primary reasons for the capacity decay are oxidization and corrosion of the alloy electrodes and pulverization of the alloy particles [40]. In this study, during charging/discharging cycles the hydrogen absorption causes lattice expansions that increase the lattice stress. The increscent lattice stress accelerates the grain fracture giving rise to the pulverization. In the two-phase alloys, the appearance of secondary phases offers phase boundary layers, which can buffer the lattice expansions during the process of charging, slowing down the pulverization rate of alloy particles. Moreover, from Table 3.12, we can see that the c/a ratios of Pr_5Co_{19}-type phase in two-phase alloys ($x = 0.75$ and 0.95) increase compared with that of the Pr_5Co_{19}-type single-phase alloy. The increased c/a ratio of the main phase facilitates hydrogen atoms from going in and out of the crystal lattice and decreases the lattice stress, thus weakens the incidence of lattice expansions and promotes a better cycling stability [41].

3.4.1.4 High-Rate Dischargeability and Electrochemical Kinetics

Figure 3.28(a) presents the *HRD* curves of the different alloy electrodes, and the discharge capacities at the current density of 1,500 mA g^{-1} (C_{1500}) are exhibited in Table 3.15. The *HRD* of all the alloy electrodes decreases with the increasing discharge current density. The HRD_{1500} of the Pr_5Co_{19}-type single-phase $La_{0.84}Mg_{0.16}Ni_{3.80}$ alloy is 51.5 %, which is much higher than that of the commercial AB_5-type alloys and other types of La–Mg–Ni-based alloys [42].

HRD is mainly affected by two factors, the hydrogen diffusion in alloy bulk and the electrochemical reaction occurring on the alloy surface. Compared with other types of La–Mg–Ni-based alloys, Pr_5Co_{19}-type alloy has more [$LaNi_5$] units which releases hydrogen firstly ahead of [$LaMgNi_4$] units in discharging [43]. And the hydrogen firstly releasing from [$LaNi_5$] units acts as the electrocatalyst to accelerate the electrochemical reaction on the surface of [$LaMgNi_4$] units [34].

When the $LaNi_5$-type phase abundance increases to 20.7 wt.%, the HRD_{1500} rises to 56 %. Also when 19.8 wt.% Ce_2Ni_7-type phase presents, the HRD_{1500} increases to 53.2 %. But when the Ce_2Ni_7-type phase abundance further increases to 40.6 wt.%, the HRD_{1500} slightly decreases (49.7 %). This implies that the Ce_2Ni_7-type or $LaNi_5$-type secondary phase is helpful for the *HRD* of the alloy with Pr_5Co_{19}-type main phase. It can be ascribed to the presence of more phase boundaries which give rise to more hydrogen diffusion paths, resulting in faster hydrogen diffusion. As seen in Table 3.12, the c/a ratios of Pr_5Co_{19}-type phase in two-phase alloys are larger than that of the Pr_5Co_{19}-type single-phase alloy. The increased c/a ratio of the main phase allows fast

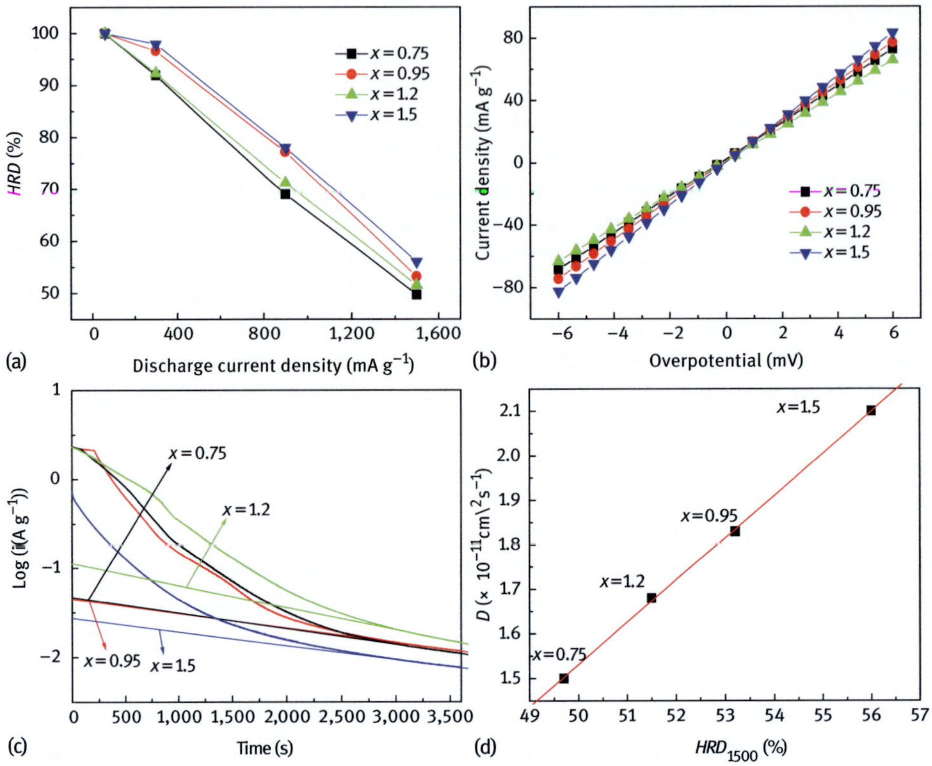

Figure 3.28: *HRD* curves of the alloy electrodes (a); linear polarization curves of the alloy electrodes at 50 % DOD and 298 K (b); semi-logarithmic plots of anodic current versus time responses of the alloy electrodes at +500 mv (c); correlation of the hydrogen diffusion coefficient versus the HRD_{1500} of the alloy electrodes (d).

Table 3.15: Electrochemical kinetic properties of the alloy electrodes.

Sample	HRD_{1500} (%)	C_{1500} (mAh g^{-1})	R_p (mΩ)	I_0 (mA g^{-1})	D ($\times 10^{-11} cm^2 s^{-1}$)
$x = 0.75$	49.7	193	85.06	301.9	1.50
$x = 0.95$	53.2	199	79.33	323.7	1.85
$x = 1.2$	51.5	174	93.20	275.5	1.68
$x = 1.5$	56.0	188	67.58	354.2	2.05

hydrogen atoms diffusion responsible for high kinetic. Moreover, $LaNi_5$ phase makes a greater contribution than Ce_2Ni_7-type phase due to the fact that the $LaNi_5$ secondary phase not only works as hydrogen absorption/desorption phase but also acts as electrocatalyst to accelerate the electrochemical reactions [32]. However, the inherent character of the secondary phase will be the dominant factor on the express of electrochemical properties when its abundance is excessive.

The *HRD* of a metal hydride electrode is mainly dominated by two factors, the charge–transfer reaction at the electrode/electrolyte interface and the hydrogen diffusion rate in the alloy bulk, which are characterized by surface exchange current density (I_0) and diffusion coefficient (D), respectively. Figure 3.28(b) shows the linear polarization curves of the electrodes. The polarization resistance (R_p) can be obtained from the slope of the curves and the exchange current density I_0 is calculated using the following equation [44]:

$$I_0 = (RTI)/(F\eta)(\eta \rightarrow 0) \tag{3.8}$$

where R is the gas constant (J mol^{-1} K^{-1}), T is the absolute temperature (K), I is the applied current (mA), F is the Faraday constant (C mol^{-1}) and η is the total over potential (mV). The results obtained are listed in Table 3.15. As shown in Table 3.15, the I_0 of the alloy with Pr_5Co_{19}-type single phase is 275 mA g^{-1}, and when a secondary phase arises in the alloy, I_0 increases in varying degrees. The highest value of I_0 is 354.16 mA g^{-1}, and it pertains to the alloy containing 20.7 wt.% $LaNi_5$ phase.

It is known that exchange current density (I_0) is generally used to measure the electrocatalytic activity and judge the speed of charge–transfer reaction on the alloy electrode surface. Accordingly, the growth of I_0 indicates that the two-phase alloys have higher charge–transfer efficiency than the single-phase alloy. In addition, from Table 3.15, it can be seen that there is no distinct linear relationship between the HRD_{1500} and the I_0. This implies that the charge–transfer reaction on the surface of the alloy electrodes is not the rate-determining factor.

The hydrogen diffusion coefficient, which has an important influence on the charging/discharge reaction, was evaluated by the potentiostatic step method. Figure 3.28(c) illustrates the semi-logarithmic plots of anodic current versus time responses of the alloy electrodes. It can be seen that after a sufficiently long discharge time the semi-logarithmic value of the anodic current density shows an approximately linear dependence on the time. At this point, the rate of charge–transfer on the surface of alloy electrodes is so fast that the concentration of hydrogen nearly reaches zero. The electrode reaction is controlled by the hydrogen diffusion velocity in the alloy bulk. When the reaction time is long enough, the hydrogen diffusion coefficient D can be calculated by the slope of the linear region of the corresponding plots according to the following equation [45]:

$$\log i = \log((C_0 - C_s)6FD/da^2) - \pi^2 Dt/2.303a^2 \tag{3.9}$$

where i, F, D, d, a, C_0, C_s and t are anodic current density (A g^{-1}), the Faraday constant, the hydrogen diffusion coefficient (cm^2 s^{-1}), the density of the alloy (g cm^{-3}), the radius of the alloy particle, the initial hydrogen concentration in the bulk of the alloy (mol cm^{-3}), the surface hydrogen concentration of the alloy (mol cm^{-3}) and the discharge time (s) respectively. The average particle radius is assumed as a = 13 μm.

The hydrogen diffusion coefficient D values calculated by eq. (3.4) are summarized in Table 3.15. The D of the alloy with Pr_5Co_{19}-type single phase is 1.68×10^{-11} $cm^2\ s^{-1}$. For the alloy containing 19.8 wt.% Ce_2Ni_7-type phase and the alloy containing 19.5 wt.% $LaNi_5$ phase, the values of D increase to 1.85×10^{-11} and 2.05×10^{-11} $cm^2\ s^{-1}$. The results suggest that both the Ce_2Ni_7-type phase and $LaNi_5$ phase have a positive effect on the hydrogen diffusion in the alloy bulk. As mentioned above, the incremental phase boundaries brought in by the secondary phases and the increased c/a ratios of Pr_5Co_{19}-type phase accelerate the hydrogen diffusion. When the amount of Ce_2Ni_7-type phase increases to 40.6 wt.%, the D of the alloy electrode decreases to 1.50×10^{-11} $cm^2\ s^{-1}$. In this case, the intrinsic property of Ce_2Ni_7-type phase dominates. Since the Ce_2Ni_7-type phase has less [$LaNi_5$] units working as catalysts to accelerate the electrochemical reaction than Pr_5Co_{19}-type phase, its D is lower than that of Pr_5Co_{19}-type phase. The increasing Ce_2Ni_7-type phase becomes the main restraining factor on hydrogen diffusion.

Figure 3.28(d) represents the relationship of the hydrogen diffusion coefficient D with the HRD_{1500} of the alloy electrodes. A linear dependence of the HRD_{1500} on the hydrogen diffusion coefficient is established which can manifest that the hydrogen diffusion of alloy electrodes is responsible for the HRD_{1500}.

3.5 Conclusions

In this section, the relationship between phase compositions and electrochemical properties has been clarified. Based on the research, great development in the electrochemical properties of La–Mg–Ni-based alloys has been made, which establishes this alloy system as a key electrode material for Ni/MH batteries. Phase composition is proved to be one of the most significant factors affecting the electrochemical performance of La–Mg–Ni-based alloys. The following conclusions can be drawn:

(1) the maximum discharge capacity of the AB_3-type alloy electrodes first mount up then decline with the growing mount of $(La,Mg)_2Ni_7$ phase. Whereas the increasing abundance of $(La,Mg)_2Ni_7$ phase which contains less [$LaMgNi_4$] (Laves-type) subunits than $(La,Mg)Ni_3$ phase, resulting in less pulverization and oxidation, favours the cycling stability of alloy electrodes. The HRD first increases and then decreases with the rising of $(La,Mg)_2Ni_7$ phase content, to which a similar tendency is obtained in hydrogen diffusion coefficient. Comparatively, alloy electrode with ratio of $(La,Mg)_2Ni_7$ to $(La,Mg)Ni_3$ phase abundance close to 1:1 exhibits the best HRD_{1440} characteristic (47.5 %).

(2) The cycling stability of the A_2B_7-type alloys is promoted by the $(La,Mg)_5Ni_{19}$ phase network distributing among the $(La,Mg)_2Ni_7$ main phase, which not only decreases the strains, alleviates the pulverization and oxidation of the alloy electrodes, but also enhances the amorphization resistance of the alloys during hydrogen absorption/desorption. Furthermore, the $(La,Mg)_5Ni_{19}$ phase with

high Ni content acts as a catalyst for the fast discharge of the $(La,Mg)_2Ni_7$ phase, leading to a remarkable improvement in the *HRD* of the alloy electrodes.

(3) The *HRD* of the A_2B_7-type alloys is improved by adjusting $LaNi_5$ phase amount. But the degradation of the alloy electrodes is aggravated due to more serious pulverization and oxidation caused by the discrepant expansion ratio between $LaNi_5$ and $(La,Mg)_2Ni_7$ phases during hydrogen absorption process.

(4) The presence of Ce_2Ni_7-type phase in the A_5B_{19}-type alloy is more favourable to the maximum discharge capacity and cycling stability while the existence of $LaNi_5$ phase has a beneficial effect on the *HRD*.

References

[1] Liu JJ, Han SM, Li Y, Zhang L, Zhao YM, Yang SQ, Liu BZ. Phase structures and electrochemical properties of La–Mg–Ni-based hydrogen storage alloys with superlattice structure. Int J Hydrogen Energy 2016;41:20261–75.

[2] Zhang JL, Han SM, Li Y, Liu JJ, Yang SQ, Zhang L, Wang JD. Effects of $PuNi_3$- and Ce_2Ni_7-type phase abundance on electrochemical characteristics of La–Mg–Ni-based alloys. J Alloys Compd 2013;581:693–8.

[3] Liu JJ, Han SM, Han D, Li Y, Yang SQ, Zhang L, Zhao YM. Enhanced cycling stability and high rate dischargeability of $(La,Mg)_2Ni_7$-type hydrogen storage alloys with $(La,Mg)_5Ni_{19}$ minor phase. J Power Sources 2015;287:237–46.

[4] Liu JJ, Han SM, Li Y, Yang SQ, Shen WZ, Zhang L, Zhou Y. An investigation on phase transformation and electrochemical properties of as-cast and annealed $La_{0.75}Mg_{0.25}Ni_x$ (x = 3.0, 3.3, 3.5, 3.8) alloys. J. Alloys Compd 2013;552:119–26.

[5] Zhao YM, Han SM, Li Y, Liu JJ, Zhang L, Yang SQ, Ke DD. Characterization and improvement of electrochemical properties of Pr_5Co_{19}-type single-phase $La_{0.84}Mg_{0.16}Ni_{3.80}$ alloy. Electrochim Acta 2015;152:265–73.

[6] Liu JJ, Han SM, Li Y, Zhao X, Yang SQ, Zhao YM. Cooperative effects of Sm and Mg on electrochemical performance of La–Mg–Ni-based alloys with A_2B_7- and A_5B_{19}-type super-stacking structure. Int J Hydrogen Energy 2015;40:1116–27.

[7] Liu JJ, Li Y, Han SM, Yang SQ, Shen WZ, Chen XC, Zhao YM. Microstructure and electrochemical characteristics of step-wise annealed $La_{0.75}Mg_{0.25}Ni_{3.5}$ alloy with A_2B_7- and A_5B_{19}-type super-stacking structure. J Electrochem Soc 2013;160:A1139–45.

[8] Liao B, Lei YO, Chen LX, Lu GL, Pan HG, Wang OD. Effect of the La/Mg ratio on the structure and electrochemical properties of $La_xMg_{3-x}Ni_9$ (x = 1.6–2.2) hydrogen storage electrode alloys for nickel–metal hydride batteriesm. J Power Sources 2004;129:358–67.

[9] Denys RV, Yartys VA. Effect of magnesium on the crystal structure and thermodynamics of the $La_{3-x}Mg_xNi_9$ hydrides. J. Alloy Compd. 2011;509:S540–S548.

[10] Denys RV, Riabov BA, Yartys AV, Sato M, Delaplane GR. Mg substitution effect on the hydrogenation behaviour, thermodynamic and structural properties of the La_2Ni_7–$H(D)_2$ system. J Solid State Chem 2008;181:812–21.

[11] Higashiyama N, Matsuura Y, Nakamura H, Kimoto M, Nogami M, Yonezu I, Nishio K. Influence of preparation methods of non-stoichiometric hydrogen-absorbing alloys on the performance of nickel–metal hydride secondary batteries. J Alloy Compd 1997;253–254:648–51.

[12] Zhang L, Han SM, Li Y, Liu JJ, Zhang JL, Wang JD, Yang SQ. Formation mechanism, phase structure and electrochemical properties of the La–Mg–Ni-based multiphase alloys by powder sintering $LaNi_5$ and $LaMgNi_4$. Int J Hydrogen Energy 2013;38:10431–7.

[13] Zhang JL, Han SM, Li Y, Liu JJ, Q S. Yang, Zhang L, Wang JD. Effects of $PuNi_3$- and Ce_2Ni_7-type phase abundance on electrochemical characteristics of La–Mg–Ni-based alloys. J Alloy Compd 2013;581:693–98.
[14] Gao J, Yan XL, Zhao ZY, Chai YJ, Hou DL. Effect of annealed treatment on microstructure and cyclic stability for La–Mg–Ni hydrogen storage alloys. J Power Sources 2012;209:257–61.
[15] Zhang FL, Luo YC, Chen JP, Yan RX, Chen JH. La–Mg–Ni ternary hydrogen storage alloys with Ce2Ni7 type and Gd2Co7-type structure as negative electrodes for Ni/Mh batteries. J Alloy Compd 2007;430:302–7.
[16] Dong XP, Lü FX, Zhang YH, Yang LY, Wang XL. Effect of La/Mg on the structure and electrochemical performance of La–Mg–Ni system hydrogen storage electrode alloy. Mater Chem Phys 2008;108:251–6.
[17] Willems JJ, Buschow KH, From permanent magnets to rechargeable hydride electrodes. J Less–Common Met 1987;129:13–30.
[18] Chung UI, Lee JY. A study on hydrogen-induced amorphization in the La–Ni system. J Non-Cryst Solids 1989;110:203–10.
[19] Huang TZ, Han JT, Zhang YH, Yu JM, Sun GX, Ren H, Yuan XX. Study on the structure and hydrogen absorption–desorption characteristics of as-cast and annealed $La_{0.78}Mg_{0.22}Ni_{3.48}Co_{0.22}Cu_{0.12}$ alloys. J Power Sources 2011;196:9585–9.
[20] Zhang QA, Fang MH, Si TZ, Fang F, Sun DL, Ouyang LZ, Zhu M. Phase stability, structural transition, and hydrogen absorption-desorption features of the polymorphic La_4MgNi_{19} compound. J Phys Chem C 2010;114:11686–11692.
[21] Zhang QA, Zhao B, Fang MH, Liu CR, Hu QM, Fang F, Sun DL, Ouyang LZ, Zhu M. $(Nd_{1.5}Mg_{0.5})Ni_7$-based compounds: Structural and hydrogen storage properties. Inorg Chem 2012;51:2976–83.
[22] Lemort L, Latroche M, Knosp B, Bernard P. Elaboration and characterization of new pseudo-binary hydride-forming phases $Pr_{1.5}Mg_{0.5}Ni_7$ and $Pr_{3.75}Mg_{1.25}Ni_{19}$: a comparison to the binary Pr_2Ni_7 and Pr_5Ni_{19} ones. J Phys Chem C 2011;115:19437–44.
[23] Férey A, Cuevas F, Latroche M, Knosp B, Bernard P. Elaboration and characterization of magnesium-substituted La_5Ni_{19} hydride forming alloys as active materials for negative electrode in Ni-MH battery. Electrochim Acta 2009;54:1710–14.
[24] Guzik MN, Hauback BC, Yvon K. Hydrogen atom distribution and hydrogen induced site depopulation for the $La_{2-x}Mg_xNi_7$–H system. J Solid State Chem 2012;186:9–16.
[25] Zhang L, Han SM, Han D, Li Y, Zhao X, Liu JJ. Phase decomposition and electrochemical properties of single phase $La_{1.6}Mg_{0.4}Ni_7$ alloy. J Power Sources 2014;268:575–83.
[26] Liu YF, Pan HG, Gao MX, Wang QD. Advanced hydrogen storage alloys for Ni/MH rechargeable batteries. J Mater Chem 2011;21:4743–55.
[27] Gao ZJ, Kang L, Luo YC. Microstructure and electrochemical hydrogen storage properties of La–R–Mg–Ni-based alloy electrodes. New J Chem 2013;37:1105–14.
[28] Li HW, Ishikawa K, Aoki K. A study on hydrogen-induced amorphization in C15 Laves phase $DyNi_2$ under different hydrogen pressures. J Alloys Compd 2005;399:69–77.
[29] Liu YF, Pan HG, Gao MX, Lei YQ, Wang QD. XRD study on the electrochemical hydriding/dehydriding behavior of the La–Mg–Ni–Co-type hydrogen storage alloys. J Alloys Compd 2005;403:296–304.
[30] Buschow KH, Van Der Goot AS. The crystal structure of rare-earth nickel compounds of the type R_2Ni_7. J Less-Commen Metals 1970;22:419–28.
[31] Liu YF, Cao YH, Huang L, Gao MX, Pan HG. Rare earth–Mg–Ni-based hydrogen storage alloys as negative electrode materials for Ni/MH batteries. J Alloys Compd 2011;509:675–86.
[32] Li Y, Han SM, Li JH, Hu L. Study on phase structure and electrochemical properties of $Ml_{1-x}Mg_xNi_{2.80}Co_{0.50}Mn_{0.10}Al_{0.10}$ (x = 0.08, 0.12, 0.20, 0.24, 0.28) hydrogen storage alloys. Electrochim Acta 2007;52:5945–5949.

[33] Zhang FL, Luo YC, Chen JP, Yan RX, Kang L, Chen JH. Effect of annealing treatment on structure and electrochemical properties of $La_{0.67}Mg_{0.33}Ni_{2.5}Co_{0.5}$ alloy electrodes. J Power Sources 2005;150:247–54.

[34] Liu ZY, Yan XL, Wang N, Chai YJ, Hou DL. Cyclic stability and high rate discharge performance of $(La,Mg)_5Ni_{19}$ multiphase alloy. Int J Hydrogen Energy 2011;36:4370–4.

[35] Si TZ, Pang G, Zhang QA, Liu DM, Liu N. Solid solubility of Mg in Ca_2Ni_7 and hydrogen storage properties of $(Ca_{2-x}Mg_x)Ni_7$ alloys. Int J Hydrogen Energy 2009;34:4833–7.

[36] Serin V, Zhang JX, Magén C, Serra R, Hytch MJ, Lemort L, Latroche M, Ibarra MR, Knosp B, Bernard P. Identification of the atomic scale structure of the $La_{0.65}Nd_{0.15}Mg_{0.20}Ni_{3.5}$ alloy synthesized by spark plasma sintering. Intermetallics 2013;32:103–8.

[37] Zhang YH, Cai Y, Zhao C, Zhai TT, Zhang GF, Zhao DL. Electrochemical performances of the as-melt $La_{0.75-x}M_xMg_{0.25}Ni_{3.2}Co_{0.2}Al_{0.1}$ (M = Pr, Zr; $x = 0$, 0.2) alloys applied to Ni/metal hydride (MH) battery. Int J Hydrogen Energy 2012;37:14590–7.

[38] Zhang D, Yamamoto T, Inui H, Yamaguchi M. Characterization of stacking faults on basal planes in intermetallic compounds La_5Ni_{19} and La_2Ni_7. Intermetallics 2000;8:391–7.

[39] Nakamura J, Iwase K, Hayakawa H, Nakamura Y, Akiba E. Structural study of La_4MgNi_{19} hydride by in situ X-ray and neutron powder diffraction. J Phys Chem C 2009;113:5853–9.

[40] Miao H, Gao MX, Liu YF, Zhu D, Pan HG. An improvement on cycling stability of Ti–V–Fe-based hydrogen storage alloys with Co substitution for Ni. J Power Sources 2008;184:627–32.

[41] Liu BZ, Hu MJ, Ji LQ, Fan YP, Wang YG, Zhang Z. Phase structure and electrochemical properties of $La_{0.7}Ce_{0.3}Ni_{3.75}Mn_{0.35}Al_{0.15}Cu_{0.75-x}(Fe_{0.43}B_{0.57})_x$ hydrogen storage alloys. J Alloys Compd. 2012;516:53–7.

[42] Wang BP, Chen YZ, Liu YN. Structure and electrochemical properties of $(La1\text{-}xDyx)_{0.8}Mg_{0.2}Ni_{3.4}Al_{0.1}$ ($x = 0.0$–0.20) hydrogen storage alloys. Int J Hydrogen Energy 2012;37:9082.

[43] Chu HL, Qiu SJ, Tian QF, Sun LX, Zhang Y, Xu F, Liu YY, Qi YN, Fan MQ. Effect of ball-milling time on the electrochemical properties of La–Mg–Ni-based hydrogen storage composite alloys. Int J Hydrogen Energy 2007;32:4925.

[44] Li Y, Han SM, Liu ZP. Effect of Mo–Ni treatment on electrochemical kinetics of La–Mg–Ni-based hydrogen storage alloys. Int J Hydrogen Energy 2010;35:12858.

[45] Li Y, Han D, Han SM, Zhu X, Hu L, Zhang Z. Effect of rare earth elements on electrochemical properties of La–Mg–Ni-based hydrogen storage alloys. Int J Hydrogen Energy 2009;34:1399.

4 Effect of Element Composition on Microstructure and Electrochemical Characteristics of RE–Mg–Ni-Based Hydrogen Storage Alloys

The overall excellent electrochemical performances of RE–Mg–Ni-based alloys are of significance for nickel-metal hydride secondary battery. Lots of efforts have been devoted to improving the electrochemical properties of RE–Mg–Ni-based alloys through different methods, such as annealing, melt-spinning and element substitution. Adjusting the elements composition or substituting elements in RE–Mg–Ni-based alloy is considered as the most effective approach due to the fact that the element's composition determines the phase structure and electrochemical characteristics.

It is well known that ternary RE–Mg–Ni alloys are originated from corresponding binary La-Ni alloys and can also be regarded as the replacement of Mg to RE at the specific location [1]. So Mg becomes indispensible in RE–Mg–Ni-based alloys and renders a significant influence on their phase compositions due to its selective occupation, limited solubility and small atomic radius. For the phases with superlattice structures, Mg and RE share the $4f$ atomic site for the hexagonal structure and share the $6c$ atomic site for rhombohedral ones [2]; both $4f$ and $6c$ atomic sites just locate in $[A_2B_4]$ subunits. Mg can hardly enter $[AB_5]$ subunit structure or $LaNi_5$ phase [3, 4]. Therefore, the addition of Mg element is beneficial for the formation of the phases with higher $[A_2B_4]$ subunits. Although Mg can enter $[A_2B_4]$ subunits, its solubility is limited. For instance, at 773 K, the solubility limit of Mg in $La_{1-x}Mg_xNi_3$, 16.8 at.% Mg, is significantly higher than that in $La_{2-x}Mg_xNi_7$, 3 at.% Mg. If Mg amount exceeds the upper limit, phases with higher Mg content will appear, and if Mg content is low, phases with lower Mg content will appear [5]. Besides, the solid solubility of Mg is temperature-dependent. The solubility of Mg in $La_{2-x}Mg_xNi_7$ increases to 6 at.% Mg at 1,073 K [6]. In terms of difference in solid solubility of Mg in various stacking structures, the phase composition of an alloy can be significantly impacted by the alternative content of Mg. Thus, the investigation that the solid solubility of Mg atoms in RE–Mg–Ni-based alloys, and the impact of various Mg amount substituting RE on structural changes and hydrogen storage characteristics are crucial adjective to the electrochemical hydrogen storage performances of superlattice RE–Mg–Ni-based alloys.

La–Mg–Ni-based alloys as a representative of RE–Mg–Ni-based alloys have been studied through element substitution. It is confirmed that electrochemical hydrogen storage performance can be improved by substituting La by Pr, Nd, Sm etc. Pr partial substitution for La was found to stabilize the AB_2-type phase structure and relive the hydrogen-induced amorphization of $La_{1-x}Pr_xMgNi_{3.6}Co_{0.4}$ (x = 0–0.4) alloys [7]. It was also reported that the cyclic hydrogen storage property of Pr–Mg–Ni Pr_4MgNi_{19} superlattice alloy was up to 1,000 absorption/desorption cycles [8].

DOI 10.1515/9783110501483-004

Besides, some of the earlier literatures also reported that partial substitution of Pr for La in La–Mg–Ni-based alloys could improve the cycling stability of the alloy electrodes [9, 10]. Moreover, Zhang et al. have found that appropriate amount of substitution of Nd for La can improve the *HRD* and cycle stability due to that Nd can improve resistance to oxidation so as to enhance cycle life of the alloys [10–14] as well as to improve surface exchange current and bulk diffusion [14]. In our group, Li et al. have studied $La_{0.60}R_{0.20}Mg_{0.20}(NiCoMnAl)_{3.5}$ (R = La, Ce, Pr, Nd) alloys and $La_{0.80-x}Nd_xMg_{0.20}Ni_{3.20}Co_{0.20}Al_{0.20}$ (x = 0.20–0.60) alloys and found that the Nd partial substitution can increase the La_2Ni_7 phase abundance and improve the cycle life and *HRD* dramatically [12, 15]. In addition, Rui et al. improved the *HRD* and the cycling stability of the $La_{1.3}CaMg_{0.7}Ni_9$ alloy using Sm to substitute for La. The *HRD* at the discharge current density of 1,200 mA g^{-1} was increased by 15.2 % and the discharge retention at the 100th cycle was improved by 10 % [16]. Zhang et al. and Li et al. found that the electrochemical properties of the alloys could be improved by the addition of Sm element [17, 18]. The above descriptions indicate that Pr, Nd and Sm are interesting and valuable elements in enhancement of cycle stability of hydrogen storage alloys. However, the existing reports have hardly expounded how Pr, Nd or Sm work in improving the cyclic performance of the alloy electrodes. On the other hand, the effect of the substitution of La by Pr, Nd, Sm etc. on the phase composition is ambiguous or divergent. As for the influence of Pr, according to the study of Zhang et al. on the as-cast and melt spinning $La_{0.75-x}Pr_xMg_{0.25}Ni_{3.2}Co_{0.2}Al_{0.1}$ (x = 0–0.4) alloys containing (La,Mg)Ni_3 and $LaNi_5$ phases and those containing $(La,Mg)_2Ni_7$ and $LaNi_5$ phases [19, 20], the increase of Pr benefits the formation of (La,Mg)Ni_3 and $(La,Mg)_2Ni_7$ phases with superlattice structure [8, 20]. However, the study of Pan et al. on the $La_{0.7-x}Pr_xMg_{0.3}Ni_{2.45}Co_{0.75}Mn_{0.1}Al_{0.2}$ (x = 0–0.3) alloy containing (La,Mg)Ni_3 and $LaNi_5$ main phases showed that the abundance of the (La,Mg)Ni_3 phase decreases from 69.48 to 40.52 wt.%, whereas the $LaNi_5$ phase abundance increases from 30.52 to 59.48 wt.% when the Pr content increases from x = 0 to 0.30 [10]. The above descriptions indicate that Pr is an interesting and valuable element in the enhancement of cycle stability of hydrogen storage alloys. However, the existing reports have hardly expounded how Pr works in improving the cyclic performance of the alloy electrodes. Understanding the above problem not only promotes the proper use of Pr in La–Mg–Ni-based alloys but also enlightens the discovery of other similar elements to further enhance the overall electrochemical performance of La–Mg–Ni-based alloys. Therefore, it is necessary to study the effect of substitution of La by Pr, Nd or Sm on phase composition and electrochemical hydrogen storage performances and explore the functioning mechanism of Pr, Nd or Sm in perspective of structural stabilization of the superlattice structure of the alloy phases during repeated charge/discharge cycles, aiming to improve the cycle life of La–Mg–Ni-based electrodes of nickel metal hydride batteries.

Besides the rare earth elements, the transition elements in B-side also affect the phase composition and electrochemical performances. Many reports have shown that partial substitution of Ni by Co, Cr, W and Si is an effective method to improve the electrochemical performances of the alloys. Co plays an important role in improving cycle stability of La–Mg–Ni-based alloys. However, the high cost of Co element prompted seeking of other substituting element with similar electrochemical functions. Al incorporation is generally believed to be one of the most effective methods to alleviate the capacity degradation of La–Mg–Ni-based alloys [21–23]. Zhao et al. [22] reported that cyclic stability of the $La_{0.7}Mg_{0.3}Ni_{2.75}Co_{0.75-x}Al_x$ (x = 0–0.4) alloys was tremendously increased from 45.9 % (x = 0) to 82.1 % (x = 0.4) at the 100th cycle. The Al_2O_3 oxide layer contributed to the improvement by protecting Co and other active materials from oxidation and corrosion. Liu et al. [23] found that addition of Al could delay the oxidation of the alloy surface layer in electrolyte, slow down the capacity degradation and prolong the cycling lifetime. Moussa et al. [24] reported that Fe partial substitution for Co in $MmNi_{3.55}Mn_{0.4}Al_{0.3}Co_{0.75-x}Fe_x$ alloys and found that the life cycle was prolonged. Partial Co substitution by Fe was also reported in rapid quenched AB_5-type alloys, and it was found that cycling stability could also be improved [25]. However, detailed work on the affecting mechanism of Al or Fe element on the cyclic stability of La–Mg–Ni-based alloys was not very sufficient. For example, how the addition of Al or Fe element can relive the pulverization of the alloys during cycling was not clear enough. Moreover, the effect of Al addition on the degradation of kinetics performance during electrochemical cycling in the alkaline solution remains unrevealed. Thus, it is necessary to investigate the effect of Al or Fe incorporation on the microstructure and electrochemical performance especially discharge capacity degradation and electrochemical kinetics degradation.

Based on the above analysis, the effect of Mg element, rare earth elements and transition elements on the phase composition and electrochemical performances will be investigated in this section. Firstly, the phase composition and electrochemical performances of $La_{2-x}Mg_xNi_7$ (x = 0.40–0.60), $Pr_{3-x}Mg_xNi_9$ (x = 0.45–1.2) and $Nd_{1-x}Mg_xNi_{2.8}$ (x = 0.10–0.50) alloys prepared by powder sintering will be discussed for clarifying the solid solution of Mg and the effect of Mg in the different RE–Mg–Ni-based alloys. Secondly, the phase composition and electrochemical performances of $La_{0.8-x}Pr_xMg_{0.2}Ni_{3.4}Al_{0.1}$ (x = 0, 0.1, 0.2 and 0.3) and $La_{0.75-x}Nd_xMg_{0.25}Ni_{3.3}$ (x = 0, 0.15) alloys will be demonstrated for explaining the mechanism of the influence of rare earth substitution on the phase structures and the improvement of cycling stability. In addition, the phase composition and electrochemical performances of $La_{0.80}Mg_{0.20}Ni_{2.95}Co_{0.70-x}Al_x$ (x = 0, 0.05, 0.10, 0.15) and $La_{0.74}Mg_{0.26}Ni_{2.55}Co_{0.65-x}Fe_x$ (x = 0, 0.10, 0.20, 0.30) alloys will be illustrated, and the emphases have been placed on the enhancement mechanism of cycling stability by Al or Fe substitution.

4.1 Effect of Mg Element on the Phase Composition and Electrochemical Performances

4.1.1 Phase Composition and Electrochemical Performances of $La_{2-x}Mg_xNi_7$ (x = 0.40–0.60) Alloy

4.1.1.1 Morphology and Crystal Structures

Figure 4.1(a–c) shows the SEM micrographs of the alloys of x = 0.40, 0.50 and 0.60. EDS analysis of $La_{1.60}Mg_{0.40}Ni_7$ alloy indicates that the two regions in Figure 4.1(a) are the magnesium region (A) and the Mg-free region (B), corresponding to A_2B_7-type and AB_5-type phases, respectively. The morphology of the $La_{1.50}Mg_{0.50}Ni_7$ alloy is uniform except for some small pores which may be caused by the volatilization of local accumulation of Mg during the sintering process. Elemental composition of area A suggests that the region is A_2B_7-type phase (Figure 4.1(b)). Figure 4.1(c) is the morphology of the $La_{1.40}Mg_{0.60}Ni_7$ alloy, in which two areas A and C belong to A_2B_7-type and AB_3-type phases, respectively. The surface of the alloy is loose, especially for the C region, due to the more severe migration and volatilization of Mg during the sintering process.

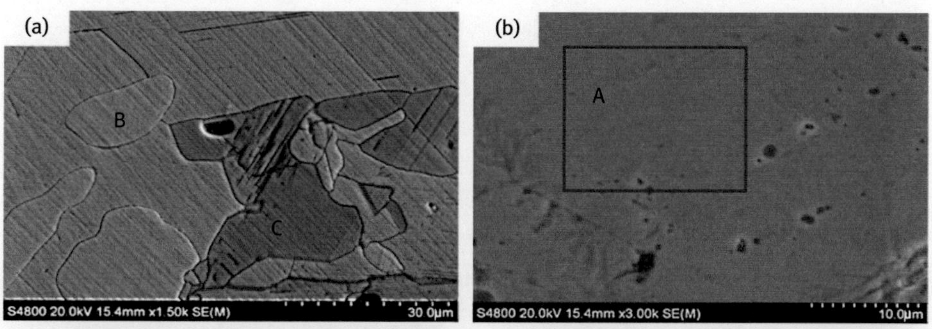

Figure 4.1: SEM backscattering electron images for alloys: (a) $La_{1.60}Mg_{0.40}Ni_7$, (b) $La_{1.50}Mg_{0.50}Ni_7$ and (c) $La_{1.40}Mg_{0.60}Ni_7$.

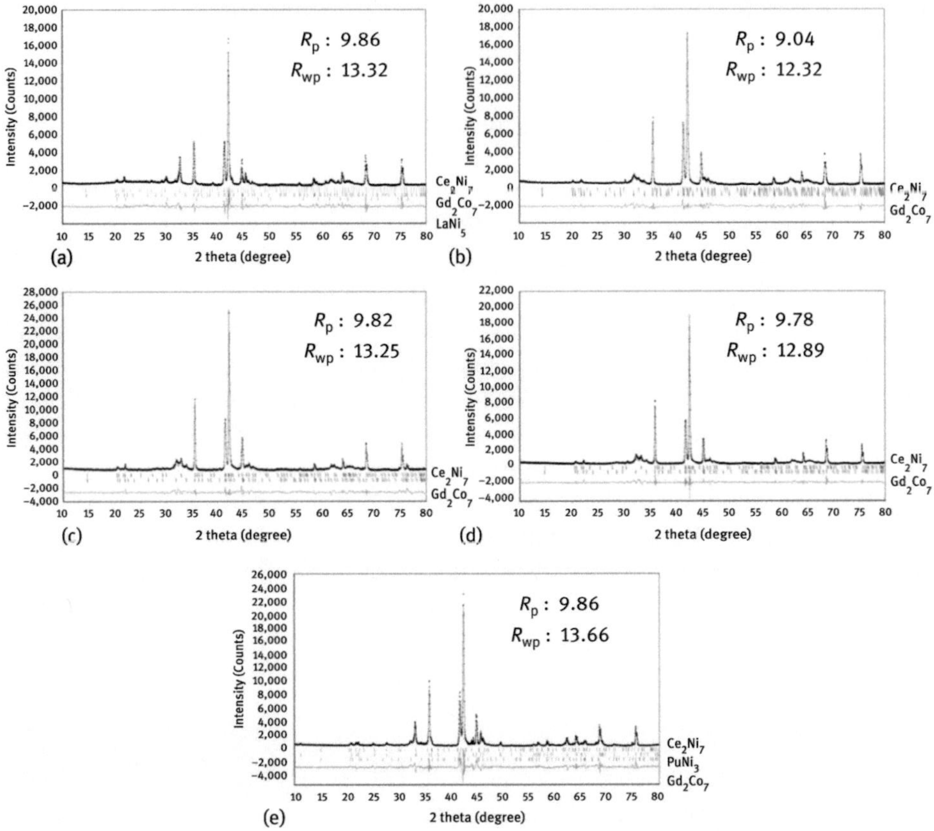

Figure 4.2: Rietveld refinements of the XRD patterns for alloys: (a) $La_{1.60}Mg_{0.40}Ni_7$, (b) $La_{1.52}Mg_{0.48}Ni_7$, (c) $La_{1.51}Mg_{0.49}Ni_7$, (d) $La_{1.50}Mg_{0.50}Ni_7$ and (e) $La_{1.40}Mg_{0.60}Ni_7$.

Figure 4.2(a–e) shows the Rietveld refinements of the XRD patterns of $La_{2-x}Mg_xNi_7$ (x = 0.40–0.60) alloys. The alloys consist of 2H- and 3R-type polymorphic structures when x is in the range of 0.48–0.50 (seen in Figure 4.2(b–d)). As shown in Figure 4.2(a), $CaCu_5$-type $LaNi_5$ phase appears when Mg content decreases to 0.40. However, when the amount of Mg increases to 0.60, superlattice $PuNi_3$-type phase turns up (Figure 4.2(e)). The XRD patterns of the alloys coincide with the SEM results. These results suggest that polymorphic structures 2H- and 3R-type $(La, Mg)_2Ni_7$ phases can exist in the concentration range of Mg from 0.48 to 0.50, and further trace increase and decrease the amount of Mg are favourable to the formations of $PuNi_3$-type surperlattice and $CaCu_5$-type structures when Mg content is beyond the allotropic scope, respectively. The maximum occupation of Mg to La is 50 % in $[A_2B_4]$ slab, yielding $[LaMgNi_4]$ slab, which can stack with two $[AB_5]$ slabs forming Ce_2Ni_7-type and Gd_2Co_7-type phases, equally, the maximum content of Mg is 0.50 for $La_{2-x}Mg_xNi_7$ (x = 0.40–0.60) alloys without formation of new phases. Upon further increase of Mg content, the extra Mg atoms have to assemble with La and Ni atoms; therefore, the superfluous Mg atoms have to take away La and Ni atoms from [AB_5] slabs which are used to stack to form Gd_2Co_7-type structure considering the markedly decreasing content from 49.94 to 2.04 wt.% (Table 4.1), and then form [$LaMgNi_4$] slabs. At that moment, the amount of [$LaMgNi_4$] slabs ascends but that descends to [AB_5] slabs, contributing to the stacking of $PuNi_3$-type superlattice structure whose [A_2B_4] slabs account for 50 % higher than other superlattice structures. Oppositely, when the amount of Mg is lower than 0.48, the [A_2B_4] slabs in Gd_2Co_7-type structure are in an unsaturated stage. Consequently, the relative redundant La and Ni atoms in [A_2B_4] slabs tend to form [AB_5] slabs and the ratio of [A_2B_4] to [AB_5] slabs reduces, thereby prompting the developing of $CaCu_5$-type $LaNi_5$ phase.

Table 4.1 summarizes the phase abundances and structure parameters of $La_{2-x}Mg_xNi_7$ (x = 0.40–0.60) alloys. The abundances of 2H-type $(La,Mg)_2Ni_7$ phase are 60.48, 54.53 and 50.06 %, while 39.52, 45.47 and 49.94 % for 3R-type phase, respectively, when x increases from 0.48 to 0.50. The rising mass of the 3R-type phase indicates that the increasing partial substitution of Mg for La is beneficial for the construction of the 3R-type phase and complies with the fact that the layered structure of an La–Ni alloy is size-dependent, with the priority of a 3R-type structure for smaller atomic radii and a 2H-type structure for larger R atomic radii [26, 27]. Structural stability of $MgZn_2$-type Laves slabs is higher than that of $MgCu_2$-type Laves slabs [28]. The increasing Mg atoms prefer to intrude into $MgZn_2$-type Laves slabs of 3R-type structure, thereby slowing down the energy of the structure. The parameter changes of a-axis, c/a ratio and unit cell volume V of the alloys with allotropic structures are depicted as Figure 4.3. With increase of Mg content from 0.48 to 0.50, the parameters of a and V of 2H-type phase rise while the c/a value is almost unchanged, but all the three parameters of 3R-type phase decrease, especially for c/a which drops rapidly. Because

Table 4.1: Phase abundances and lattice parameters of $La_{2-x}Mg_xNi_7$ (x = 0.40–0.60) alloys.

Alloy	Str. type	Phase abundance (%)	a (Å)	c (Å)	c/a	V(Å^3)
x = 0.40	Ce_2Ni_7	74.33	5.0392	24.282	4.819	533.99
	Gd_2Co_7	18.16	5.0380	36.341	7.213	798.81
	$CaCu_5$	7.51	4.9726	3.989	0.802	85.42
x = 0.48	Ce_2Ni_7	60.48	5.0188	24.154	4.813	526.89
	Gd_2Co_7	39.52	5.0282	36.254	7.211	793.80
x = 0.49	Ce_2Ni_7	54.53	5.0190	24.156	4.813	526.98
	Gd_2Co_7	45.47	5.0278	36.229	7.206	793.13
x = 0.50	Ce_2Ni_7	50.06	5.0197	24.151	4.811	527.01
	Gd_2Co_7	49.94	5.0270	36.204	7.190	792.32
x = 0.60	Ce_2Ni_7	77.62	5.0173	24.126	4.809	525.96
	$PuNi_3$	20.34	5.0312	24.273	4.824	532.10
	Gd_2Co_7	2.04	5.0308	36.113	7.178	791.53

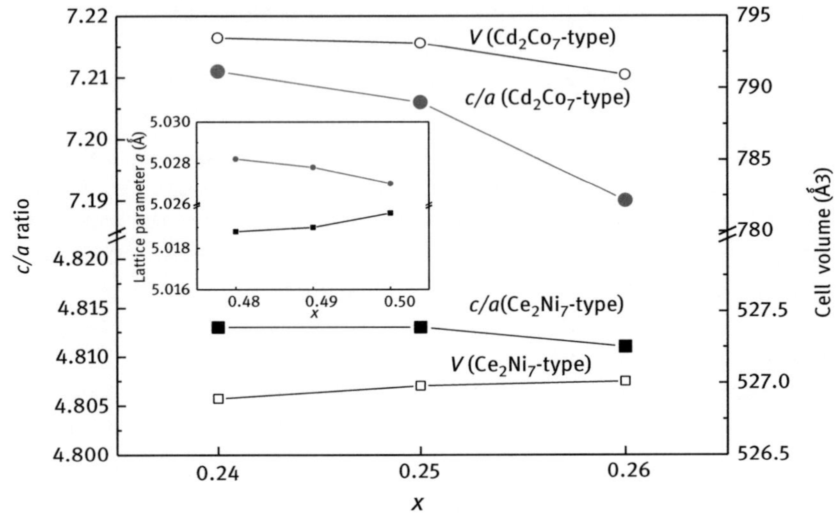

Figure 4.3: Parameters of c/a ratio, cell volumes V and lattice parameter a as functions of x for $La_{2-x}Mg_xNi_7$ ($x = 0.48–0.50$) alloys.

of the large differences in atomic radii $R_{Mg} = 1.602$ Å and $R_{La} = 1.897$ Å, increase in Mg/La ratio will lead to a decrease of a, c and V parameters [4]. Therefore, we can deduce that the increasing Mg atoms selectively enter into 3R-type phase. Simultaneously, some Mg atoms transfer into 3R-type phase from 2H-type phase allowing for its incremental a and V with the reduced phase content, while that is the secondary factor considering the nearly unchanged c/a ratio of 2H-type phase.

4.1.1.2 Electrochemical *P–C* Isotherms

The effect of different phase compositions and different contents of the 2H- and 3R-type structures on the discharge plateau pressures of the *P–C* isotherms of the alloys is shown in Figure 4.4. It can be seen that the alloys with only superlattice structures show single discharge plateau, while two discharge plateaus appear in the *P–C* curves of the $x = 0.40$ alloy which contains $LaNi_5$ phase in addition to the superlattice structure phases. The presentation of $LaNi_5$ phase causes the appearance of the second plateau with a plateau pressure of 0.1 MPa. For the three alloys with allotropic structures, the plateau pressure of the alloys increases by 0.008 MPa from 0.010 MPa with increasing of Mg, suggesting that the decreasing cell volume V of 3R-type phase caused by selective entrance of Mg makes a higher discharge pressure for 3R- than 2H-type phases, while the pressure gap between the two phases is difficult to be identified from the measured *P–C* isotherms. In comparison, the plateau pressure of the alloy comprising a $PuNi_3$-type phase is slightly lower than the other alloys, about 0.007 MPa, indicating that the plateau pressure of $PuNi_3$-type phase is lower than that of $(La,Mg)_2Ni_7$ and $LaNi_5$ phases. The reason is that its larger cell volume V provides

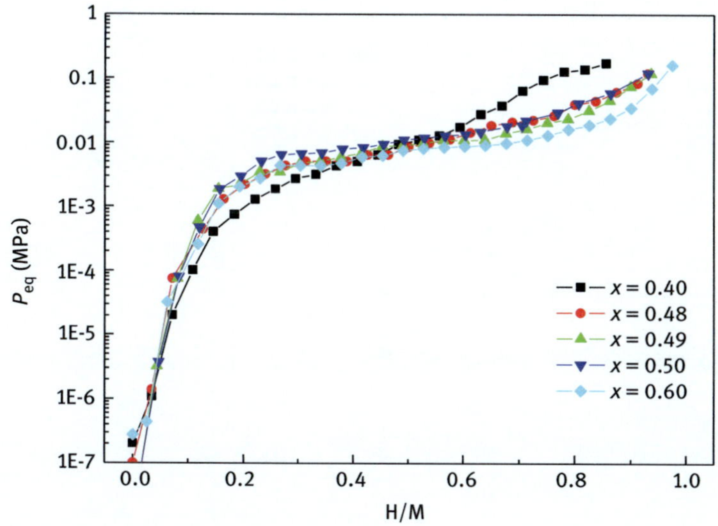

Figure 4.4: *P–C* isotherms of the $La_{2-x}Mg_xNi_7$ (x = 0.40–0.60) alloy electrodes.

Table 4.2: Electrochemical properties of $La_{2-x}Mg_xNi_7$ (x = 0.40–0.60) alloy electrodes.

Samples	N	C_{max} (mAh g^{-1})	HRD_{1440} (%)	S_{100} (%)	H/M
x = 0.40	2	360.1	52.1	76.0	0.855
x = 0.48	2	384.5	45.8	78.1	0.930
x = 0.49	2	387.7	45.7	78.2	0.936
x = 0.50	2	386.1	45.8	78.0	0.931
x = 0.60	1	378.5	46.2	74.5	0.975

more spaces for hydrogen atoms to deliberate, thereby lowering the discharge plateau pressure.

The electrochemical hydrogen storage capacities of each alloy obtained from the electrochemical *P–C* isotherms are listed in Table 4.2. The hydrogen storage capacity of the x = 0.60 alloy is 0.975 H/M. And that is 0.855 H/M for the x = 0.40 alloy because of the cost of 3R-type $(La,Mg)_2Ni_7$ phase by the $LaNi_5$ phase whose discharge capacity is relatively lower. Within the Mg concentration range of the existence of 2H- and 3R-type $(La,Mg)_2Ni_7$ phases, the H/M of the two allotropes are similar about 0.93. The *P–C* isotherms show that the alloys with $(La,Mg)_2Ni_7$ phase show one plateau and similar H/M values, stating similar discharge behaviour of the 2H- and 3R-type phases [29].

4.1.1.3 Electrochemical Performances

Figure 4.5 shows the relationship between discharge capacity and cycle number of the alloy electrodes at the charge–discharge current density of 0.2°C, and the electrochemical characteristics of the alloy electrodes are also summarized in Table 4.3. It can be

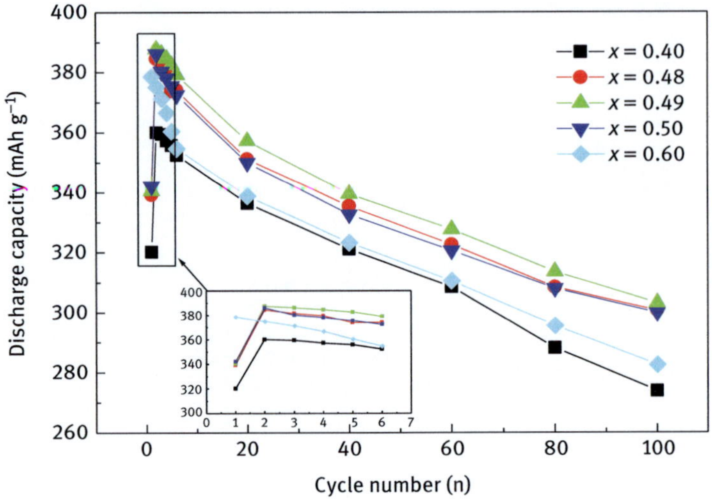

Figure 4.5: Discharge capacity versus cycle number of the $La_{2-x}Mg_xNi_7$ ($x = 0.40–0.60$) alloy electrodes.

Table 4.3: Kinetic properties of $La_{2-x}Mg_xNi_7$ (x = 0.40–0.60) alloy electrodes.

Samples	R_{ct} **(mΩ)**	I_0 **(mA g^{-1})**	D**(×10^{-10} cm^2 s^{-1})**
x = 0.40	258.5	323.2	1.87
x = 0.48	280.8	252.1	1.52
x = 0.49	286.8	251.2	1.58
x = 0.50	282.5	253.0	1.54
x = 0.60	240.3	281.2	1.79

seen that all the alloy electrodes display good activation performance and can be fully activated within two cycles, especially for the alloy with $PuNi_3$-type $(La,Mg)Ni_3$ phase and allotropic $(La,Mg)_2Ni_7$ phase which can be activated at the first cycle.

The maximum discharge capacity of the alloys with only allotropic structure 2H- and 3R-type $(La,Mg)_2Ni_7$ phases is very similar ranging from 384.5 to 387.7 mAh g^{-1}, and that decreases to 378.5 and 360.1 mAh g^{-1} with the appearance of superlattice $PuNi_3$-type $(La,Mg)Ni_3$ and $CaCu_5$-type $LaNi_5$ phases, respectively. Ignoring the phase interaction, the maximum discharge capacity of an alloy is decided by its phase composition and the intrinsic phase structural characteristic of each phase. The maximum hydrogen capacities of three kinds of single-phase structural alloys are in the order: $(La,Mg)Ni_3$ [30] > $(La,Mg)_2Ni_7$ [3] > $LaNi_5$ [31], with the values decreasing from 13.2 to 7 H, which may identically reflect the fact of the electrochemical capacities of those alloys. Therefore, the maximum discharge capacities of the alloys with phases of $(La,Mg)Ni_3$ and $(La,Mg)_2Ni_7$, and with $(La,Mg)_2Ni_7$ and $LaNi_5$ are higher and lower than the alloys with allotropic $(La,Mg)_2Ni_7$ phase in view of their phase compositions, respectively. The allotropic structures of $(La,Mg)_2Ni_7$ phase own parallel stacking structures except for $[A_2B_4]$ slabs, resulting in small gap (within 4 mAh g^{-1}) on the maximum discharge capacities among the alloys of $x = 0.48–0.50$.

Figure 4.5 also shows the cycling stabilities of the samples with 100 % depth of discharge (DOD) at the 100th cycle. The alloys with only $(La,Mg)_2Ni_7$ phase in different amounts of 2H- and 3R-type structures own very similar degradation behaviour which can be found from the nearly S_{100} (calculated as $C_{100}/C_{max} \times 100$ %, where C_{100} is the discharge capacity at 100 cycles) values (78.0–78.2 %), and the cycling stabilities of the alloy electrodes with only $(La,Mg)_2Ni_7$ phase are superior than the alloy electrodes of $x = 0.40$ and 0.60. Figure 4.6(a) shows the XRD pattern of the $La_{1.5}Mg_{0.50}Ni_7$ alloy after charge–discharge for 40 cycles. The peaks of the $LaNi_5$ phase appear, while that of the allotropic structures disappear with only some peaks of 2H-type $(La,Mg)_2Ni_7$ phase still existing for $x = 0.50$. That indicates the $(La,Mg)_2Ni_7$ phase has decomposed

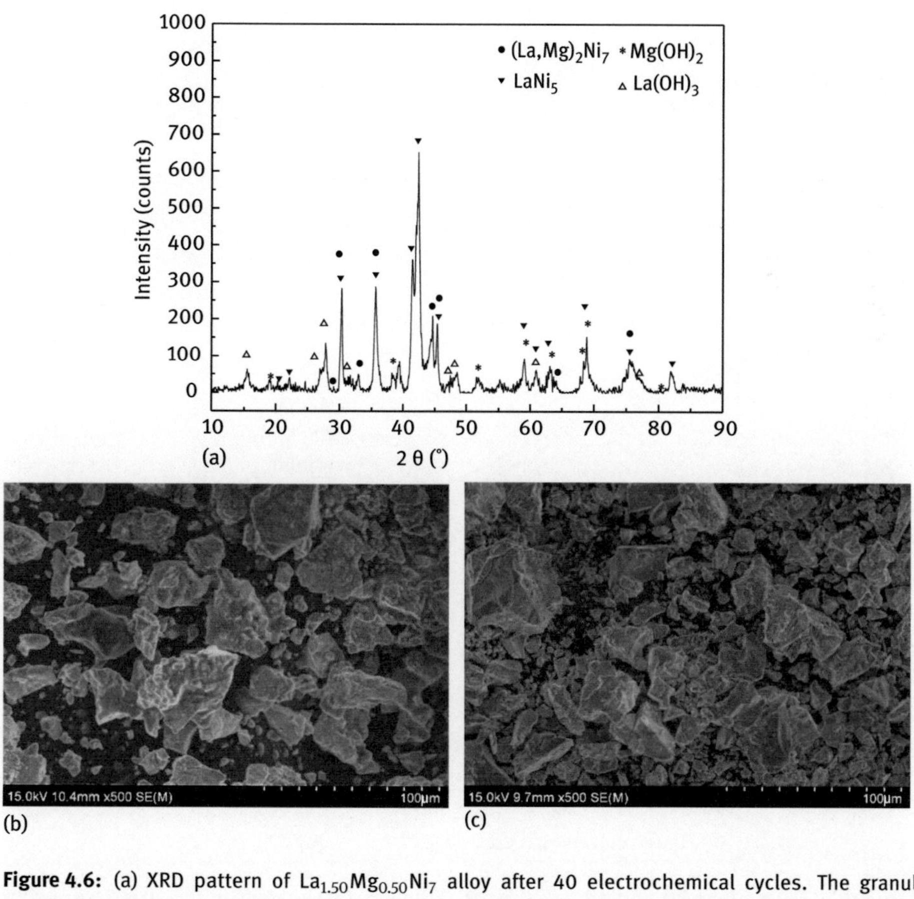

Figure 4.6: (a) XRD pattern of $La_{1.50}Mg_{0.50}Ni_7$ alloy after 40 electrochemical cycles. The granular morphologies after 40 electrochemical cycles for alloys: (b) $La_{1.50}Mg_{0.50}Ni_7$and (c) for $La_{1.60}Mg_{0.40}Ni_7$.

to $CaCu_5$-type $LaNi_5$ phase during cycling. Meanwhile, peaks of $Mg(OH)_2$ and $La(OH)_3$ are detected which can be owed to the oxidation of Mg and La, leading to the discharge capacity decay. It can be deduced that $Mg(OH)_2$ and $La(OH)_3$ are from the oxidation of elements on the alloy surface as well as amorphous La and Mg in the alloy bulk after the pulverization. Therefore, the capacity degradation of the alloy electrodes can be ascribed to the phase decomposition on a structural perspective, and the oxidation-pulverization of the alloy particles. But the alloys with allotropic structures have better cycling stability than the alloys with multiphase structures for the following reasons: on the one hand, the difference in the expansion rate between multiphase structure during hydrogen absorption/desorption forces the pulverization of the alloy particles which can be proved in Figure 4.6(b) and (c), showing that the particle size of the allotropic structure alloy is much bigger than that of the corresponding multiphase structure alloy after 40 electrochemical cycles, thereby bringing more fresh surfaces exposed to alkali, giving rise to oxidative damage. On the other hand, the phase boundaries exited in multiphase structure alloy increase the inconsistency of the expansion of the multiphase structure and offer pathways to alkali which aggravates the pulverization and oxidation of the alloy [32], thereby leading to worse cycling stability of the alloy electrode.

4.1.1.4 Electrochemical Kinetic Characteristics

Figure 4.7 shows the *HRD* curves of those alloy electrodes. It can be seen that alloy of $x = 0.40$ has outstanding *HRD* at 1,440 mA g^{-1} among the five alloy electrodes. That can be owed to the $LaNi_5$ phase which is believed to be good at the hydrogen diffusion

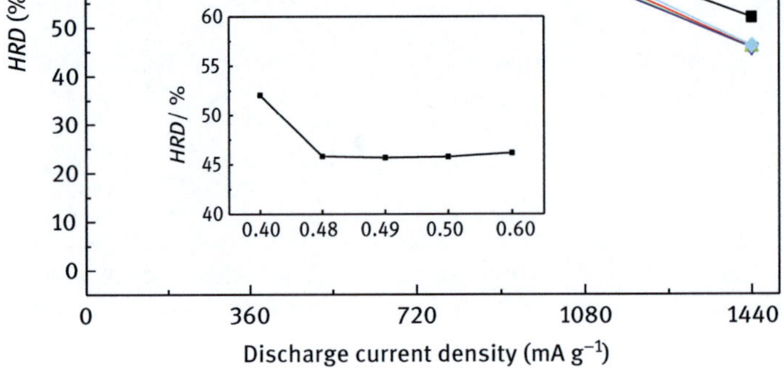

Figure 4.7: *HRD* as a function of discharge current density of the $La_{2-x}Mg_xNi_7$ (x = 0.40–0.60) alloy electrodes.

in alloy bulk and the electrochemical reaction on the alloy surface [15]. For one thing, $LaNi_5$ phase has a high catalytic effect on the hydrogen absorption/desorption process due to its high amount of Ni. For another, the metal hydrides are less stability in $LaNi_5$ phase shown up as higher plateau pressure in *P–C* isotherms. So, the existence of $LaNi_5$ phase is conducive for hydrides liberation at large discharge current densities. It also can be seen that the *HRDs* of the alloys of x = 0.48–0.50 are close to each other, meaning that there are no differences on *HRD* between the 2H- and 3R-type $(La,Mg)_2Ni_7$ phases. With the appearance of $(La,Mg)Ni_3$ phases, the *HRD* increased slightly to 46.2 % form 45.8 %. It turns out that the crystal boundaries between different phases provide pathways for hydrogen diffusion, while that is unobvious to allotropes considering the similar stacking model.

Figure 4.8 presents the linear polarization curves of the $La_{2-x}Mg_xNi_7$ (x = 0.40–0.60) alloy electrodes. The exchange current density (I_0) can be calculated according to the following equation [33]. The values of I_0 are listed in Table 4.3. The I_0 of the alloy electrodes with the $LaNi_5$ phase or $(La,Mg)Ni_3$ phase are superior than those alloys with allotropic structure $(La,Mg)_2Ni_7$ phases which are close to each other for the allotropes.

The semi-logarithmic curves of anodic current versus time responses of the $La_{2-x}Mg_xNi_7$ (x = 0.40–0.60) alloy electrodes are shown in Figure 4.9. Assuming that the alloy particles are uniform spheres with radius of 13 μm and the reaction time is long enough, *D* value is calculated according to the literature [34] and the results are also tabulated in Table 4.3. It can be seen that the alloy electrodes with 2H- and 3R-type $(La,Mg)_2Ni_7$ phases gain similar *D* values about 1.50×10^{-10} $cm^2\ s^{-1}$, and the

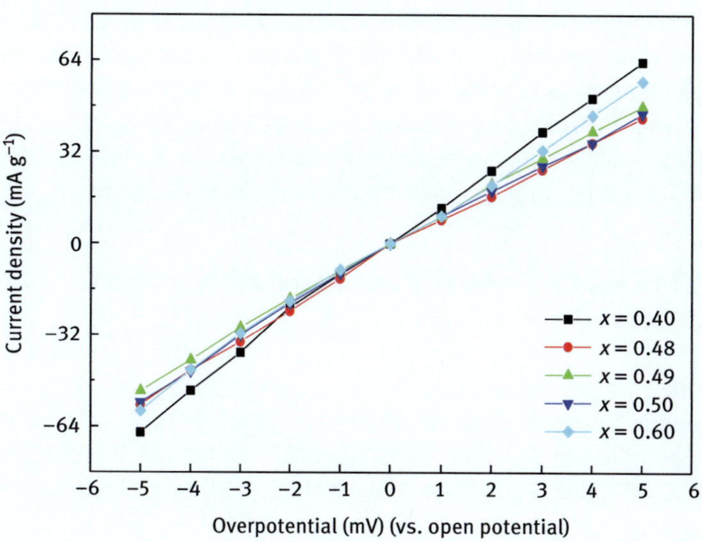

Figure 4.8: Linear polarization curves of the $La_{2-x}Mg_xNi_7$ (x = 0.40–0.60) alloy electrodes at 50 % DOD and 298 K.

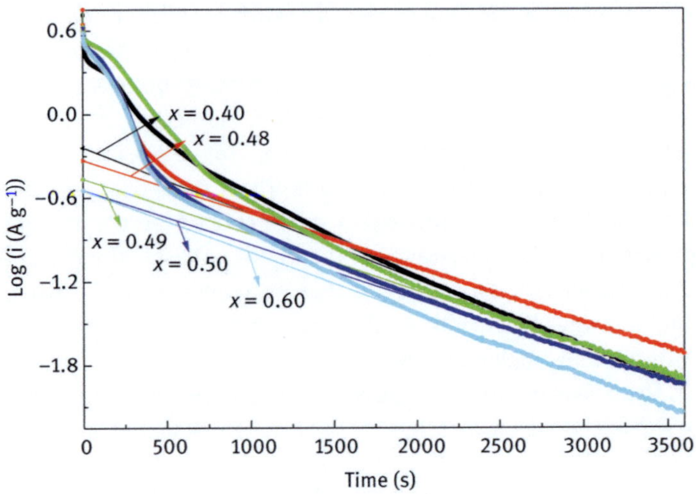

Figure 4.9: Correspondence of the anodic current density versus discharge time of the $La_{2-x}Mg_xNi_7$ (x = 0.40–0.60) alloy electrodes at 50 % DOD and 298 K.

appearance of $LaNi_5$ phase or (La,Mg)Ni_3 phase is beneficial to the hydrogen diffusion in the alloys, calculated to be 1.87×10^{-10} and 1.79×10^{-10} $cm^2\ s^{-1}$, suggesting that crystal boundaries can accelerate hydrogen diffusion in the alloys.

The kinetic properties of I_0 of the allotropic alloy electrodes are in accordance with the *HRD* properties, while *D* values show the disparities. The similar structures of the 2H- and 3R-type lead to the almost accordant hydrogen diffusion in two allotropes when discharge at large current density, therefore, *HRD* values cannot be decided by the *D* values. However, kinetic properties of the allotropic alloy electrodes can be estimated by the I_0. Assuming the allotropic alloys as one kind electrode and dividing the $La_{2-x}Mg_xNi_7$ (x = 0.40–0.60) alloys electrodes into three kinds, the results imply that hydrogen diffusion in alloy bulk and hydrogen reaction on the surface of alloy electrodes work jointly in discharging process with high-rate current density.

4.1.2 Phase Composition and Electrochemical Performances of $Pr_{3-x}Mg_xNi_9$ (x = 0.45–1.2) Alloys

4.1.2.1 Microstructure

The XRD patterns of the alloy samples are shown in Figure 4.10. As shown in Figure 4.10, $PuNi_3$-type phase is the main constituent for all studied $Pr_{3-x}Mg_xNi_9$ (x = 0.45–1.2) alloys, and Pr_2MgNi_9 alloy (x = 1.0) is with a $PuNi_3$-type single phase. The secondary phase for x = 0.45 and 0.6 is Gd_2Co_7-type $(Pr,Mg)_2Ni_7$ phase, but for the alloy of x = 1.0, the secondary phase changes to $MgCu_4Sn$-type $(Pr,Mg)_2Ni_4$ phase. Because most of the peaks of Ce_2Ni_7-type and $PuNi_3$-type phases are overlaps, it is difficult to determine whether Ce_2Ni_7-type phase is included for alloys of x = 0.45

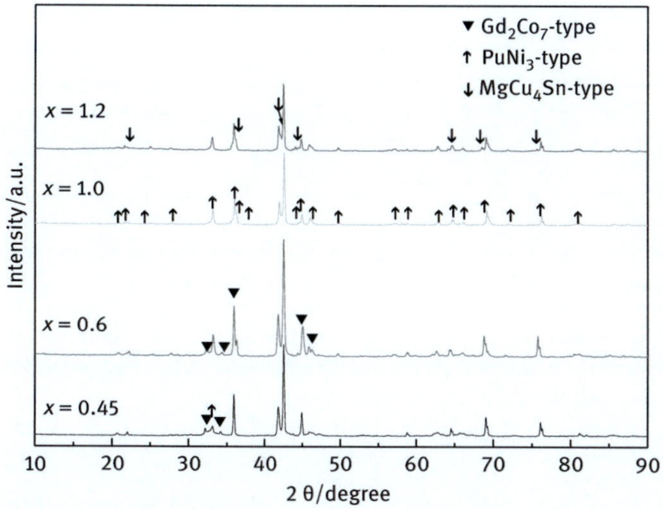

Figure 4.10: X-ray diffraction patterns for $Pr_{3-x}Mg_xNi_9$ (x = 0.45–1.2) alloys.

and 0.6. Therefore, Ce_2Ni_7-type structure was firstly introduced to the Rietveld refinement process. However, the calculated pattern did not fit well with the observed pattern, with S were about 2.3, indicating the absence of Ce_2Ni_7-type phase in the two alloys. Therefore, a two-phase model of Cd_2Co_7-type and $PuNi_3$-type was applied. The phase abundances and cell parameters analysed by the Rietveld refinement of the XRD data are listed in Table 4.4 and images for x = 1.0 and 1.2 alloys are shown in Figure 4.11 as typical examples. The results show that as Mg increases from 0.45 to 0.6, $PuNi_3$-type phase abundance increases from 52.31 to 72.54 wt.% and reaches 100 % when x rises to 1.0. But when Mg content exceeds 1.0, $MgCu_4Sn$-type phase forms with a content of 26.32 wt.%. Moreover, with the increase of Mg content, the cell parameters a, c and V of Gd_2Co_7-type phase show little changes, but those of $PuNi_3$-type phase decrease until the content of Mg reaches 1.0, and no more changes occur with further increasing of Mg. The cell parameter changes are due to the fact that atomic radius of Mg

Table 4.4: Phase abundances and lattice parameters of $Pr_{3-x}Mg_xNi_9$ (x = 0.45–1.2) alloys.

Alloy	Str. type	Space group	Phase abundance (%)	a (nm)	c (nm)	V (nm^3)
x = 0.45	PuN_3	R-3m	52.31	0.5020	2.4332	0.5310
	Gd_2Co_7	R-3m	47.69	0.5022	3.6146	0.7895
x = 0.6	PuN_3	R-3m	72.54	0.5014	2.4320	0.5295
	Gd_2Co_7	R-3m	27.46	0.5022	3.6133	0.7892
x = 1.0	PuN_3	R-3m	100.00	0.4993	2.4308	0.5248
x = 1.2	$PuNi_3$	R-3m	73.68	0.4998	2.4306	0.5258
	$MgCu_4Sn$	Fd-3m	26.32	0.7167	–	0.3682

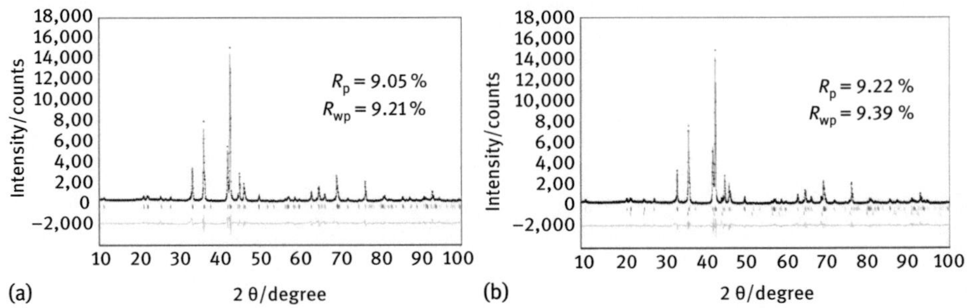

Figure 4.11: Rietveld refinement of the observed XRD pattern for alloys: (a) Pr_2MgNi_9 and (b) $Pr_{1.8}Mg_{1.2}Ni_9$. Vertical bars below the patterns show the positions of all possible reflection peaks: (a) $PuNi_3$-type phase and (b) $PuNi_3$- and $MgCu_4Sn$-type phases.

(0.1602 nm) is rather smaller than that of Pr (0.1824 nm). Additional Mg constantly goes into $PuNi_3$-type phase and reaches the maximum solid solubility at content of 1.0 for $PuNi_3$-type phase. When Mg exceeds its maximum solid solubility in $PuNi_3$-type phase, the exceeding Mg has to intrude into the $MgCu_4Sn$-type $(Pr,Mg)_2Ni_4$ phase, in which Mg has larger solid solubility.

In fact, Mg content can significantly impact the phase structure of one ternary RE–Mg–Ni-based compound. On one hand, the solid solubility of Mg in ternary RE–Mg–Ni-based compounds is decided by its formative mount of $[A_2B_4]$ slabs in the super-stacking structures. In the view of the $[A_2B_4]$ slabs in AB_2, AB_3 and A_2B_7-type super-stacking structures, the ratios of $[A_2B_4]$ in the structures are 1, 1/2 and 1/3 respectively, indicating that the solid solubility of Mg in those structures also decreases. And when Mg reaches a maximum solid solubility value for one phase, the increasing Mg will contribute to a new phase's formation that has more $[A_2B_4]$ slabs. On the other hand, when increasing the content of Mg to form more $[A_2B_4]$ slabs, RE and Ni atoms, which have been used to form other phases, have to be applied now, leading to the phase transformation. In this study, the content of Mg (0.45 and 0.6) for Cd_2Co_7-type phase at that phase content is in full solid solubility, but for $PuNi_3$-type phase it is unsaturated. Therefore, the additional Mg continually occupies Pr_2 sites in $PuNi_3$-type phase with the Pr and Ni atoms which are used to stack to Cd_2Co_7-type phase, and reaches the maximum value at 1.0 for $PuNi_3$-type with a decreasing content of Cd_2Co_7-type phase. With further increasing Mg content, the redundant Mg atoms take away Pr and Ni atoms from $PuNi_3$-type phase, forming $(Pr,Mg)_2Ni_4$ phase. In conclusion, the phase transformation can be made via changing Mg content of alloys.

The backscattered SEM images of the alloys of $x = 0.6$ and 1.0 are shown in Figure 4.12(a) and (b). Apparent phase boundaries can be seen in the morphology of the $Pr_{2.40}Mg_{0.6}Ni_9$ alloy; and EDS analysis indicates that regions of #A and #B of $Pr_{2.40}Mg_{0.6}Ni_9$ alloy in Figure 4.12(a) correspond to AB_3-type and A_2B_7-type phases with compositions of $Pr_{2.1}Mg_{0.9}Ni_{9.0}$ and $Pr_{3.04}Mg_{0.96}Ni_{13.8}$, respectively. And the small

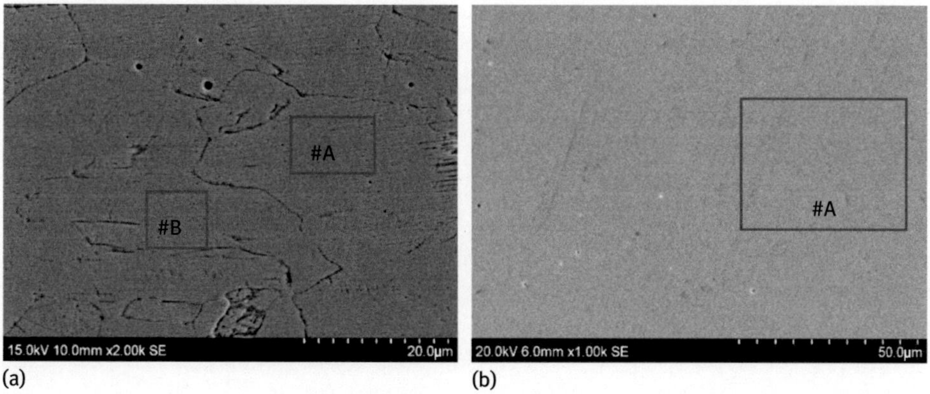

Figure 4.12: SEM backscattering electron images for alloys: (a) $Pr_{2.40}Mg_{0.6}Ni_9$ and (b) Pr_2MgNi_9.

pores might be caused by the volatilization of local accumulation of Mg in the sintering process [35]. Figure 4.12(b) is the morphology of Pr_2MgNi_9 alloy, of which the homogenous morphology can be determined. Elemental composition of area #A suggests that the composition of the region is $Pr_{2.01}Mg_{0.99}Ni_{9.03}$. The composition result from EDS corresponds well with the XRD and ICP results, indicating the single-phase characteristic.

4.1.2.2 Charge and Discharge Property

Figure 4.13(a) and (b) shows the relationship between discharge capacity and cycle number of $Pr_{3-x}Mg_xNi_9$ ($x = 0.44–1.2$) and three single-phase alloy electrodes. The electrochemical properties of each alloy electrode are summarized in Table 4.5, respectively. As shown in Figure 4.13(a), all $Pr_{3-x}Mg_xNi_9$ ($x = 0.45–1.2$) alloy electrodes are fully activated at four or five cycles to reach the maximum discharge capacities. When x is 0.45–1.0, the maximum discharge capacity of the alloys of $x = 0.45–1.0$ with super-stacking structures are all little higher than the alloy with minor $(Pr,Mg)_2Ni_4$ phase, because of its relatively lower discharge capacity.

The cycling stability of the alloy electrodes is a major factor that needs to be improved for RE–Mg–Ni-based-based alloys before applied as Ni/MH battery electrodes.

Table 4.5: Electrochemical properties of $Pr_{3-x}Mg_xNi_9$ ($x = 0.45–1.2$) alloy electrodes.

Sample	N	C_{max} (mAh g^{-1})	HRD_{1500} (%)	S_{100} (%)
$x = 0.45$	4	349	62.2	82.8
$x = 0.6$	4	347	57.6	83.6
$x = 1.0$	4	342	56.7	86.3
$x = 1.2$	5	338	58.8	62.1

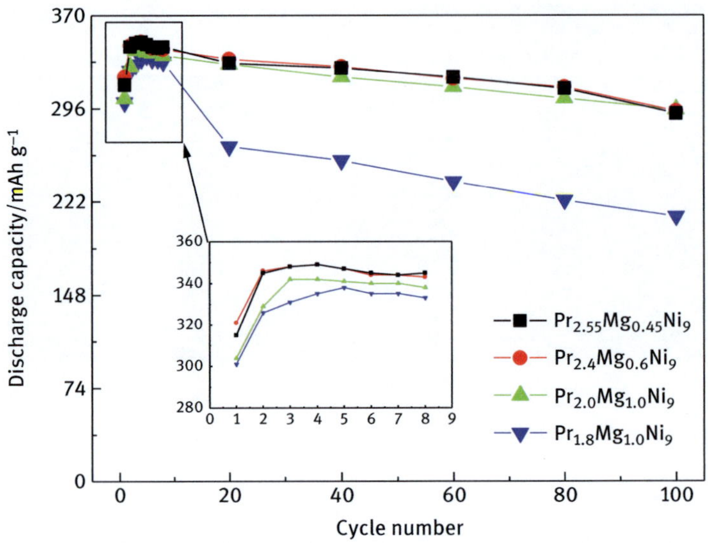

Figure 4.13: Discharge capacity versus cycle number for $Pr_{3-x}Mg_xNi_9$ (x = 0.45–1.2) alloys.

The capacity retaining rate (S_{100}) is used to evaluate the cycling stability of an alloy electrode, and it is defined as $S_{100} = C_{100}/C_{max} \times 100$ %, where C_{max} is the maximum discharge capacity and C_{100} is the discharge capacity at the 100th charge–discharge cycle. The capacity-retaining rate of single-phase Pr_2MgNi_9 alloy is excellent, up to 86.3 %, but existence of a secondary phase is adverse to the cycling stability, whether it is Gd_2Co_7-type or $MgCu_4Sn$-type phases. Generally, the capacity decay of an alloy electrode is due to the pulverization and oxidation of alloy electrodes [36]. The pulverization of single-phase alloy can be reduced in terms of its synchronous volume expansion/contraction during charge/discharge cycles [37]. Correspondingly, the suppressed pulverization further decreases oxidation of active elements. While, because of the appearance of the secondary phases, difference in the expansion rates between different phase structures during hydrogen absorption/desorption happens, aggravates the pulverization of the alloy electrodes and further deteriorates the oxidation of Pr and Mg elements on the new cracked surface of the alloys [29, 38]. Besides, the negative effect on cycling stability of $MgCu_4Sn$-type phase is much more noticeable than Gd_2Co_7-type phase, and this may be due to the wide divergence between $PuNi_3$-type and $MgCu_4Sn$-type phases. In addition, higher Mg content in $MgCu_4Sn$-type phase also accelerates the pulverization-oxidation.

The calculated HRD_{1500} for $Pr_{3-x}Mg_xNi_9$ (x = 0.45 – 1.2) alloy electrodes and three single-phase alloy electrodes are also listed in Table 4.5, respectively. From Table 4.5, it can be seen that the HRD of $Pr_{3-x}Mg_xNi_9$ (x = 0.45 – 1.2) alloy firstly decreases to 56.7 % (the single-phase alloy) but then slightly increases as Mg increases to 1.2. The changes can be attributed to following reasons: on one hand, the gradual disappearance of Gd_2Co_7-type phase reduces the phase boundaries which lower the pathways for the hydrogen diffusion in discharging process; on the other hand, the hydride of $MgCu_4Sn$-type phase is relatively unstable in regard to its higher discharge plateau in comparison with AB_3- or A_2B_7-type phases, and its hydrides can be easily released hydrogen from the alloy's bulk, thereby improving the *HRD*. The *HRD* of Nd_2MgNi_9 and Pr_2MgNi_9 alloys are almost twice as that of La_2MgNi_9 alloy. The reason is that Nd_2MgNi_9 and Pr_2MgNi_9 alloys have higher dehydriding pressures and the enthalpy changes of the double alloys are also lower, therefore, the hydrogen atoms are relatively easy to break loose from their phase structure, so Nd_2MgNi_9 and Pr_2MgNi_9 alloys exhibit excellent HRD properties.

4.1.3 Phase Composition and Electrochemical Performances of $Nd_{1-x}Mg_xNi_{2.8}$ (x = 0.10–0.50) Alloys

4.1.3.1 Phase Structure

Figure 4.14 shows the XRD patterns of the obtained $Nd_{1-x}Mg_xNi_{2.8}$ (x = 0.10, 0.20, 0.36, 0.50) alloys. All the alloys contain $PuNi_3$-type (Nd,Mg)Ni_3 as the main phase (3R-type, SG: *R-3m*). When the Mg content are 0.1 and 0.2, the alloys contain minor phases of the Ce_2Ni_7-type $(Nd,Mg)_2Ni_7$ and $MgCu_4Sn$-type $NdMgNi_4$ phases [39]. But as x increased to 0.36, the alloy only possesses $PuNi_3$-type single-phase characteristic. And with further increase in x to 0.50, minor $MgNi_2$ phase present in the alloy.

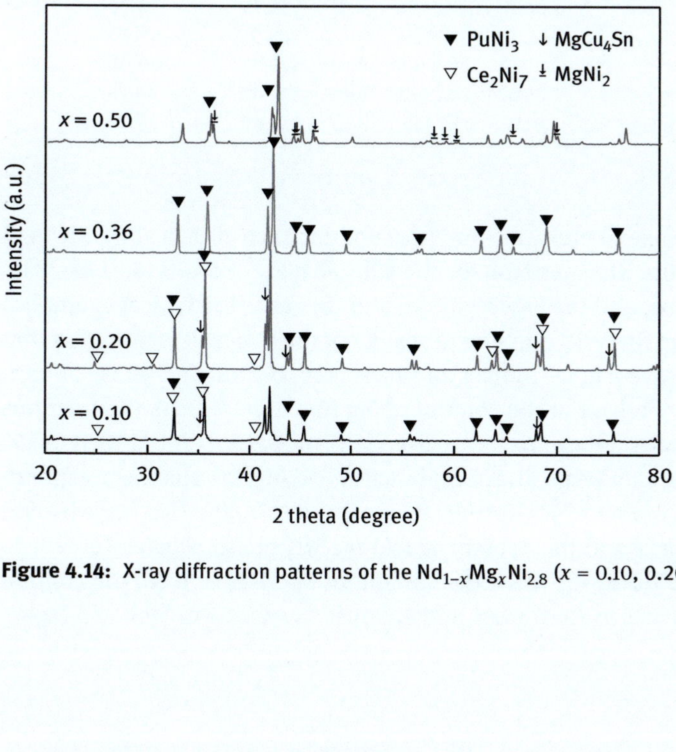

Figure 4.14: X-ray diffraction patterns of the $Nd_{1-x}Mg_xNi_{2.8}$ (x = 0.10, 0.20, 0.36, 0.50) alloys.

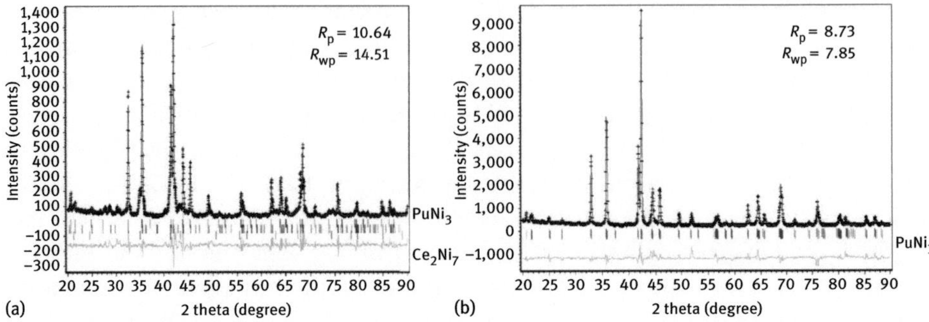

Figure 4.15: Rietveld analysis of the $Nd_{1-x}Mg_xNi_{2.8}$ (x = 0.10, 0.20, 0.36, 0.50) alloys of x = 0.10 (a) and x = 0.36 (b). s is used to judge the degree of the refinement: $s = R_{wp}/R_e$, where R_{wp} is the weighed sum of residuals of the least square fit and R_e is the value statistically expected.

Table 4.6: Phase abundances and lattice parameters of the $Nd_{1-x}Mg_xNi_{2.8}$ (x = 0.10, 0.20, 0.36, 0.50) alloys.

Sample (x)	Phase	Space group	Lattice parameter a (Å)	Lattice parameter c(Å)	c/a	Cell volume (Å^3)	Phase abundance (wt.%)
x = 0.10	$PuNi_3$	*R-3m*	5.0294	24.6819	4.91	540.68	68.32
	Ce_2Ni_7	*P63/mmc*	5.0120	24.6427	4.92	536.09	11.17
	$NdMgNi_4$	*Fd-3m*	7.2087	7.2087	1	374.60	20.51
x = 0.20	$PuNi_3$	*R-3m*	5.0114	24.5239	4.90	533.37	76.77
	Ce_2Ni_7	*P63/mmc*	5.0030	24.5408	4.91	531.95	10.70
	$NdMgNi_4$	*Fd-3m*	7.2519	7.2519	1	381.81	12.53
x = 0.36	$PuNi_3$	*R-3m*	5.0019	24.2748	4.85	525.95	100.00
x = 0.50	$PuNi_3$	*R-3m*	5.0002	24.2689	4.85	525.46	70.76
	$MgNi_2$	*P63/mmc*	4.8115	15.7453	3.27	315.68	29.24

To further identify the phase composition and phase abundance of each alloy, Rietveld refinement analysis on the XRD patterns for the $Nd_{1-x}Mg_xNi_{2.8}$ (x = 0.10, 0.20, 0.36, 0.50) alloys are performed, and the Figure 4.15(a) and (b) are as the typical examples. Table 4.6 summarizes the Rietveld refinement results of the alloys. It is apparent that changes in Mg content have a great impact on phase composition and phase abundance of the alloys. With increase of the content of Mg from 0.10 to 0.20, the fractions of $(Nd,Mg)_2Ni_7$ and $NdMgNi_4$ phases decrease from 11.17 to 10.70 wt.% and 10.51 to 7.53 wt.%, respectively. At the same time, the abundance of $(Nd,Mg)Ni_3$ phase correspondingly increases. As the content of Mg further increases to 0.36, the $(Nd,Mg)_2Ni_7$ and $NdMgNi_4$ phases disappear and the content of $(Nd,Mg)Ni_3$ phase reaches 100 wt.%, namely one pure phase. However, when the content of Mg rises to 0.50, the content of $(Nd,Mg)Ni_3$ phase reduces to 70.76 wt.% and emerges as an unhydrogenated $MgNi_2$

phase. Table 4.6 shows that the cell parameters of each phase of the $Nd_{1-x}Mg_xNi_{2.8}$ (x = 0.10, 0.20, 0.36, 0.50) alloys, in which the cell parameters of Ce_2Ni_7-type phase for each alloy are very close, but for $PuNi_3$-type (Nd,Mg)Ni_3 phase, parameters of a, c and V all significantly decrease before unchanged (about 526 $Å^3$). This is caused by the smaller radius of Mg (1.72 Å) than Nd (1.82 Å). When Mg enters into one phase, the cell volume reduction would be an inevitable trend. Therefore, from the above results, it can be determined that the Mg has been getting into $PuNi_3$-type phase when x is not more than 0.36 and reaches the maximum value at 0.36. Further increases Mg content, the extra Mg has to exist in precursor form in the $MgNi_2$ phase as a secondary phase with the $PuNi_3$-type main phase. The solubility limit of Mg is lower than that previously reported by Hu et al. [40], where its solubility limit of Mg reached 0.5 at 1,123 K. The reason is closely related to the deviation of the element composition for the Ni in this study, where the Ni content is ca. 0.2 lower than the ideal 3.0 value. Besides, the higher prepared temperature of 1,183 K can be also ascribed to lower solubility limit of Mg in this study.

In general, Mg atoms only enter into the [A_2B_4] subunits in the RE–Mg–Ni-based-based superstacking alloys. Therefore the solid solubility of Mg for ternary Nd–Mg–Ni alloys is decided by the quantity of [A_2B_4] subunits. In views of the proportion of [A_2B_4] subunits in AB_3-type and A_2B_7-type structures are 1/2 and 1/3, respectively, the solid solubility of Mg in AB_3-type structure is higher than A_2B_7-type structure. Hence, when the Mg content increases, it has a trend to form the phase endowed more [A_2B_4] subunits. Besides, the stability of [A_2B_4] subunits mainly depends on the Nd–Ni and Ni–Ni bonds with strong bonding ability. When the cell volume contracts after Mg excess in, the Ce_2Ni_7-type phase stacking structure will be destroyed, generating more $NdMgNi_4$ subunits and $NdNi_5$ subunits [41], this contributes to the double subunits in 1:1 ratio forming $PuNi_3$-type phase. Consequently, increasing content of Mg facilitates occurrences of the series of phase changes during the sintering process.

The backscattered SEM image of $Nd_{1-x}Mg_xNi_{2.8}$ (x = 0.2, 0.36, 0.50) alloys is shown in Figure 4.16. The alloys of the x = 0.20 (Figure 4.16(a)) and 0.50 (Figure 4.16(c)) are multiphase morphologies. The EDS on elemental composition from Figure 4.16(a) indicates that the areas A, B and C comprise of A_2B_7, AB_3 and AB_2 phases, respectively, and areas A and D in Figure 4.16(c) are $Nd_{0.66}Mg_{0.33}Ni_{3.10}$ and $MgZn_{2.11}$, respectively. While the morphology of the alloy of x = 0.36 is homogenous (Figure 4.16(b)) and the elemental composition of area A is $Nd_{0.64}Mg_{0.36}Ni_{3.10}$ (Figure 4.16(d)). The composition result from the EDS corresponds well with the XRD and ICP results.

4.1.3.2 Electrochemical *P–C* Isotherms

The *P–C* isotherms of the $Nd_{1-x}Mg_xNi_{2.8}$ (x = 0.10, 0.20, 0.36, 0.50) alloys are shown in Figure 4.17. No obvious discharge plateaus for the alloys of x = 0.10 and 0.20 can be observed, but single discharge plateau appears in the *P–C* curve of the x = 0.36 alloy and the platform shows a wide and flat one. Moreover, single platform can be observed in the *P–C* curve of the x = 0.50 alloy. However, the length of the plateau

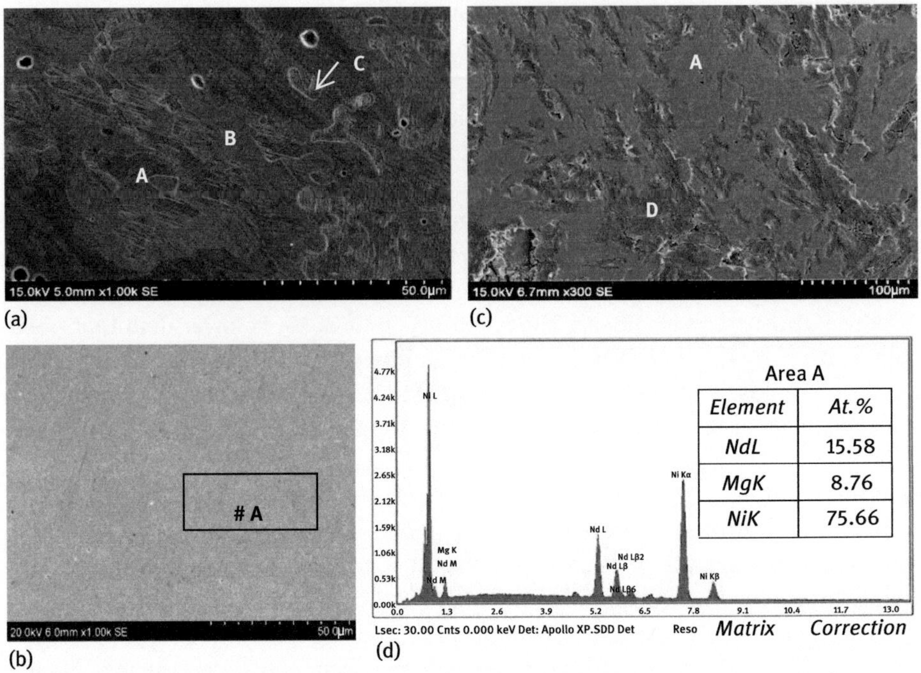

Figure 4.16: Scanning electron microscopy micrographs for $Nd_{0.90}Mg_{0.10}Ni_{2.8}$ (a), $Nd_{0.64}Mg_{0.36}Ni_{2.8}$ (b) and $Nd_{0.50}Mg_{0.50}Ni_{2.8}$ (c) alloy. (d) EDS pattern for $Nd_{0.64}Mg_{0.36}Ni_{2.8}$ alloy.

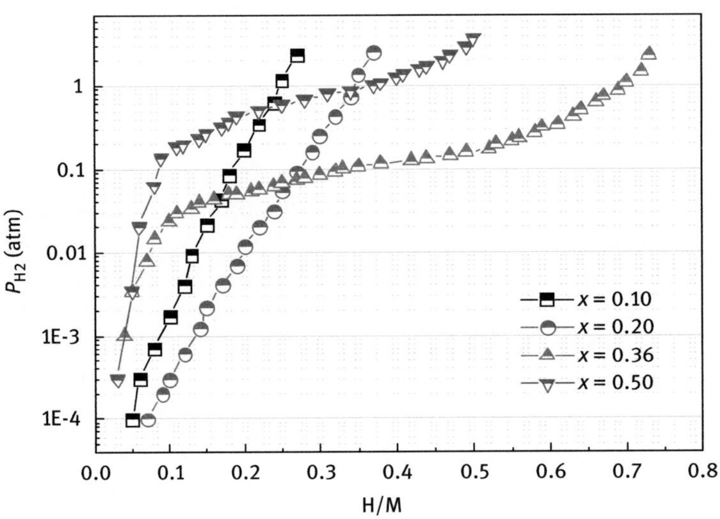

Figure 4.17: *P–C* isotherms of the $Nd_{1-x}Mg_xNi_{2.8}$ (x = 0.10, 0.20, 0.36, 0.50) alloy electrodes.

becomes narrow in comparison with that of the x = 0.36 alloy and the discharge plateau pressure of the alloy is elevated to 0.63 atm., which may be attributed to $MgNi_2$ phase that cannot absorb hydrogen. Besides, the hydrogen storage capacity of the single-phase $Nd_{0.64}Mg_{0.36}Ni_{2.8}$ alloy is 0.73 H/M, while that decreases to 0.50 H/M for the x = 0.50 alloy because of the existence of the unhydrogenated $MgNi_2$ phase. The hydrogen storage capacities of the alloys of x = 0.10 and 0.20 are relatively low, which is about 0.27 and 0.37 H/M, respectively. This is closely related to the phenomenon of hydrogen-induced non-crystallization of $MgCu_4Sn$-type structure, which occurs in the fast hydrogen-absorbing process [42]. Therefore, the discharge plateaus are unobvious when the content of Mg is low in the alloys. Besides, the dehydrogenated plateau pressure of $MgCu_4Sn$-type $NdMgNi_4$ phase is higher than the superlattice phases. The two factors make the alloy with low content of Mg possess titling platform. However, when Mg increases to 0.36, the $PuNi_3$-type single phase exhibits the structural stability during dehydrogenated process, which also means appropriate Mg can prevent the hydride of the alloy amorphization. In order to prove whether the alloy has been non-crystallized, the $Nd_{0.80}Mg_{0.20}Ni_{2.8}$ alloy was selected as an example and the contrastive XRD patterns of the alloy before and after hydrogenated are as shown in Figure 4.18. It can be seen the diffraction peaks of the alloy become wider and the intensity of the peaks is greatly reduced, therefore, we can deduce that the alloy transformed to amorphous state during hydrogenation.

4.1.3.3 Electrochemical Properties

Figure 4.19 shows the discharge capacity of the $Nd_{1-x}Mg_xNi_{2.8}$ (x = 0.10, 0.20, 0.36, 0.50) alloy electrodes within 100 charge/discharge cycles. The activation property, maximum discharge capacity and capacity retention rate of the alloy electrodes are summarized in Table 4.7. As shown in Figure 4.19 and Table 4.7, the alloys all possess excellent activation capability and can be completely activated at the second cycle. The maximum discharge capacity (C_{max}) of the alloy electrodes first increases from

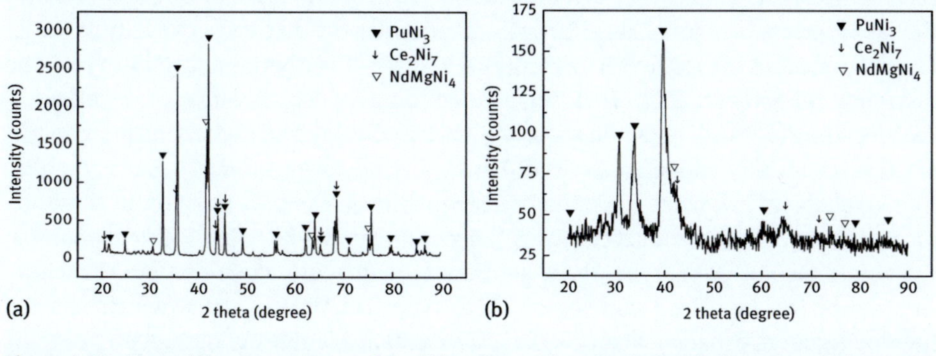

Figure 4.18: X-ray diffraction patterns of the $Nd_{0.80}Mg_{0.20}Ni_{2.8}$ alloy before hydrogenated (a) and after hydrogenated (b).

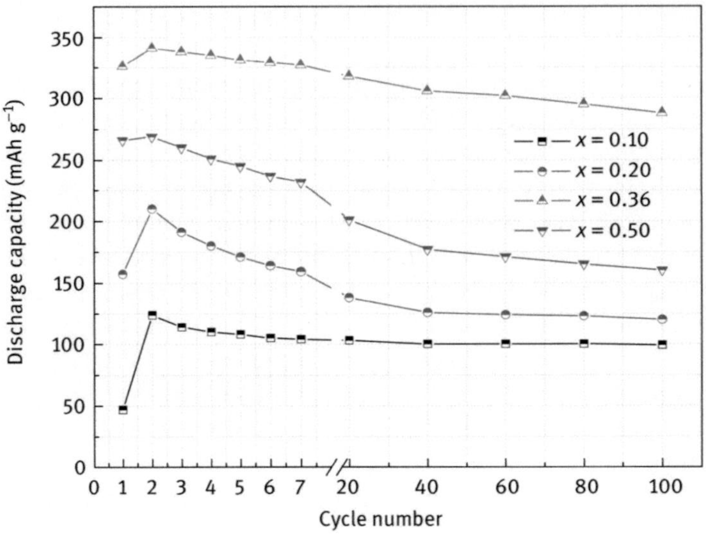

Figure 4.19: Discharge capacity versus cycle number of the $Nd_{1-x}Mg_xNi_{2.8}$ (x = 0.10, 0.20, 0.36, 0.50) alloys.

Table 4.7: Summary of electrochemical properties of $Nd_{1-x}Mg_xNi_{2.8}$ (x = 0.10, 0.20, 0.36, 0.50) alloys.

Sample (x)	N	C_{max} (mAh g^{-1})	HRD_{1800} (%)	S_{100} (%)	R_p (mΩ)	I_0 (mA g^{-1})	D (×10^{-10} cm^2 s^{-1})
x = 0.10	2	124	17.0	65.7	145.3	176.6	0.66
x = 0.20	2	210	18.5	79.8	120.3	213.4	0.82
x = 0.36	2	341	55.7	84.5	77.4	331.9	1.29
x = 0.50	2	216	42.3	59.5	108.7	236.2	1.17

124 to 341 mAh g^{-1}, reaching the maximum value with increase of Mg from 0.10 to 0.36, then decreases to 216 mAh g^{-1} with further increase of Mg to 5.0. This result indicates that appropriate content of Mg effectively ameliorates the discharge capacity of the alloy electrode and the result is in accordance to the *P–C* isotherms analysis. In fact, the discharge capacity variation is closely related to the phase alteration of the alloys. It has been found $PuNi_3$-type phase La–Mg–Ni-based alloy has higher electrochemical capacity especially which is about 400 mAh g^{-1}; however, for $MgCu_4Sn$-type phase that is relatively low where the study on single-phase $NdMgNi_4$ compound shows its discharge capacity is about 200 mAh g^{-1}. According to the XRD and Rietveld analysis, the abundance of $PuNi_3$-type phase shows a tendency of increase before reduction. In spite of the increased content of $PuNi_3$-type (Nd,Mg)Ni_3 phase which has the higher discharge capacity when x < 0.36, the existence of the $MgCu_4Sn$-type $NdMgNi_4$ phase yet lowers the discharge capacity of the alloy electrodes. But the discharge capacity of the alloy electrode is significantly improved with the decrease of $NdMgNi_4$

phase. When Mg content reaches 0.36, the disappearance of $NdMgNi_4$ phase with the lasting remaining pure $PuNi_3$-type $(Nd,Mg)Ni_3$ phase makes the $Nd_{0.64}Mg_{0.36}Ni_{2.8}$ show the highest discharge capacity value. And as x further increases to 0.50, the unhydrogenated $MgNi_2$ phase drops the discharge capacity of the alloy electrode again.

Figure 4.19 also shows a comparison of the electrochemical cycling stability of the $Nd_{1-x}Mg_xNi_{2.8}$ (x = 0.10, 0.20, 0.36, 0.50) alloys within 100 cycles. The capacity retention (S_{100}) is used to represent the cycling stability of the alloy electrode and can be calculated as $C_{100}/C_{max} \times 100$ % (where C_{100} is the discharge capacity at 100 cycles), with the results listed in Table 4.7. With the increase of content of Mg from 1.0 to 2.0, the cycling stability of the alloy electrodes is enhanced from 65.7 to 79.8 %, and the single-phase alloy with x = 0.36 has apparently improved cycle life as compared to the other samples, reaching 84.5 %. However, the capacity retention ability of the alloy markedly deteriorates as further increases Mg to 0.50. This can be mainly explained by the phase changes with the increase of Mg via pulverization and oxidation mechanism as the following: the pulverization of single-phase alloy can be reduced on account of its synchronous volume expansion/contraction during charge/discharge cycles. Correspondingly, the suppressed pulverization further decreases oxidation of active elements. While, with appearance of secondary phases, the different expansion rates between different phase structures during hydrogen absorption/desorption happen, aggravating the pulverization of the alloy electrodes and further deteriorating the oxidation of Nd and Mg elements from the new cracked surface of the alloys [38]. Besides, although the alloys of x = 0.10 and 0.20 contain similar secondary phases of Ce_2Ni_7-type and $MgCu_4Sn$-type phases, the improvement of the compositional homogeneity of x = 0.20 than x = 0.10 is the cause of the increasing cycling stability. Furthermore, negative effect on cycling stability of $MgNi_2$ phase is much noticeable than other phases, which may be due to its widely divergent from the $PuNi_3$-type structure in regards to its unhydrogenated property during the charge/discharge process, accelerating the pulverization-oxidation.

4.1.3.4 Electrochemical Kinetics

The *HRD* of the $Nd_{1-x}Mg_xNi_{2.8}$ (x = 0.10, 0.20, 0.36, 0.50) alloy electrodes at 298 K are shown in Figure 4.20. The discharge retention at the discharge current density of 1,800 mA g^{-1} (HRD_{1800}) is also summarized in Table 4.7. *HRD* of all the alloys reduces with the increasing discharge current density. The HRD_{1800} value firstly increases then decreases with the increasing content of Mg, and the single-phase alloy (x = 0.36) exhibits the best HRD when the discharge current density exceeds 360 mA g^{-1}, whose capacity remains 140 mAh g^{-1} at the current density of 1,800 mA g^{-1} and the corresponding *HRD* is 55.7 %.

The linear polarization curves of the $Nd_{1-x}Mg_xNi_{2.8}$ (x = 0.10, 0.20, 0.36, 0.50) are drawn in Figure 4.21. The exchange current density I_0 can be calculated according to

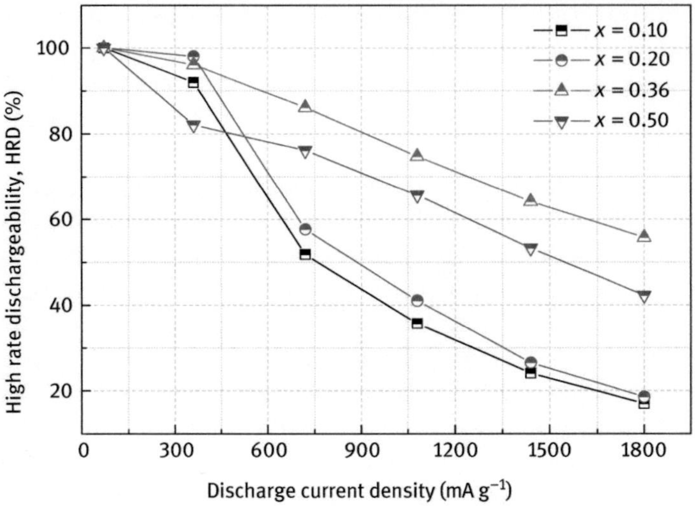

Figure 4.20: *HRD* as a function of discharge current density of the $Nd_{1-x}Mg_xNi_{2.8}$ alloy electrodes.

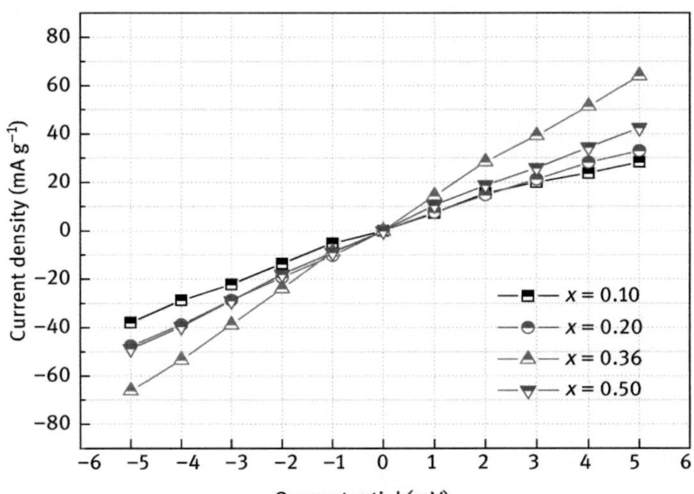

Figure 4.21: Linear polarization curves of the $Nd_{1-x}Mg_xNi_{2.8}$ alloy electrodes at 50 % DOD and 298 K.

the equation [33]. The values of the polarization resistance R_p and the exchange current density I_0 of the alloy electrodes are calculated and listed in Table 4.7. As shown in Table 4.7, I_0 increases from 176.6 mA g^{-1} ($x = 0.10$) to 331.9 mA g^{-1} ($x = 0.36$) and decreases to 236.2 mA g^{-1} ($x = 0.50$), which is caused by transformation of the $MgCu_4Sn$-type minor phases to $PuNi_3$-type phase and appearance of $MgNi_2$ phase.

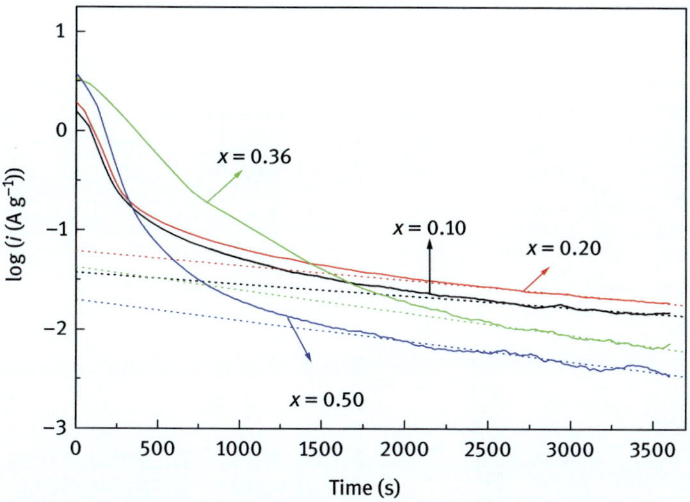

Figure 4.22: Correspondence of the anodic current density versus discharge time of the $Nd_{1-x}Mg_xNi_{2.8}$ (x = 0.10, 0.20, 0.36, 0.50) alloy electrodes at 50 % DOD and 298 K.

For hydrogen diffusion coefficient test, the potential static step method can be used. Each semi-logarithmic curve of anode current density versus time can be divided into two time domains from Figure 4.22. Assuming that the alloy particles are uniform spheres with radius of 13 μm and the reaction time is long enough, the diffusion coefficient D can be calculated [34] and listed in Table 4.7. The hydrogen diffusion rate improves from 0.66×10^{-10} cm^2 s^{-1} (x = 0.10) to 1.29×10^{-10} cm^2 s^{-1} (x = 0.36) and reduces to 1.17×10^{-10} cm^2 s^{-1} (Table 4.7). The order of D value is also consistent with that of the HRD_{1800} value, which indicates that hydrogen diffusion in alloy bulk is a conclusive factor than hydrogen reaction on surface of alloy electrodes in discharging process at high rate current density.

4.2 Effect of Rare Earth Elements on the Phase Composition and Electrochemical Performances

4.2.1 Phase Structure and Electrochemical Performances of $La_{0.8-x}Pr_xMg_{0.2}Ni_{3.4}Al_{0.1}$ (x = 0, 0.1, 0.2 and 0.3) Alloys

4.2.1.1 Microstructure

Figure 4.23(a–d) shows the Rietveld refinements of the XRD patterns of $La_{0.8-x}Pr_xMg_{0.2}Ni_{3.4}Al_{0.1}$ alloy samples and the phase abundances from the refinements are shown in Figure 4.23(e). It can be seen that all the alloys are mainly composed of $(La,Mg)_2Ni_7$ and $(La,Mg)_5Ni_{19}$ superlattice phases. Besides, $LaNi_5$ phase

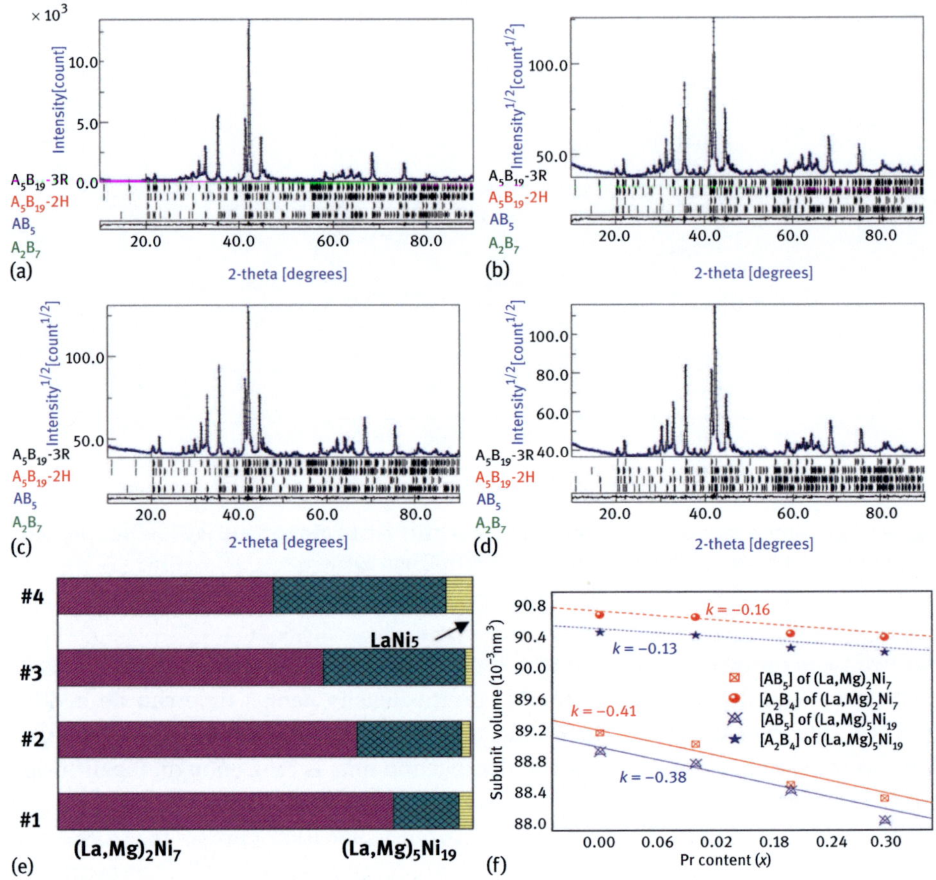

Figure 4.23: Rietveld refinements of the $La_{0.8-x}Pr_xMg_{0.2}Ni_{3.4}Al_{0.1}$ alloy samples (a)–(d), phase abundance of the $La_{0.8-x}Pr_xMg_{0.2}Ni_{3.4}Al_{0.1}$ alloy phases (c) and cell volumes of [$LaMgNi_4$] and [$LaNi_5$] subunits of (La,Mg)$_2$Ni$_7$ and (La,Mg)$_5$Ni$_{15}$ superlattice phases.

also exists due to the inclusion of Al element which quietly prefers to enter $LaNi_5$ phase in replacement of Ni. The addition of Pr exerts an unconscious influence on the phase compositions but alters the relative contents of the phases. It can be seen that the (La,Mg)$_5$Ni$_{19}$ phase abundance gradually increases in sacrifice of (La,Mg)$_2$Ni$_7$ phase with Pr substitution for La, indicating that Pr addition leads to the phase transformation from (La,Mg)$_2$Ni$_7$ phase to (La,Mg)$_5$Ni$_{19}$ phase. Moreover, it is noticeable that when x reaches 0.3, the $LaNi_5$ phase content tremendously increases.

To explore the reason for phase transformations caused by the substitution of Pr for La, the distribution of Pr element in the superlattice structure is investigated by analysing the structures of the constituent [AB_5] and [A_2B_4] subunits. It can be seen from Figure 4.23(f) that both [AB_5] and [A_2B_4] subunit volumes contract with increasing Pr substitution for La due to the smaller atomic radius of Pr than that of La. However, the slopes of the curves for $[AB_5]$ subunits are larger than those of the curves for $[A_2B_4]$ subunits in both $(La,Mg)_2Ni_7$ and $(La,Mg)_5Ni_{19}$ phases in Figure 4.23(f), indicating that the volumes of $[AB_5]$ subunits decrease faster than those of $[A_2B_4]$ subunits with higher Pr content. This means that most of the added Pr element with smaller atomic radius than La forms $[AB_5]$ subunit instead of $[A_2B_4]$ subunit. Resultantly, the content of the $(La,Mg)_5Ni_{19}$ phase with higher $[AB_5]/[A_2B_4]$ ratio increases in sacrifice of the $(La,Mg)_2Ni_7$ phase with relatively lower $[AB_5]/[A_2B_4]$ ratio. However, when x reaches 0.3, Pr content in $[AB_5]$ subunit of the superlattice phases may be saturated, and the extra Pr is left to form $LaNi_5$ phase, leading to the surge of $LaNi_5$ phase content once x reaches 0.3. But for all the studied alloys, the abundance of $LaNi_5$ non-superlattice phase content stays at a very low level. This is likely due to the stepwise annealing method which is believed to be in favour of the peritectic reaction between $LaNi_5$ phase and liquid phase, thus promoting the formation of superlattice phases [43].

The phase distribution of the $La_{0.6}Pr_{0.2}Mg_{0.2}Ni_{3.4}Al_{0.1}$ alloy is shown in Figure 4.24(a) and (b) and the EDS spectrum is presented in Figure 4.24 (c) and (d).

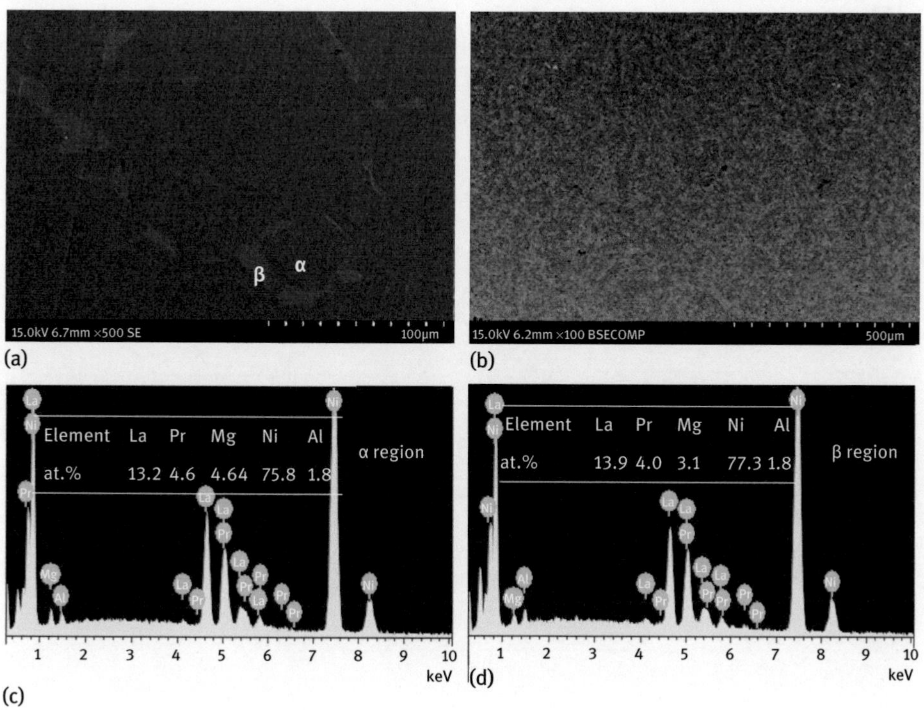

Figure 4.24: SEM and EDS of the alloy samples: (a)–(b) SEM images of the $La_{0.6}Pr_{0.2}Mg_{0.2}Ni_{3.4}Al_{0.1}$ alloy sample, (c) EDS of the area of Figure 4.24(a) A and (d) EDS of the area of Figure 4.24(a) B.

Two areas can be identified from Figure 4.24(a), which are labelled as α and β. The EDS result exhibited in the inset of the figure demonstrates that the compositions of area A and B are $(La,Mg)_2Ni_7$ phase and $(La,Mg)_5Ni_{19}$ phase with compositions of $La_{0.58}Pr_{0.21}Mg_{0.21}Ni_{3.38}Al_{0.08}$ ($AB_{3.46}$) and $La_{0.66}Pr_{0.19}Mg_{0.15}Ni_{3.68}Al_{0.09}$ ($AB_{3.77}$), respectively. It can be seen that areas A and B have very close colours because of their similar compositions and corrosion degrees by the acid pretreatment. $LaNi_5$ phase is missing in the presented photograph, which may be due to its relatively small amount.

4.2.1.2 Hydrogen Absorption and Desorption

The *P–C* isotherms of the $La_{0.8-x}Pr_xMg_{0.2}Ni_{3.4}Al_{0.1}$ alloy samples at 30°C are shown in Figure 4.25. It can be seen that the *P–C* curves have clear plateau regions corresponding to the co-existence of α and β phases. Moreover, the curves exhibit one hydrogen absorption/desorption plateau, indicating that the $(La,Mg)_2Ni_7$ and $(La,Mg)_5Ni_{19}$ phases have very close equilibrium pressures due to their similar super-stacking structures [44]. The plateau pressure of the alloy samples during hydrogen absorption (P_{abs}) and desorption (P_{ads}) increases with increasing amount of Pr substitution for La. This observation is in accordance with the decreasing trend of the cell parameters of the alloy phases with increasing Pr content.

The hydrogen storage capacity (C_{abs}) of the alloy samples decreases with the addition of Pr content, which is caused by the following structural changes: on one hand, increasing Pr content decreases the cell volumes of the alloy phases, leaving less room for hydrogen atoms. Moreover, the theoretical hydrogen storage capacity of the $(La,Mg)_5Ni_{19}$ phase is less than that of the $(La,Mg)_2Ni_7$ phase. Therefore, the hydrogen storage capacity shows a decreasing trend with more Pr substitution for La. But it can be noticed that the capacity decrease is not very considerable, which may be due to the catalytic effect of the $(La,Mg)_5Ni_{19}$ phase with high-Ni content on the hydrogen absorption/desorption of the alloys samples. This catalytic effect can also be

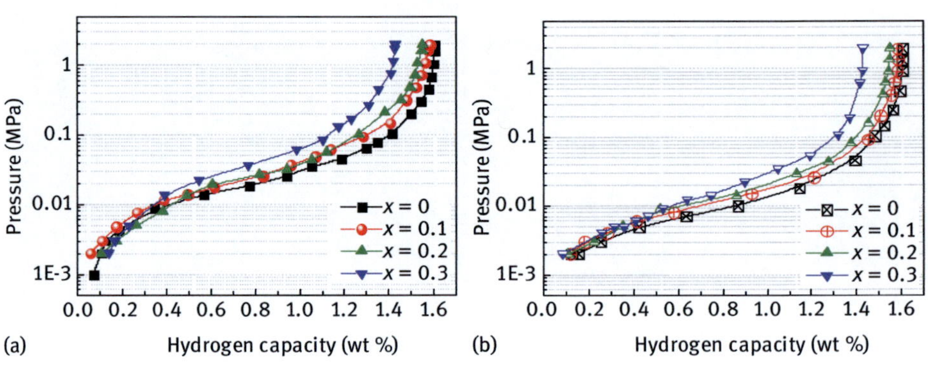

Figure 4.25: *P–C* isotherms of the $La_{0.8-x}Pr_xMg_{0.2}Ni_{3.4}Al_{0.1}$ alloy samples: (a) hydrogen absorption and (b) hydrogen desorption.

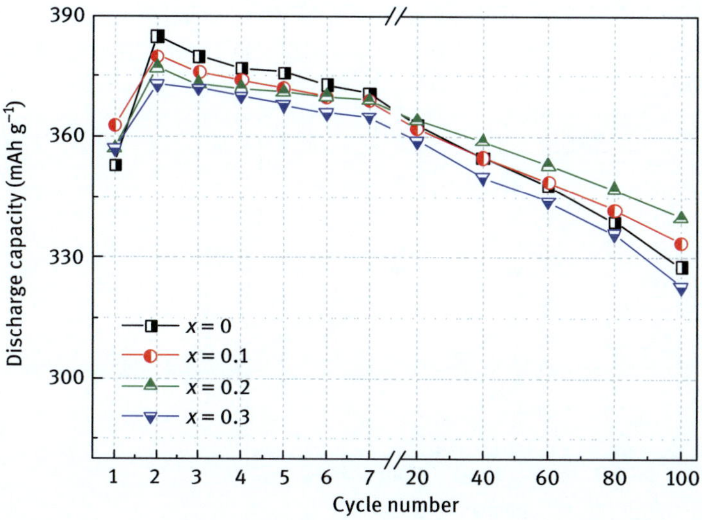

Figure 4.26: Discharge capacity of the $La_{0.8-x}Pr_xMg_{0.2}Ni_{3.4}Al_{0.1}$ alloy electrodes within 100 cycles.

reflected by the increasing hydrogen absorption/desorption reversibility (C_{ads}/C_{abs}) which increases from 0.923 ($x = 0$) to 0.943 ($x = 0.2$).

4.2.1.3 Electrochemical Properties

The discharge capacity of the alloy electrodes within 100 charge/discharge cycles is shown in Figure 4.26. It can be seen that the alloy electrodes have good activation property and can be fully activated after two cycles. The maximum discharge capacity (C_{max}) of the alloy electrodes decreases with increasing amount of Pr substitution for La, but the decreasing trend is gentle. The evolution trend in the electrode discharge capacity essentially follows the same trend as the hydrogen storage capacity that is demonstrated by *P–C* isotherms.

The *HRD* curves at various discharge capacities are shown in Figure 4.27. It can be seen that the *HRD* at different discharge current densities first increases and then decreases with increasing Pr content. This evolution trend in *HRD* is closely related to the phase composition of the alloys. The addition of Pr increased the abundance of $(La,Mg)_5Ni_{19}$ phase which has superior discharge performance at large current densities, thus increasing the *HRD* of the alloy electrodes. However, when x exceeds 0.2, the $LaNi_5$ phase sharply increased in sacrifice of $(La,Mg)_2Ni_7$ and $(La,Mg)_5Ni_{19}$ super-stacking phases with good *HRD* property, leading to the slight decrease in *HRD*.

By comparing the slopes of the curves in Figure 4.27, it can be seen that the capacity degradation of the alloy electrodes is first alleviated and then aggravated with increasing Pr substitution for La, and the cycling stability at the 100th cycle (S_{100}) first increases from 86.3 % ($x = 0.1$) to 90.2 ($x = 0.2$) and then decreases to 86.6 ($x = 0.3$).

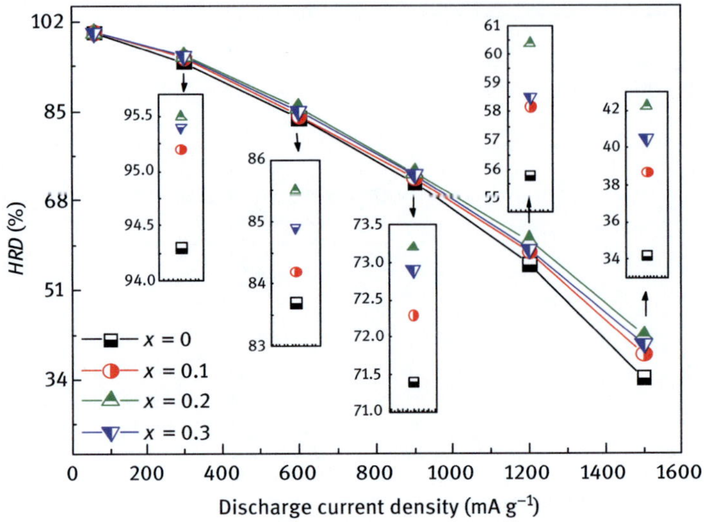

Figure 4.27: *HRD* of the $La_{0.8-x}Pr_xMg_{0.2}Ni_{3.4}Al_{0.1}$ alloy electrodes at various discharge current densities.

The decrease in the discharge retention is due to the uprush of the $LaNi_5$ phase. It was reported that the expansion rate of the cell volume for $LaNi_5$ non-stacking phase is quite different from those of the $(La,Mg)_2Ni_7$ and $(La,Mg)_5Ni_{19}$ superstacking phases, and the discrete expansion leads to serious pulverization and capacity degradation. The reason for the improvement in the cycling stability of the alloy electrodes by adding appropriate amount of Pr is discussed in detail below.

The XRD patterns of the $La_{0.8}Mg_{0.2}Ni_{3.4}Al_{0.1}$ and $La_{0.6}Pr_{0.2}Mg_{0.2}Ni_{3.4}Al_{0.1}$ alloy powders after 100 cycles are shown in Figure 4.28 (red curves) and those before cycling are also presented (blue curves) for comparison. It can be seen that the alloy powders still keep their original peak patterns after cycling, but the intensity of diffraction peaks decreases and the width of the diffraction peaks becomes broader, indicating the pulverization and amorphization of the alloy powders. Moreover, it is noticeable that the main peak of the $La_{0.8}Mg_{0.2}Ni_{3.4}Al_{0.1}$ alloy after cycling is much broader than that of $La_{0.6}Pr_{0.2}Mg_{0.2}Ni_{3.4}Al_{0.1}$ alloy and the calculated FWHM of $La_{0.8}Mg_{0.2}Ni_{3.4}Al_{0.1}$ alloy is apparently higher, indicating more serious pulverization and amorphization of $La_{0.8}Mg_{0.2}Ni_{3.4}Al_{0.1}$ alloy powder than $La_{0.6}Pr_{0.2}Mg_{0.2}Ni_{3.4}Al_{0.1}$ alloy powder after 100 cycles.

To study the pulverization degree of the alloy electrodes, the particle size distribution curves of the $La_{0.8}Mg_{0.2}Ni_{3.4}Al_{0.1}$ and $La_{0.6}Pr_{0.2}Mg_{0.2}Ni_{3.4}Al_{0.1}$ alloy powders before and after 100 cycles are presented in Figure 4.29. It can be seen that before cycling, the mean particle size of the $La_{0.8}Mg_{0.2}Ni_{3.4}Al_{0.1}$ and $La_{0.6}Pr_{0.2}Mg_{0.2}Ni_{3.4}Al_{0.1}$ alloy powders are similar, being 51.4 μm. However, the mean particle size of the $La_{0.8}Mg_{0.2}Ni_{3.4}Al_{0.1}$ alloy powders decreases to 26.1 μm after 100 cycles and that of

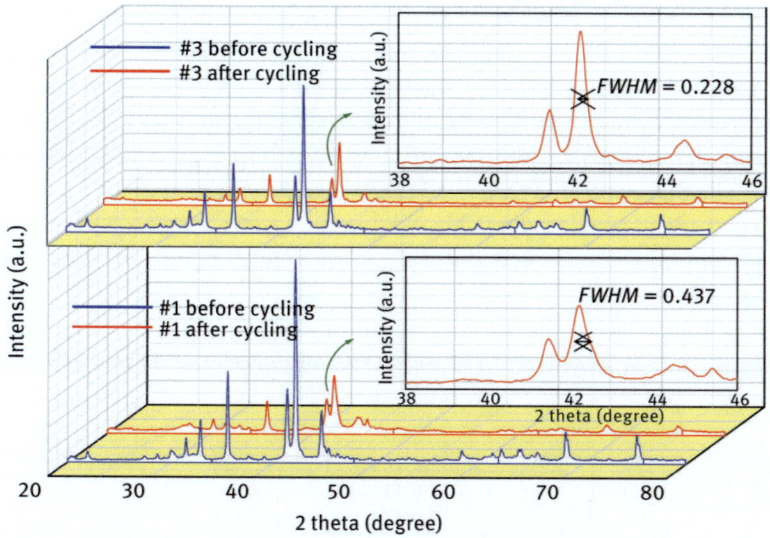

Figure 4.28: Comparison of the XRD patterns between the $La_{0.8}Mg_{0.2}Ni_{3.4}Al_{0.1}$ and $La_{0.6}Pr_{0.2}Mg_{0.2}Ni_{3.4}Al_{0.1}$ alloy powders before and after 100 cycles. (a) $x = 0$ (b) $x = 0.2$.

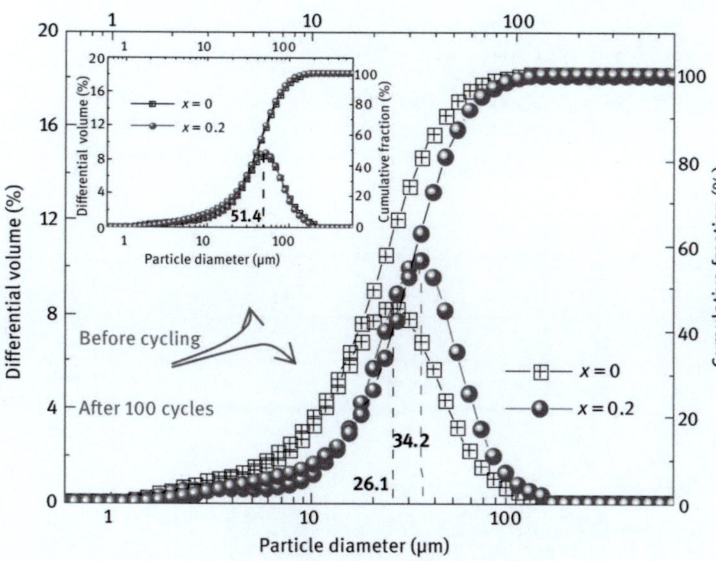

Figure 4.29: Particle size distribution of $La_{0.8}Mg_{0.2}Ni_{3.4}Al_{0.1}$ and $La_{0.6}Pr_{0.2}Mg_{0.2}Ni_{3.4}Al_{0.1}$ alloy powders before and after 100 charge/discharge cycles.

$La_{0.6}Pr_{0.2}Mg_{0.2}Ni_{3.4}Al_{0.1}$ alloy powders is much larger, being 34.2 μm. The results suggest that the pulverization of the $La_{0.6}Pr_{0.2}Mg_{0.2}Ni_{3.4}Al_{0.1}$ alloy powders after cycling is slighter than that of $La_{0.8}Mg_{0.2}Ni_{3.4}Al_{0.1}$ alloy powders. Larger alloy particles are usually less sensitive to the corrosion of alkaline electrolyte due to their lower activity. The oxygen contents of the $La_{0.8}Mg_{0.2}Ni_{3.4}Al_{0.1}$ and $La_{0.6}Pr_{0.2}Mg_{0.2}Ni_{3.4}Al_{0.1}$ alloy powders after 100 electrochemical cycles in 6 M KOH alkaline solution are measured to be 3.23 mass% and 2.62 mass%, respectively. The lower oxygen content of the cycled $La_{0.6}Pr_{0.2}Mg_{0.2}Ni_{3.4}Al_{0.1}$ alloy electrode is due to its lower activity in alkaline solution thanks to the larger particle size.

The Rietveld refinements of the XRD patterns of the $La_{0.8}Mg_{0.2}Ni_{3.4}Al_{0.1}$ and $La_{0.6}Pr_{0.2}Mg_{0.2}Ni_{3.4}Al_{0.1}$ alloy powders after 100 cycles are shown in Figure 4.30(a)

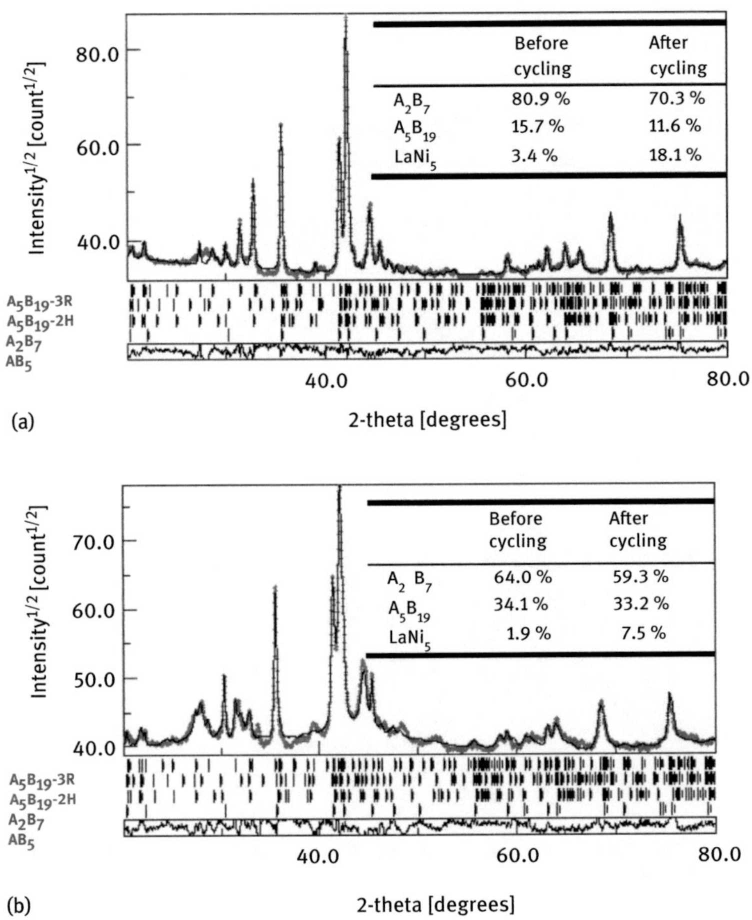

Figure 4.30: Rietveld refinements of the $La_{0.8}Mg_{0.2}Ni_{3.4}Al_{0.1}$ and $La_{0.6}Pr_{0.2}Mg_{0.2}Ni_{3.4}Al_{0.1}$ alloy powders after 100 cycles. (a) $x = 0$ (b) $x = 0.2$.

and (b) with phase abundance presented in the insets of the figure. In comparison with the phase abundances of the as-prepared alloys, the $LaNi_5$ phase abundance of $La_{0.8}Mg_{0.2}Ni_{3.4}Al_{0.1}$ alloy after cycling apparently increases at the cost of $(La,Mg)_2Ni_7$ phase abundance, and $(La,Mg)_5Ni_{19}$ phase abundance also slightly declined. But as for $La_{0.6}Pr_{0.2}Mg_{0.2}Ni_{3.4}Al_{0.1}$ alloy, the $(La,Mg)_2Ni_7$ phase abundance does not drop as much, and the $(La,Mg)_5Ni_{19}$ phase abundance almost keep unchanged. Figure 4.31(a) and (b) shows the TEM electron diffraction patterns of $La_{0.8}Mg_{0.2}Ni_{3.4}Al_{0.1}$ and $La_{0.6}Pr_{0.2}Mg_{0.2}Ni_{3.4}Al_{0.1}$ alloy particles after 100 gaseous hydrogen absorption/desorption cycles, and the images of the alloy powders are presented in the insets of the figure. It is can be observed from the insets that the $La_{0.8}Mg_{0.2}Ni_{3.4}Al_{0.1}$ alloy powder is irregular after cycling, while the $La_{0.8}Mg_{0.2}Ni_{3.4}Al_{0.1}$ alloy powder shows a regular shape after 100 cycles. Moreover, from the electron diffractions, it is seen that the cycled $La_{0.8}Mg_{0.2}Ni_{3.4}Al_{0.1}$ alloy exhibits a large extent of amorphization while most part of the $La_{0.8}Mg_{0.2}Ni_{3.4}Al_{0.1}$ alloy still keeps in the form of crystallization. Our previous study [45] reported that the $(La,Mg)_2Ni_7$ phase would decompose into $LaNi_5$ phase and amorphous phase due to the consecutive cell volume change of $[LaMgNi_4]$ subunits during hydrogen absorption/desorption. The above result indicates that Pr can stabilize the structural stability of the super-stacking phases. Actually, Pr was found to be able to relive the hydrogen-induced amorphization of the AB_2-type $La_{1-x}Pr_xMgNi_{3.6}Co_{0.4}$ alloys [7]. Moreover, the amount of the $(La,Mg)_5Ni_{19}$ phase network with good structural stability increases with increasing Pr content, which also benefits the anti-amorphization ability of the alloys. However, in comparison with that of AB_5-type alloys [46], the cycling stability

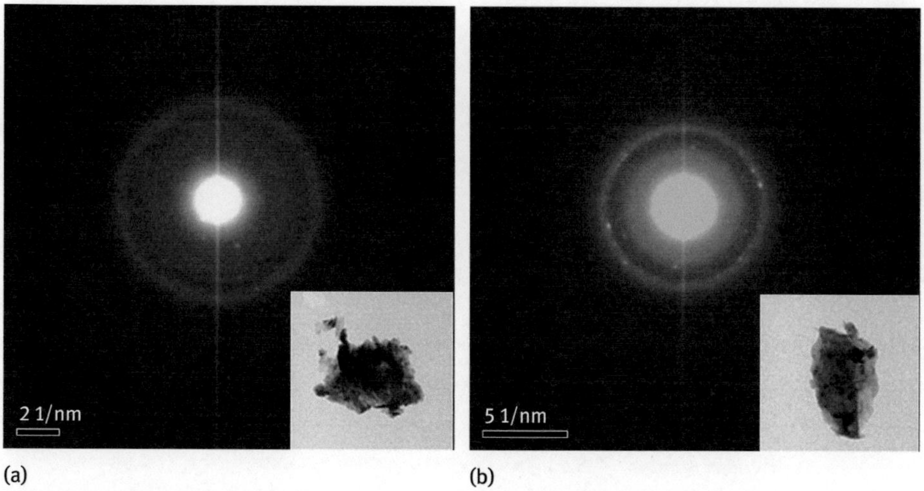

Figure 4.31: TEM diffraction patterns of $La_{0.8}Mg_{0.2}Ni_{3.4}Al_{0.1}$ and $La_{0.6}Pr_{0.2}Mg_{0.2}Ni_{3.4}Al_{0.1}$ alloy powders after 100 cycles. (a) $La_{0.8}Mg_{0.2}Ni_{3.4}Al_{0.1}$ (b) $La_{0.6}Pr_{0.2}Mg_{0.2}Ni_{3.4}Al_{0.1}$.

of $La_{0.8-x}Pr_xMg_{0.2}Ni_{3.4}Al_{0.1}$ (x = 0, 0.1, 0.2 and 0.3) alloys in this study is relatively lower. Further improvement is expected to be made.

4.2.2 Phase Structure and Electrochemical Performances of $La_{0.75-x}Nd_xMg_{0.25}Ni_{3.3}$ (x = 0, 0.15) Alloys

4.2.2.1 Crystal Structure

Figure 4.32 represents the Rietveld refinement of the XRD patterns of the $La_{0.75-x}Nd_xMg_{0.25}Ni_{3.3}$ (x = 0, 0.15) alloys. The $La_{0.75}Mg_{0.25}Ni_{3.3}$ alloy contains Ce_2Ni_7-type hexagonal (2H-A_2B_7) phase and Gd_2Co_7-type rhombohedral (3R-A_2B_7) phase accompanied with minor $CaCu_5$-type $LaNi_5$ phase. While the $La_{0.60}Nd_{0.15}Mg_{0.25}Ni_{3.3}$ alloy is composed of 2H-A_2B_7 phase and 3R-A_2B_7 phase. The crystallographic data and phase compositions determined by the Rietveld analysis are summarized in Table 4.8.

Compared with $La_{0.75}Mg_{0.25}Ni_{3.3}$ alloy consisting of 72.96 wt.% 2H-A_2B_7 phase, 15.68 wt.% 3R-A_2B_7 phase and 11.37 wt.% $LaNi_5$ phase, the $La_{0.60}Nd_{0.15}Mg_{0.25}Ni_{3.3}$ alloy only contains 2H- and 3R-A_2B_7 phases, suggesting that the partial substitution Nd for

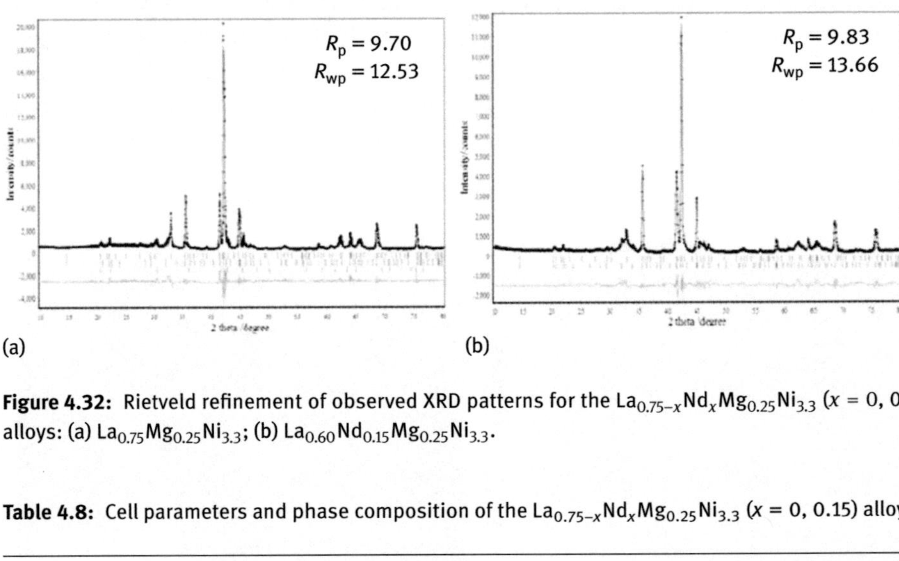

Figure 4.32: Rietveld refinement of observed XRD patterns for the $La_{0.75-x}Nd_xMg_{0.25}Ni_{3.3}$ (x = 0, 0.15) alloys: (a) $La_{0.75}Mg_{0.25}Ni_{3.3}$; (b) $La_{0.60}Nd_{0.15}Mg_{0.25}Ni_{3.3}$.

Table 4.8: Cell parameters and phase composition of the $La_{0.75-x}Nd_xMg_{0.25}Ni_{3.3}$ (x = 0, 0.15) alloys.

Sample (x)	Phase	Space group	Lattice parameter		c/a	Cell volume (Å)	Phase abundance (wt.%)
			A (Å)	C (Å)			
x = 0	$LaNi_5$	$P6/mmm$	4.9689	3.9900	0.803	85.32	11.37
	Ce_2Ni_7	$P63/mmc$	5.0367	24.2833	4.821	533.49	72.96
	Gd_2Co_7	R-$3m$	5.0337	36.2944	7.210	796.43	15.68
x = 0.15	Ce_2Ni_7	$P63/mmc$	4.9909	24.0437	4.817	518.67	79.84
	Gd_2Co_7	R-$3m$	5.0089	36.1702	7.221	785.90	20.16

La benefits the formation of A_2B_7 phase while suppresses that of the $CaCu_5$-type phase. The phase abundances calculated by the Rietveld refinement show the disappearance of the $CaCu_5$-type phase with partial substitution of Nd for La accompanied with the consumption of $LaNi_5$ alloy. According to the binary phase diagram, NdNi alloy becomes molten state firstly at 540°C (melting point). The molten NdNi wraps up the rest alloys, generating the subsequent peritectic reaction at sintering temperature.

Tables 4.9 and 4.10 display the refined atomic coordinates and occupation numbers of the 2H-A_2B_7, 3R-A_2B_7 and $CaCu_5$-type phases in the $La_{0.75-x}Nd_xMg_{0.25}Ni_{3.3}$ (x = 0, 0.15) alloys. The results indicate that the Mg atoms occupy $4f1$ site in the 2H-A_2B_7 phase and take up $6c2$ sites in the 3R-A_2B_7 phase. Due to the fact that the Mg–Ni bonds are shorter than the La–Ni bonds in the A_2B_4 subunits [27], the Mg–Ni ionic bond strength is stronger than the La–Ni ionic bond strength. Therefore the Nd atom prefers to enter the La site rather than the Mg site. On the other hand, the variation of the Mg occupation rate on the La site brings almost no change on the reliability during the Rietveld refinement. So the Nd atom can only enter the La site in the A_2B_4 subunits. With the temperature rising, the molten NdNi and $LaMgNi_4$ parcelled

Table 4.9: Atomic coordinates and occupation number for the 2H-type A_2B_7 structure in the $La_{0.60}Nd_{0.15}Mg_{0.25}Ni_{3.3}$.

Ce_2Ni_7	Site	X	y	z	Occupy
La_1	4f	0.3333	0.6667	0.0217	0.6724
Mg	4f	0.3333	0.6667	0.0324	0.3276
La_2	4f	0.3333	0.6667	0.1724	1
Ni_1	2a	0	0	0	1
Ni_2	4e	0	0	0.1775	1
Ni_3	4f	0.3333	0.6667	0.8393	1
Ni_4	6h	0.8337	1.6674	0.2500	1
Ni_5	12k	0.8333	1.6667	0.0866	1

Table 4.10: Atomic coordinates and occupation number for the 3R-type A_2B_7 structure in the $La_{0.60}Nd_{0.15}Mg_{0.25}Ni_{3.3}$.

Gd_2Co_7	Site	X	y	z	Occupy
La_1	6c	0	0	0.0559	1
La_2	6c	0	0	0.1639	0.5127
Mg_2	6c	0	0	0.1298	0.4873
Ni_1	3b	0	0	0.5000	1
Ni_2	6c	0	0	0.2791	1
Ni_3	6c	0	0	0.3867	1
Ni_4	9e	0.5000	0	0	1
Ni_5	18h	0.5032	−0.5032	0.1099	1

around the $LaNi_5$, so that the replaced La atom in the $LaMgNi_4$ subunits or the unreacted NdNi transform with the rest Ni to be more A_2B_4 subunits, which crystallize with the AB_5 subunits sufficiently.

As reported in reference [27], the 2H- and 3R-A_2B_7 structures are almost equally stable. Like the Mg–Ni ionic bonds, the Nd–Ni ionic bonds are shorter in length than La–Ni bonds, which could strengthen the ionic bonds of the A_2B_7-type structure. Thus it is more stable for the 2H- and 3R-type A_2B_7 structure by stacking the A_2B_4 subunits containing Nd–Ni bonds with AB_5 subunits. So in the sintering temperature, that is, 980°C, the AB_5 subunits and A_2B_4 subunits are stacking in regular sequence to form the 2H- and 3R-A_2B_7 structures when the peritectic reaction happens. But the formation of the 3R-A_2B_7 phase is not much more than the 2H-type caused by the substitution of Nd by La during the refinements.

We know that the substitution of Mg on the La site in the A_2B_4 subunits lowers the mismatch between the A_2B_4 and AB_5 units because of the smaller radius of the Mg than the La [17]. As mentioned above, the Nd atoms occupy the La site in the A_2B_4 subunits. For the smaller radii of Nd(1.82 Å) than the La(1.88 Å), the substitution of Nd for La can also induce the mismatch between the AB_5 and A_2B_4 subunits as the effect of Mg. Hence, the Nd substitution for the La in the $La_{0.75}Mg_{0.25}Ni_{3.3}$ alloy increases the 2H- and 3R-A_2B_7 phases and strengthens stability of the stacking structure.

The shrink percentages of the axes and cells for $La_{0.75-x}Nd_xMg_{0.25}Ni_{3.3}$ (x = 0, 0.15) are listed in Table 4.11. Lattice parameters a, c and cell volume in the Ce_2Ni_7-type phase shrink by 0.91, 0.99 and 2.78 % respectively, and the same tendency occurs in the Gd_2Co_7-type phase with $\Delta a/a = -0.49$ %, $\Delta c/c = -0.34$ %, $\Delta V/V = -1.32$ %. The result above has shown that Nd entering the A_2B_4 subunits causes linear shrinkage in the lattice parameters and cell volume due to the smaller Nd radius and promotes formation of the 2H- and 3R-A_2B_7 phases. The different rates of change in the lattice parameters identify with the different occupancies of Nd substitution for La.

4.2.2.2 *P–C* Isotherms

The electrochemical P–C isotherms of the $La_{0.75-x}Nd_xMg_{0.25}Ni_{3.3}$ (x = 0, 0.15) alloys at room temperature are drawn in Figure 4.33. It can be obviously observed that the *P–C* isotherm of the $La_{0.75}Mg_{0.25}Ni_{3.3}$ shows the distinct two plateaus, while the other isotherm for the $La_{0.60}Nd_{0.15}Mg_{0.25}Ni_{3.3}$ presents only one plateau. Li et al. [12] has reported the plateau pressure of the $LaNi_5$ phase is a slightly higher than the La_2Ni_7

Table 4.11: Shrink of the lattice axes, unit cell volume of the Ce_2Ni_7 and Gd_2Co_7 cell.

	Ce_2Ni_7 (%)	Gd_2Co_7 (%)
$\Delta a/a$	0.91	0.49
$\Delta c/c$	0.99	0.34
$\Delta V/V$	2.78	1.32

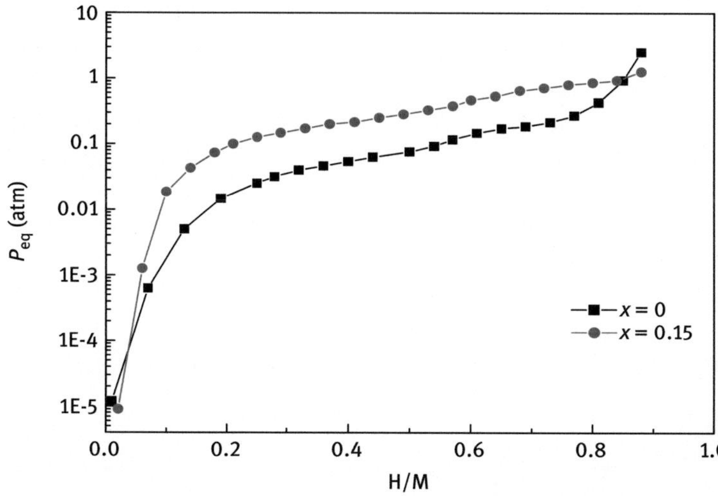

Figure 4.33: Electrochemical desorption P–C isotherms of the $La_{0.75-x}Nd_xMg_{0.25}Ni_{3.3}$ (x = 0, 0.15) alloy electrodes.

phase. And the Ce_2Ni_7 phase and Gd_2Co_7 phase have a similar plateaus pressure for the similar stacking structure [27]. We consider the higher plateaus corresponding to the $LaNi_5$ phase, and the lower one standing for the 2H- and 3R-type phases when x is 0. When x is 0.15, the plateau only represents the coexistence of the Ce_2Ni_7 phase and Gd_2Co_7 phase. What we have found is corresponding to the Rietveld refinement.

When x is 0, the plateau pressure of A_2B_7 phase and $LaNi_5$ phase are 0.006 MPa and 0.018 MPa, respectively. At the same time, the hydrogen capacity of the A_2B_7 phase is ~0.47 H/M which is higher than the AB_5 phase (~0.16 H/M). With the Nd substitution, the plateau pressure improves to 0.022 MPa and has a longer plateau, which indicates the Nd substitution increases the dehydrogenation pressure and capacity (~0.70 H/M) apparently. The cause for the change in the equilibrium pressure is the shrink of the lattice parameter due to the smaller radii of Nd than La. The shrink of the subunits results the dehydrogenation of H atoms needing more pressure to escape from the alloy bulk, which is identical with the P–C isotherms and the change of the lattice parameters.

4.2.2.3 Electrochemical Properties

The electrochemical properties of the $La_{0.75-x}Nd_xMg_{0.25}Ni_{3.3}$ (x = 0, 0.15) alloy electrodes are summarized in Table 4.12. Figure 4.34 shows the activation cycle number (N_a), maximum discharge capacity (C_{max}) and the cycle stability (S_n) of the $La_{0.75-x}Nd_xMg_{0.25}Ni_{3.3}$ (x = 0, 0.15) alloy electrodes. It can be seen that both of the electrodes can be activated within two cycles and reach their maximum capacity of 365 mAh g^{-1} (x = 0) and 376 mAh g^{-1} (x = 0.15), respectively. However the 100th cycle stability (S_{100}) of the electrodes decreases from 81.10 % (x = 0) to 76.74 % (x = 0.15).

Table 4.12: Electrochemical properties of the $La_{0.75-x}Nd_xMg_{0.25}Ni_{3.3}$ (x = 0, 0.15) alloy electrodes.

Sample	N_a	C_{max} (mAh g^{-1})	HRD_{1440} (%)	S_{100} (%)
x = 0	2	365	42.98	81.10
x = 0.15	2	376	67.70	76.74

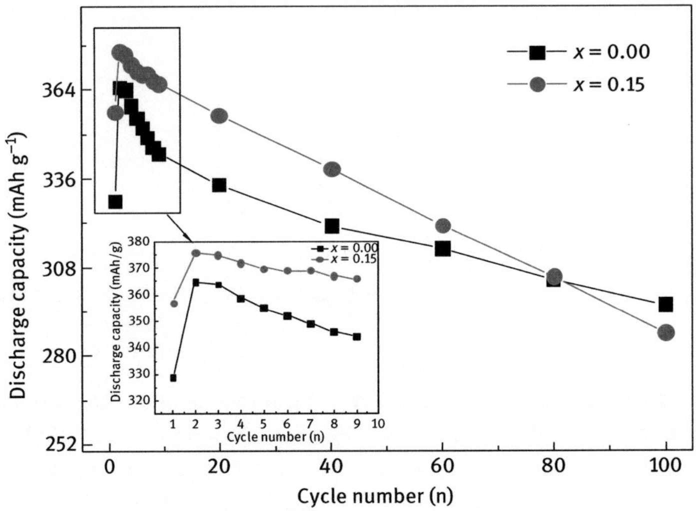

Figure 4.34: Discharge capacity of the $La_{0.75-x}Nd_xMg_{0.25}Ni_{3.3}$ (x = 0, 0.15) alloy electrodes versus cycle number.

Due to the fact that La_2Ni_7 phase has a higher capacity than $LaNi_5$ phase [1–3], the maximum capacity of the $La_{0.60}Nd_{0.15}Mg_{0.25}Ni_{3.3}$ alloy electrodes is higher than the $La_{0.75}Mg_{0.25}Ni_{3.3}$ alloy. The electrochemical discharge capacities of the alloys are in good agreement with the phase component and the hydrogen storage capacity observed in the *P–C* isotherm curves in Figure 4.33.

It is known that the decrease of the cyclic stability of the alloy electrodes can be caused mainly by the oxidation of active components in alkali liquor from the surface to interior of the electrodes and pulverization of the alloy particles resulting from the volume expansion [11]. Shen et al. [47] have reported the Nd addition suppress the oxidation of the hydrogen-absorbing elements such as La and Mg effectively. So it can be concluded the pulverization should be the dominant factor for the reduction of the cyclic stability. As shown in Table 4.8, the changes of a, c and V of the 2H- and 3R-A_2B_7 structure could result in more serious expansion after hydrogen-absorbing than the based alloy electrodes and therefore causes the deterioration in the cyclic stability of the $La_{0.60}Nd_{0.15}Mg_{0.25}Ni_{3.3}$ alloy electrode. Meanwhile $LaNi_5$ phase as a hydride storer can also be a catalyst to promote the absorbing/desorbing process of the La_2Ni_7 phase

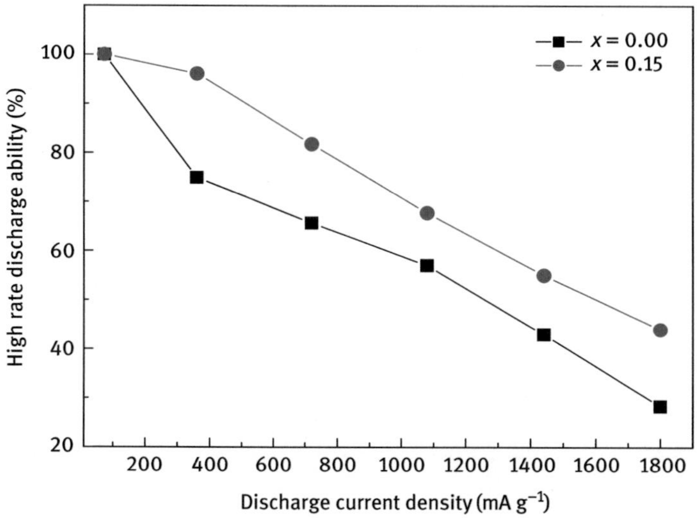

Figure 4.35: *HRD* of the $La_{0.75-x}Nd_xMg_{0.25}Ni_{3.3}$ (x = 0, 0.15) alloy electrodes.

so that the electrodes of the $La_{0.75}Mg_{0.25}Ni_{3.3}$ alloy have a higher cyclic stability than the other one.

4.2.2.4 Kinetics Characteristics

The *HRD* of the alloy electrodes at 298 K are illustrated in Figure 4.35, and the discharge retention at the discharge current density 1,440 mA g^{-1} (HRD_{1440}) are summarized in Table 4.12. It is indicated that the substitution of Nd for La has drastically improved *HRD* and the HRD_{1440} is increased from 42.98 % (x = 0) to 67.70 % (x = 0.15). It has been reported the better kinetic characteristics of the electrodes accompany with the higher hydrogen desorption plateau pressure usually [48]. As shown in Figure 4.35, the HRD_{1440} of the $La_{0.60}Nd_{0.15}Mg_{0.25}Ni_{3.3}$ alloy electrodes increases with the elevated plateau pressure.

The linear polarization curves of the $La_{0.75-x}Nd_xMg_{0.25}Ni_{3.3}$ (x = 0, 0.15) alloy electrodes are drawn in Figure 4.36(a). And the kinetics characteristics of the $La_{0.75-x}Nd_xMg_{0.25}Ni_{3.3}$ (x = 0, 0.15) alloy electrodes are summarized in Table 4.13. The polarization resistance R_p and exchange current density I_0 can be calculated as described in Notten's work [49]. Table 4.13 shows that the R_p calculated from the equation decreases from 645.02 mΩ (x = 0) to 472.66 mΩ (x = 0.15). Moreover the exchange current density I_0 of the alloy electrodes increases from 261.37 mA g^{-1} (x = 0) to 358.14 mA g^{-1} (x = 0.15). We have already known that the Nd addition accelerates the pulverization of the alloy electrodes so that it provides much more channels for the hydrogen transfer. Therefore the Nd substitution for La improves the electrocatalytic performance of the alloy electrodes significantly judging from the exchange current density. Figure 4.36(b) presents the anodic current versus discharge time of

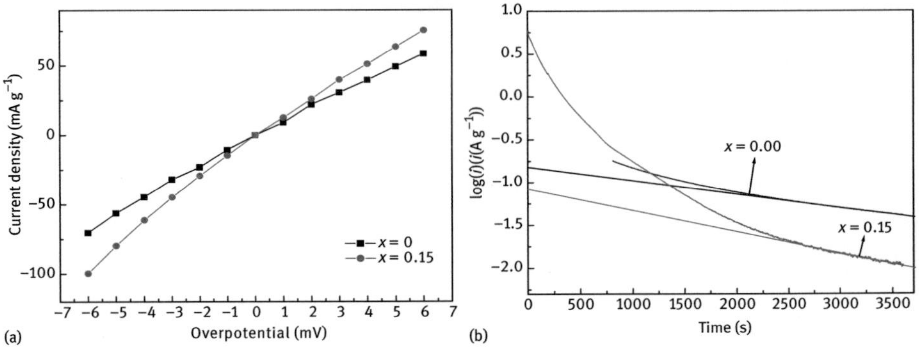

Figure 4.36: Linear polarization curves and correspondence of the anodic current density versus discharge time of the $La_{0.75-x}Nd_xMg_{0.25}Ni_{3.3}$ (x = 04, 0.15) alloy electrodes: (a) linear polarization curves; (b) correspondence of the anodic current density versus discharge time.

Table 4.13: Kinetics characteristics of the $La_{0.75-x}Nd_xMg_{0.25}Ni_{3.3}$ (x = 0, 0.15) alloy electrodes.

Sample	I_0 (mA g^{-1})	R_p (mΩ)	D (×10^{-11} cm^2 s^{-1})
x = 0	261.39	645.02	6.30
x = 0.15	358.14	472.66	9.89

$La_{0.75-x}Nd_xMg_{0.25}Ni_{3.3}$ (x = 0, 0.15) alloy electrodes. Supposing that the alloy particles have a similar size with the average particle radius of 13 μm, the coefficient of hydrogen diffusion can be calculated as explored by Zheng et al. [34]. The method of potential step is a characterization for the hydrogen diffusion rate in the alloy bulk. We give a greater anodic potential to the full-charged electrodes so that the hydrogen diffusion rate in the alloy bulk become the rate-controlling step. D values are summarized in Table 4.13. It can be seen that the hydrogen diffusion rate improves from 6.30×10^{-11} cm^2 s^{-1} (x = 0) to 9.89×10^{-11} cm^2 s^{-1} (x = 0.15).

4.3 Effect of Transition Elements on the Phase Composition and Electrochemical Performances

4.3.1 Phase Structure and Electrochemical Performances of $La_{0.80}Mg_{0.20}Ni_{2.95}Co_{0.70-x}Al_x$ (x = 0, 0.05, 0.10, 0.15) Alloys

4.3.1.1 Microstructure

Figure 4.37(a–d) shows the Rietveld refinements of the XRD patterns for the $La_{0.80}Mg_{0.20}Ni_{2.95}Co_{0.70-x}Al_x$ alloys. The XRD data were well fitted judging by s. It can be seen that the alloys mainly contain $(La,Mg)_2Ni_7$ phase in 2H- and 3R-types and

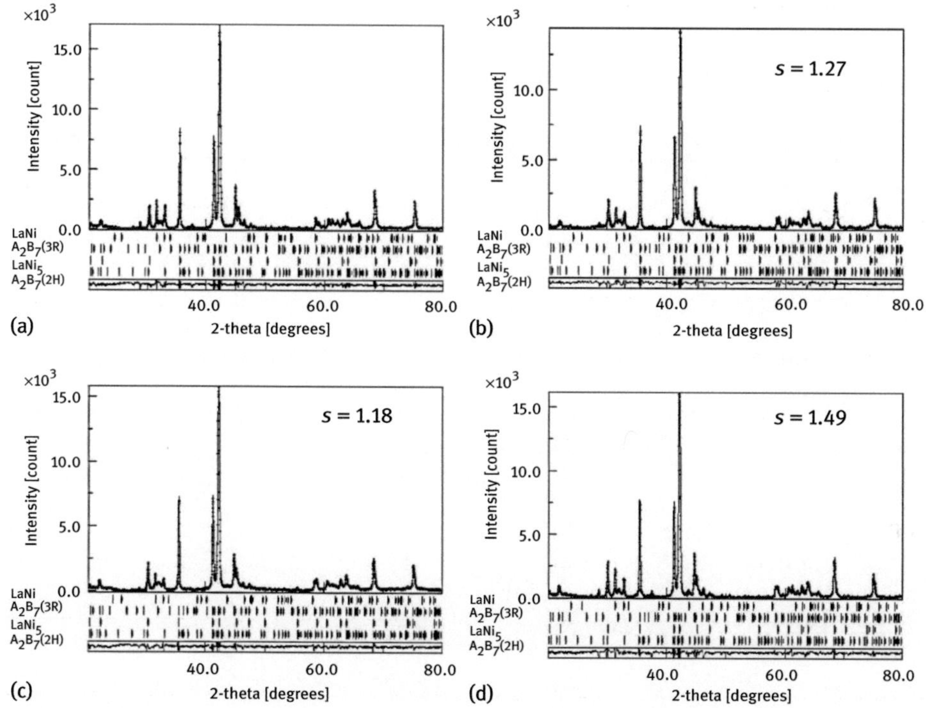

Figure 4.37: Rietveld refinements of the alloy samples, s is used to judge the degree of the refinement: $s = R_{wp}/R_e$, where R_{wp} is the weighed sum of residuals of the least square fit and R_e is the value statistically expected.

$LaNi_5$ phase in $CaCu_5$-type. The phase abundance is summarized in Table 4.14, and the variation trend of the phase abundance with increasing x is shown in Figure 4.38(a). As Al content increases, the $LaNi_5$ phase abundance increases monotonously while the $(La,Mg)_2Ni_7$ phase abundance decreases. We consider the phase transformation of the $(La,Mg)_2Ni_7$ phase to the $LaNi_5$ phase is related to the uneven distribution of the added Al element, which will be discussed later. Figure 4.38(b) shows the variation trends of the 2H- and 3R-type $(La,Mg)_2Ni_7$ phase abundances as well as the ratio of the 2H-type to 3R-type phase. It is observed that as Al content increases the ratio of the 2H-type to 3R-type phase abundance increases. Zhang et al. [27] reported that the phase crystal structure of the A_2B_7 compounds was size-dependent. For larger atomic radii, 2H-type was preferred while for smaller atomic radii, 3R-type was preferred. Similar result was also reported in our previously study that the ratio of 2H/3R-type phase decreased as the cell lattice shrank [37].

The lattice parameters of the alloy samples are presented in Table 4.14. It can be seen that the lattice parameters and the cell volumes of both $(La,Mg)_2Ni_7$ and $LaNi_5$ phases increase with increasing Al content due to the larger atomic radius of

Table 4.14: Phase abundance and cell parameters of the alloy samples.

Sample	Phase	Space group (no.)	Phase abundance (%)	*a* (nm)	*c* (nm)	*V* (nm^3)
$x = 0$	$(La,Mg)_2Ni_7$ (2H)	*P63/mmc* (194)	32.5	5.0467	24.198	533.73
	$(La,Mg)_2Ni_7$ (3R)	*R-3m* (166)	55.6	5.0455	36.301	800.32
	$LaNi_5$	*P6/mmm* (191)	10.9	5.0411	3.9901	87.814
	LaNi	*Cmcm* (63)	1.0			
$x = 0.05$	$(La,Mg)_2Ni_7$ (2H)	*P63/mmc* (194)	38.5	5.0480	24.221	534.51
	$(La,Mg)_2Ni_7$ (3R)	*R-3m* (166)	31.9	5.0480	36.344	802.05
	$LaNi_5$	*P6/mmm* (191)	26.6	5.0466	3.9943	88.099
	LaNi	*Cmcm* (63)	3.0			
$x = 0.10$	$(La,Mg)_2Ni_7$ (2H)	*P63/mmc* (194)	36.8	5.0498	24.265	535.88
	$(La,Mg)_2Ni_7$ (3R)	*R-3m* (166)	38.3	5.0532	36.398	804.90
	$LaNi_5$	*P6/mmm* (191)	20.8	5.0513	4.0063	88.529
	LaNi	*Cmcm* (63)	4.1			
$x = 0.15$	$(La,Mg)_2Ni_7$ (2H)	*P63/mmc* (194)	14.4	5.0584	24.267	537.74
	$(La,Mg)_2Ni_7$ (3R)	*R-3m* (166)	60.6	5.0544	36.453	806.51
	$LaNi_5$	*P6/mmm* (191)	24.2	5.0519	4.0155	88.751
	LaNi	*Cmcm* (63)	0.8			

Al than that of Co element. The expansion ratios of the cell volumes ($\Delta V\%$) calculated from the cell parameters are linearly fitted with the Al content (x) in Figure 4.39. It is found that the $\Delta V\%$ of all the phases augments with increasing Al content. Noticeably, the increasing slopes (k) of $(La,Mg)_2Ni_7$ and $LaNi_5$ phases labelled in Figure 4.39 are quite different. The k values of the 2H- and the 3R-type $(La,Mg)_2Ni_7$ phase are 5.02 and 5.31, respectively, which are very close, while that of the $LaNi_5$ phase is as large as 7.38, indicating that the expansion in the cell volumes of the $LaNi_5$ phase grows faster with increasing Al content than that of the $(La,Mg)_2Ni_7$ phase. One possible reason for this phenomenon is that the increased Al has a tendency to enter the $LaNi_5$ phase so that the influence of increasing Al content on the volume expansion of $LaNi_5$ phase is more obvious than that of the $(La,Mg)_2Ni_7$ phase. This Al uneven distribution in $(La,Mg)_2Ni_7$ and $LaNi_5$ phases is also in accordance with the phenomenon that the $(La,Mg)_2Ni_7$ phase abundance increases while the $LaNi_5$ phase decreases with increasing Al content. This will be further proved by the SEM-EDS.

The SEM image of $La_{0.80}Mg_{0.20}Ni_{2.95}Co_{0.55}Al_{0.15}$ is presented in Figure 4.40(a) and the EDS spectra are shown in Figure 4.40(b) and (c). It can be seen that there are two different areas in Figure 4.40(a) which are light grey (α) and dark grey (β). EDS analysis shows that α and β regions in Figure 4.40(a) are $(La,Mg)_2Ni_7$ phase with a composition of $La_{0.76}Mg_{0.24}Ni_{1.94}Co_{0.51}Al_{0.07}$ and $LaNi_5$ phase with a composition of $LaNi_{4.18}Co_{0.63}Al_{0.25}$, respectively. No difference between 2H- and 3R-type $(La,Mg)_2Ni_7$ phase is observed in the image due to their close compositions [2, 37]. The SEM result is

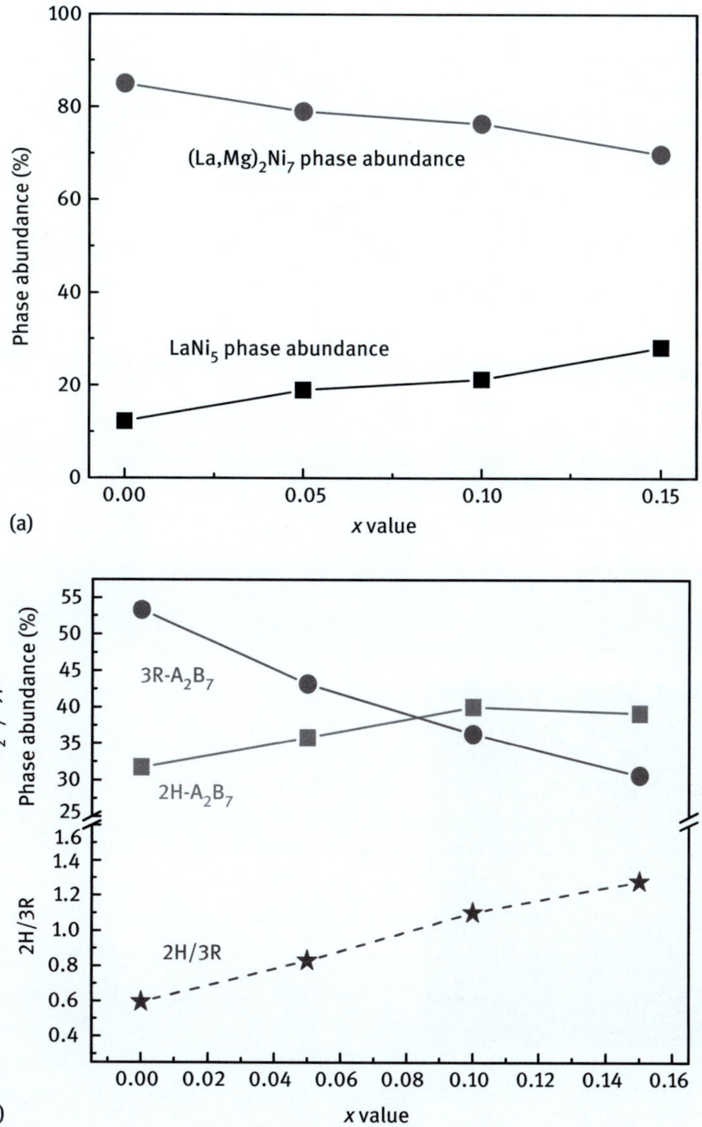

Figure 4.38: $LaNi_5$ and $(La,Mg)_2Ni_7$ phase abundance of the alloy samples (a), 2H- and 3R-type $(La,Mg)_2Ni_7$ phase abundance and their ratios (b).

in good agreement with the XRD analysis. Remarkably, it is noted that the atomic percentage of Al element in the $LaNi_5$ phase is almost four times that in the $(La,Mg)_2Ni_7$ phase, indicating that most of the introduced Al element forms the $LaNi_5$ phase rather than the $(La,Mg)_2Ni_7$ phase. This result, together with the XRD analysis, proves that the introduced Al element is preferable to enter $LaNi_5$ phase rather than $(La,Mg)_2Ni_7$

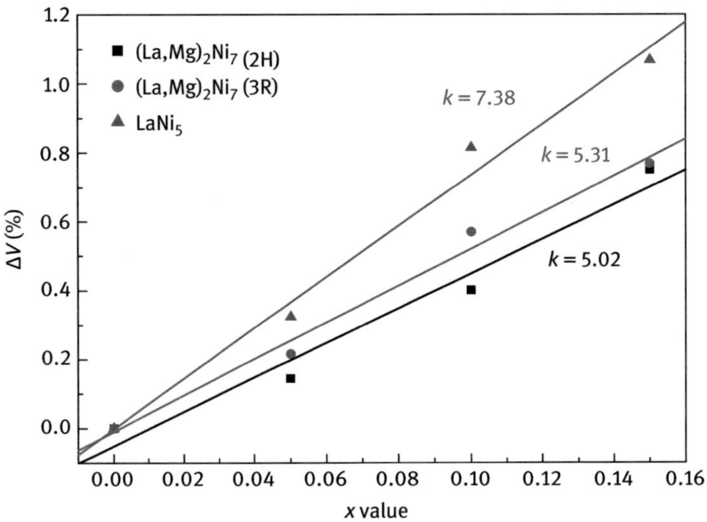

Figure 4.39: Relationship between cell volume expansion (ΔV) after hydrogenation and Al content (x).

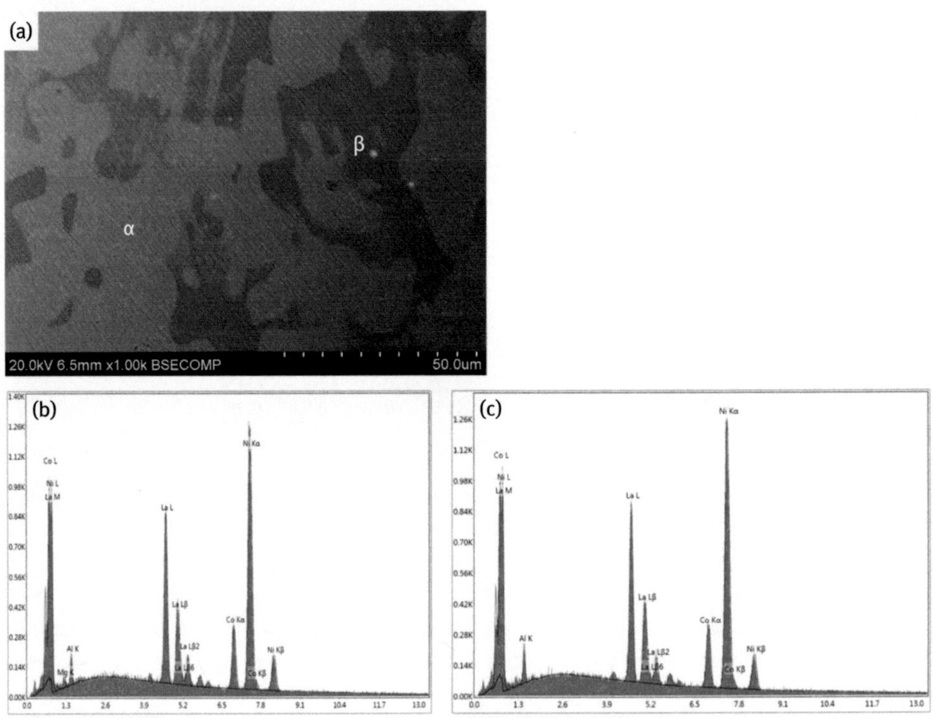

Figure 4.40: SEM and EDS of $La_{0.80}Mg_{0.20}Ni_{2.95}Co_{0.54}Al_{0.15}$ alloy: (a) SEM backscattering electron image, (b) EDS of the areas of Figure 4.40(a) α and (c) EDS of the areas of Figure 4.40(a) β.

phase, which is the reason for the phase transformation of the (La,Mg)$_2$Ni$_7$ phase to the LaNi$_5$ phase with increasing Al content. Besides, Mg also shows selective phase distribution and it only enters the (La,Mg)$_2$Ni$_7$ phase due to the fact that Mg exclusively occupies La site in the [La$_2$Ni$_4$] unit of the (La,Mg)$_2$Ni$_7$ phase, but it cannot enter [LaNi$_5$] unit or LaNi$_5$ phase where the La–La distance is longer than that of the [La$_2$Ni$_4$] unit [50].

4.3.1.2 Activation and Cycling Stability

The discharge capacity of the alloys within 150 cycles is shown in Figure 4.41(a) and (b), and the information is summarized in Table 4.15. Figure 4.41(a) shows that the alloys exert good activation performance and can reach their maximum discharge capacity (C_{max}) within two cycles. The good activation property is due to the multiphase structures whose interfaces are helpful to the diffusion of hydrogen atoms. The C_{max} of the alloys decreases with Al substitution for Co due to the increase of the LaNi$_5$ phase in sacrifice of the (La,Mg)$_2$Ni$_7$ phase whose hydrogen storage capacity is higher than LaNi$_5$ phase. The discharge capacity of the alloy electrodes decreases monotonously after activation. Figure 4.41(b) shows that the decreasing trend of alloys becomes gentler as x value increases from 0 to 0.15. From Table 4.15 it is seen that the capacity retention of the alloy electrodes at the 100th cycle (S_{100}) increases from 73.1 % (x = 0) to 86.7 (%) (x = 0.15), and the capacity retention at the 150th cycle (S_{150}) increases from 61.9 % (x = 0) to 81.3 % (x = 0.15). The alleviation in the discharge capacity degradation is closely related to the alleviated pulverization and oxidation of the cycled alloys thanks to Al incorporation.

4.3.1.3 HRD and *HRD*$_{1500}$ Cycling Retention

The HRD_n of the alloy electrodes at various discharge current densities were measured after initial activation (four cycles) and are shown in Figure 4.42(a). The HRD_{1500} values are listed in Table 4.15. It can be seen that the *HRD* of the alloy electrodes

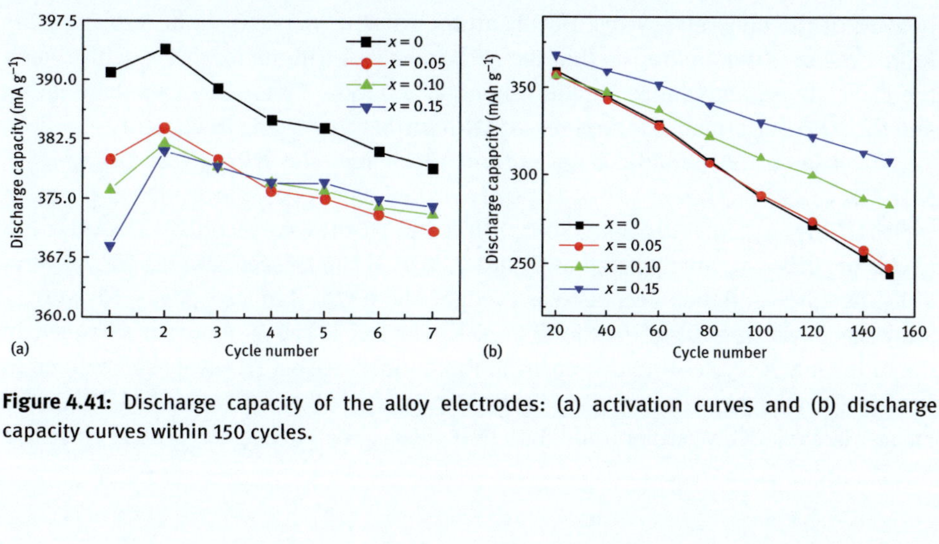

Figure 4.41: Discharge capacity of the alloy electrodes: (a) activation curves and (b) discharge capacity curves within 150 cycles.

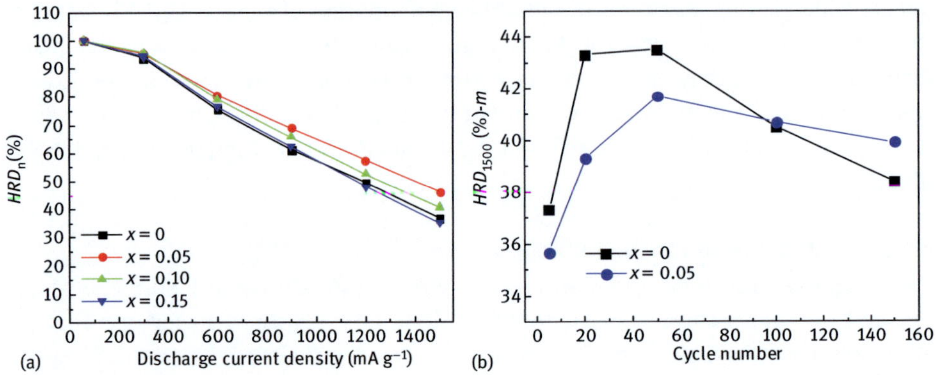

Figure 4.42: *HRD* of the alloy electrodes: (a) *HRD* at various discharge current densities and (b) HRD_{1500} curves within 150 cycles.

Table 4.15: Electrochemical properties of the alloy electrodes.

Sample	C_{max} (mA h g^{-1})	S_{100} (%)	S_{150} (%)	$HRD_{1500\text{-initial}}$ (%)	$HRD_{1500\text{-max}}$ (%)	$HRD_{1500\text{-}150}$ (%)	$S_{HRD1500\text{-}150}$ (%)
$x = 0$	394	73.1	61.9	36.59	43.5	38.4	95.7
$x = 0.05$	389	75.2	64.4	46.12	–	–	–
$x = 0.10$	382	81.2	77.1	40.53	–	–	–
$x = 0.15$	381	86.7	81.3	35.12	41.7	39.9	97.6

decreases monotonously with increasing discharge current density. The highest *HRD* values at various discharge current densities are obtained when x is 0.05. The increase of $LaNi_5$ catalytic phase and the expansion in cell volumes are the two main factors affecting *HRD*. The $LaNi_5$ catalytic phase accelerates the hydrogen desorption reaction while the expansion in the cell volumes is unfavourable to the fast discharge process of the alloy electrodes. In this study, when x increases from 0 to 0.05, the former factor is dominating so that the *HRD* increases with increasing x, while when the $LaNi_5$ phase abundance is relatively high, the latter factor becomes dominating and the *HRD* performance is suppressed with further increasing Al content.

To observe the periodic large current discharge, the HRD_{1500} of $La_{0.80}Mg_{0.20}Ni_{2.95}Co_{0.70}$ and $La_{0.80}Mg_{0.20}Ni_{2.95}Co_{0.55}Al_{0.15}$ alloy electrodes, electrodes at different cycles ($HRD_{1500\text{-}m}$, m = 5th, 20th, 50th, 100th and 150th) were recorded. The variation trends in $HRD_{1500\text{-}m}$ are depicted in Figure 4.42(b). It can be seen that the HRD_{1500} initially increases and then decreases as cycling. The HRD_{1500} of $La_{0.80}Mg_{0.20}Ni_{2.95}Co_{0.70}$ alloy electrode rises faster in the first 20 cycles but it keeps almost unchanged in the following 30 cycles and decreases monotonously with a relatively steep slope in the last 100 cycles. As for $La_{0.80}Mg_{0.20}Ni_{2.95}Co_{0.55}Al_{0.15}$ alloy electrode, its HRD_{1500} increases in a relatively gentler trend than that of $La_{0.80}Mg_{0.20}Ni_{2.95}Co_{0.70}$ alloy electrode in the initial 20 cycles, but in the following 30 cycles, it increases more apparently than that of $La_{0.80}Mg_{0.20}Ni_{2.95}Co_{0.70}$ alloy electrode and it decreases much slower in the last 100 cycles. The $S_{HRD1500}$ denoted as the ratio of the HRD_{1500} at the 150th cycle ($HRD_{1500\text{-}150}$) to the maximum HRD_{1500} ($HRD_{1500\text{-}max}$) with in 150 cycles is displayed Table 4.15. It can be seen that the $S_{HRD1500\text{-}150}$ of $La_{0.80}Mg_{0.20}Ni_{2.95}Co_{0.70}$ alloy electrode decreases to 88.3 % while that of $La_{0.80}Mg_{0.20}Ni_{2.95}Co_{0.55}Al_{0.15}$ alloy electrode remains as high as 95.7 %.

4.3.1.4 Electrochemical Kinetics of the Alloys within 150 Cycles

The exchange current density (I_0) of the alloy electrodes characterizes the electrocatalytic activity for charge–transfer at the metal/electrolyte interface and it can be obtained [51]. The variation trend of I_0 for the $La_{0.80}Mg_{0.20}Ni_{2.95}Co_{0.70}$ and $La_{0.80}Mg_{0.20}Ni_{2.95}Co_{0.55}Al_{0.15}$ alloy electrodes at the 5th, 20th, 50th, 100th and 150th cycles are shown in Figure 4.43(a). It can be seen that the I_0 of the alloy electrodes is initially increased and then decreased within 150 cycles, which is in accordance with the changing trend of HRD_{1500}. The I_0 of both alloys reaches the peak value at the 50th cycle. Although the Al incorporation relives the degradation of charge transfer, the effect is not as outstanding as that on the HRD_{1500}.

To monitor the hydrogen diffusion behaviour of the alloy electrodes within 150 cycles, the anode polarization test of $La_{0.80}Mg_{0.20}Ni_{2.95}Co_{0.70}$ and $La_{0.80}Mg_{0.20}Ni_{2.95}Co_{0.55}Al_{0.15}$ alloy electrodes was carried out at the 5th, 20th, 50th, 100th and 150th electrochemical cycles and the variation trends as cycling are presented in Figure 4.43(b). It is found that similar to the HRD_{1500} and I_0, the limiting current density (I_L) of both alloys initially rises and then decreases within 150 cycles. In comparison of the two alloy electrodes, it is found that $La_{0.80}Mg_{0.20}Ni_{2.95}Co_{0.70}$ alloy electrode shows higher initial I_L value and more obvious trend of rising in the first 5–20 cycles, while the increasing rate slows down in the following

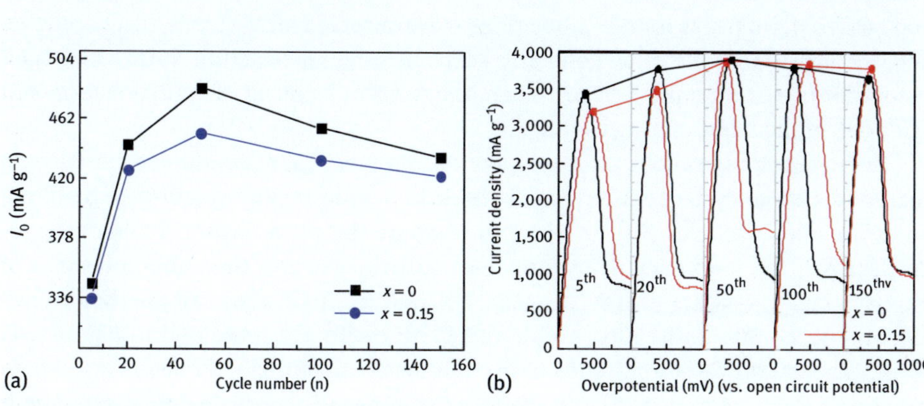

Figure 4.43: Variation trend of I_0 (a) and I_L (b) for $La_{0.80}Mg_{0.20}Ni_{2.95}Co_{0.70}$ and $La_{0.80}Mg_{0.20}Ni_{2.95}Co_{0.55}Al_{0.15}$ alloy electrodes within 150 cycles.

30 cycles. The I_L of $La_{0.80}Mg_{0.20}Ni_{2.95}Co_{0.55}Al_{0.15}$ alloy electrode increases slower than that of $La_{0.80}Mg_{0.20}Ni_{2.95}Co_{0.70}$ alloy electrode in the 5–20 cycles but faster in the 20–50 cycles. In the last 100 cycles, $La_{0.80}Mg_{0.20}Ni_{2.95}Co_{0.55}Al_{0.15}$ alloy electrode exhibits gentler degradation slope than that of $La_{0.80}Mg_{0.20}Ni_{2.95}Co_{0.70}$ alloy electrode. Noticeably, the I_L value of $La_{0.80}Mg_{0.20}Ni_{2.95}Co_{0.55}Al_{0.15}$ alloy electrode has exceeded that of $La_{0.80}Mg_{0.20}Ni_{2.95}Co_{0.70}$ alloy electrode during 50th–100th cycles.

It can be seen from the above discussion that the *HRD*, the charge transfer on the alloy surface and the hydrogen diffusion in the alloy bulk are all initially promoted and then suppressed with cycling in the alkaline solution. Al addition decreases the degradation rate of the HRD_{1500} as well as electrochemical kinetics of the alloy electrodes. Resultantly, the HRD_{1500} and hydrogen diffusion speed of the $La_{0.80}Mg_{0.20}Ni_{2.95}Co_{0.55}Al_{0.15}$ alloy electrode exceeds those of $La_{0.80}Mg_{0.20}Ni_{2.95}Co_{0.70}$ alloy electrode during the 100–150 cycles and 50–100 cycles, respectively, due to their slower degradation. The improvement in the *HRD* and kinetics retention of the Al containing alloy is closely related to the effect of Al on the microstructure evolution of the alloys with cycling which will be discussed in the next section.

4.3.1.5 Influence Mechanism of Al Incorporation

The degradation in both discharge capacity and kinetics performance of the alloy electrodes is closely related to the structure changes of the alloys during electrochemical cycling in the alkaline solution. The XRD pattern of the $La_{0.80}Mg_{0.20}Ni_{2.95}Co_{0.55}Al_{0.15}$ alloy powder after 150 electrochemical charge/discharge cycles in 6 M KOH electrolyte is shown in Figure 4.44, and the pattern of $La_{0.80}Mg_{0.20}Ni_{2.95}Co_{0.55}Al_{0.15}$ alloy before cycling is also included for comparison. It can be seen that the XRD pattern of the cycled alloy remains similar to its original pattern except for some degree of peak broadening caused by the refinement of the particle grains and some small peaks of oxidation/corrosion products (hydroxides and oxides such as $Al(OH)_3$, Al_2O_3, $Mg(OH)_2$ and $La(OH)_3$). No peaks for new alloy phases are observed after charge/discharge cycling, indicating that the alloy generally keeps its original structure during hydrogen absorption/desorption process [52]. Therefore, we will focus on the pulverization and oxidation of the alloys during cycling.

Hydrogen storage alloys undergo large cell volume expansion/contraction during charge/discharge cycles, which would result in a serious pulverization or breaking up of the alloy particles [53]. In order to compare the pulverization rates of the alloy samples, the frequency distributions of particle size and the cumulative size of the $La_{0.80}Mg_{0.20}Ni_{2.95}Co_{0.70}$ and $La_{0.80}Mg_{0.20}Ni_{2.95}Co_{0.55}Al_{0.15}$ alloy samples before cycling and at the 5th, 20th, 50th, 100th and 150th cycles are presented in Figure 4.45 (a–e). The general variation trends of the volume mean particle diameters are shown in Figure 4.45(f). It can be seen that the mean volume mean particle diameters of both alloys decrease intensively in the first five cycles and the decreasing trend becomes much gentler in the following cycles. After 50 cycles, very slight particle size decrease

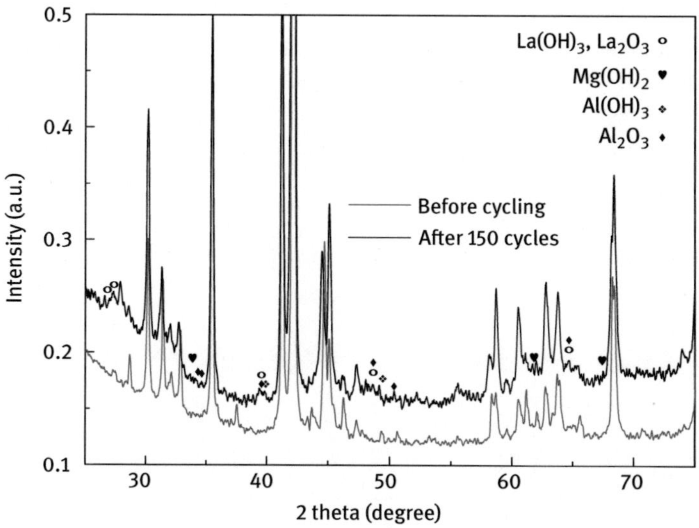

Figure 4.44: XRD patterns of $La_{0.80}Mg_{0.20}Ni_{2.95}Co_{0.55}Al_{0.15}$ alloy electrode before cycling and after 150 charge/discharge cycles.

can be observed for both alloys. In comparison of the two alloys, from Figure 4.46(a) it can be seen that the volume mean particle diameters of $La_{0.80}Mg_{0.20}Ni_{2.95}Co_{0.70}$ and $La_{0.80}Mg_{0.20}Ni_{2.95}Co_{0.55}Al_{0.15}$ alloy electrode before cycling are very close, about 50.6 μm, but the particle size of $La_{0.80}Mg_{0.20}Ni_{2.95}Co_{0.70}$ alloy electrode becomes much smaller than that of $La_{0.80}Mg_{0.20}Ni_{2.95}Co_{0.55}Al_{0.15}$ alloy electrode after the first five cycles. The initial decrease in the alloy particles results in more active interfaces, which are beneficial to the fast hydrogen absorption/desorption process. So that the kinetics performance of $La_{0.80}Mg_{0.20}Ni_{2.95}Co_{0.70}$ alloy electrode mounts up to its maximum value faster than that of $La_{0.80}Mg_{0.20}Ni_{2.95}Co_{0.55}Al_{0.15}$ alloy electrode in very early cycles. However, in the 6–50 cycles, the particle size of $La_{0.80}Mg_{0.20}Ni_{2.95}Co_{0.70}$ alloy presents a steeper decreasing slope than that of $La_{0.80}Mg_{0.20}Ni_{2.95}Co_{0.55}Al_{0.15}$ alloy, which is in accordance with the more apparent increase in the kinetics of the $La_{0.80}Mg_{0.20}Ni_{2.95}Co_{0.55}Al_{0.15}$ alloy than that of $La_{0.80}Mg_{0.20}Ni_{2.95}Co_{0.70}$ alloy during 6–50 cycles. After 50 cycles, the particle size hardly changes for both alloys.

In general, the $La_{0.80}Mg_{0.20}Ni_{2.95}Co_{0.55}Al_{0.15}$ sample exhibits far less pulverization than $La_{0.80}Mg_{0.20}Ni_{2.95}Co_{0.70}$ sample during the first five cycles and the gap lasts in the following cycles, indicating that Al incorporation relives the alloy pulverization induced by hydrogen absorption/desorption. The XRD patterns of the fully electrochemically hydrogenated $La_{0.80}Mg_{0.20}Ni_{2.95}Co_{0.70}$ and $La_{0.80}Mg_{0.20}Ni_{2.95}Co_{0.55}Al_{0.15}$ alloy samples are presented in Figure 4.46, and the unit cell parameters and the expansion rates after hydrogenation are shown in the insets of Figure 4.46. It can be seen that the expansion rate of the hydrogenated $La_{0.80}Mg_{0.20}Ni_{2.95}Co_{0.55}Al_{0.15}$

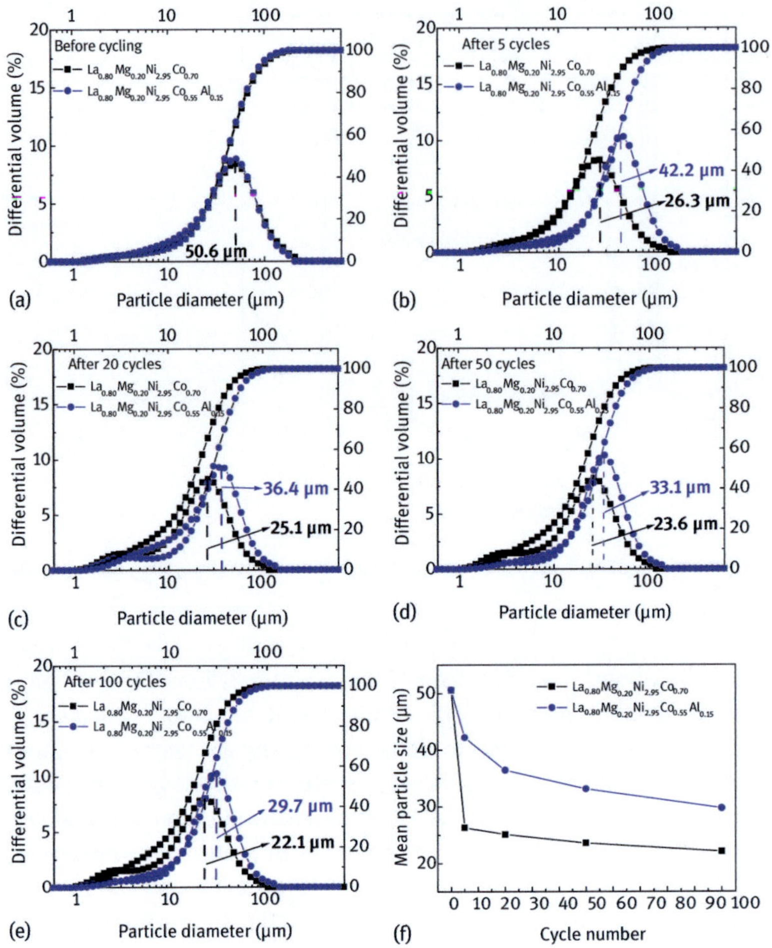

Figure 4.45: Particle size distributions of $La_{0.80}Mg_{0.20}Ni_{2.95}Co_{0.70}$ and $La_{0.80}Mg_{0.20}Ni_{2.95}Co_{0.55}Al_{0.15}$ alloys after various charge/discharge cycles.

alloy sample is smaller than that of the Al-free alloy, which means that the $La_{0.80}Mg_{0.20}Ni_{2.95}Co_{0.55}Al_{0.15}$ alloy sample undergoes smaller rate of expansion contraction during hydrogen absorption/desorption than $La_{0.80}Mg_{0.20}Ni_{2.95}Co_{0.70}$ alloy sample, indicating that larger cell volumes can accommodate more volume expansion and constriction caused by the transformation between α-phase and β-phase during hydrogen absorption process, which is in agreement with Zhao et al. [22]. From the above analysis it is concluded that Al addition can relive constriction during hydrogen absorption/desorption by reducing the expansion rate of the cell volumes after hydrogenation so that the pulverization is alleviated.

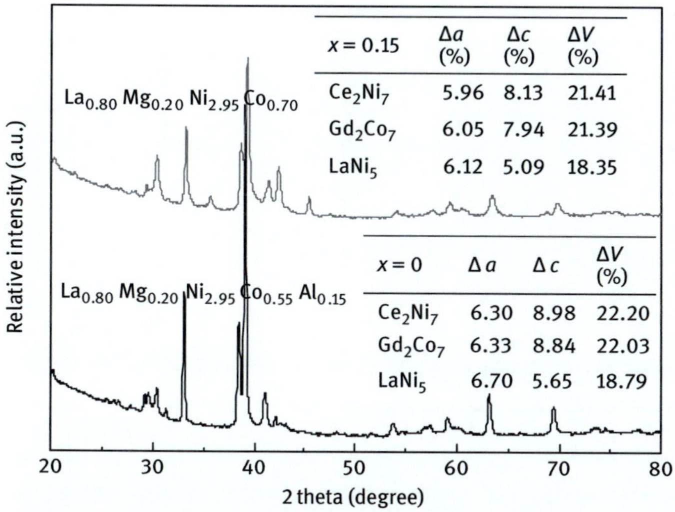

Figure 4.46: XRD of the hydrogenated $La_{0.80}Mg_{0.20}Ni_{2.95}Co_{0.70}$ and $La_{0.80}Mg_{0.20}Ni_{2.95}Co_{0.55}Al_{0.15}$ alloy powders.

The oxygen content of the $La_{0.80}Mg_{0.20}Ni_{2.95}Co_{0.70}$ and $La_{0.80}Mg_{0.20}Ni_{2.95}Co_{0.55}Al_{0.15}$ alloys before soaked in alkaline (Figure 4.47, point A), soaked in alkaline for 24 h (Figure 4.47, point B) and those at the 5th, 20th, 50th, 100th and 150th charge/discharge cycles are depicted in Figure 4.47(a). It can be seen from the inset of Figure 4.47(a) that the oxygen content of the alloys before soaked in alkaline solution is similar, while after being soaked in alkaline for 24 h, the oxygen content of $La_{0.80}Mg_{0.20}Ni_{2.95}Co_{0.55}Al_{0.15}$ alloy increases apparently, indicating that the Al oxide layer forms within relatively short time because of the stronger activity of Al element in alkaline solution. From the SEM images of $La_{0.80}Mg_{0.20}Ni_{2.95}Co_{0.70}$ and $La_{0.80}Mg_{0.20}Ni_{2.95}Co_{0.55}Al_{0.15}$ alloys after soaking in alkaline for 24 h (Figure 4.47(b-c)), it can be seen that the surface of $La_{0.80}Mg_{0.20}Ni_{2.95}Co_{0.55}Al_{0.15}$ alloy is completely covered with corrosion/oxidation product which are much denser than that of $La_{0.80}Mg_{0.20}Ni_{2.95}Co_{0.70}$ alloy. The denser corrosion/oxidation products of $La_{0.80}Mg_{0.20}Ni_{2.95}Co_{0.55}Al_{0.15}$ alloy after being immersed in alkaline for 24 h are in agreement with its higher oxygen content, which is supposed to be contributed by Al oxide layer. For the $La_{0.80}Mg_{0.20}Ni_{2.95}Co_{0.70}$ alloy sample after immersing in alkaline for 24 h, there is still some exposed area (e. g., left bottom of Figure 4.47(b)) showing a nanometric roughness. This roughness should be caused by the corrosion of the active material such as La, Mg and Al, leaving the Ni-Co-rich area which catalyses the fast charge/discharge reaction. This is also the reason that the $La_{0.80}Mg_{0.20}Ni_{2.95}Co_{0.70}$ alloy sample presents better initial kinetics performance, especially charge transfer performance. After five cycles, it is shown in Figure 4.47(a) that the oxygen content

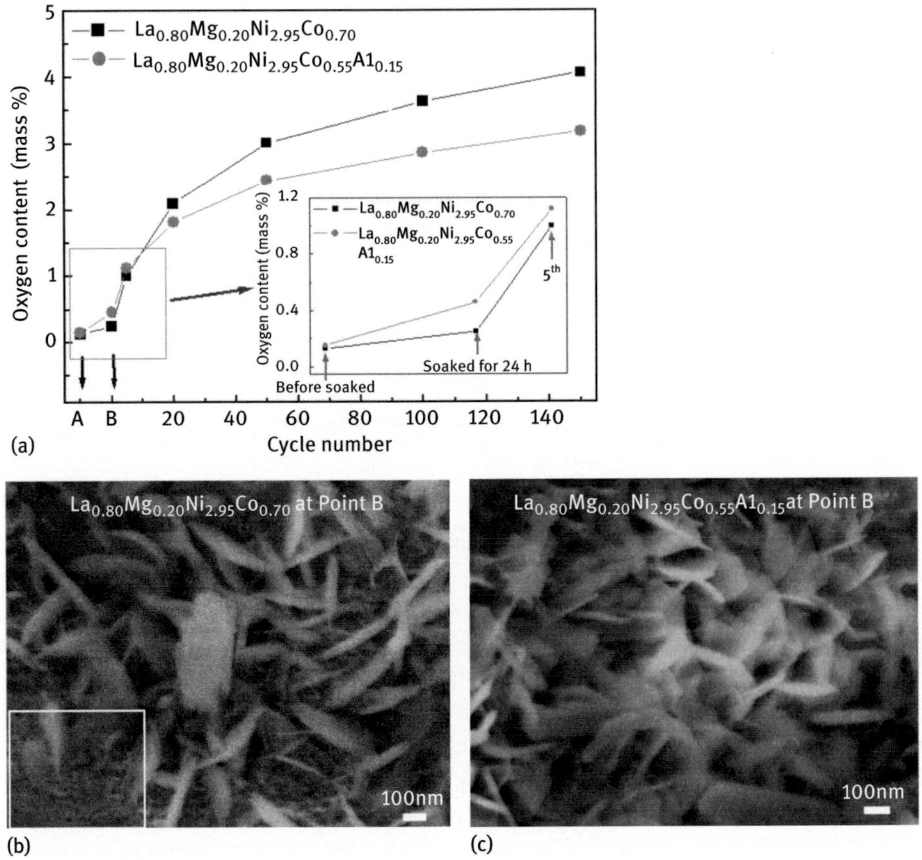

Figure 4.47: Oxygen content of the $La_{0.80}Mg_{0.20}Ni_{2.95}Co_{0.70}$ and $La_{0.80}Mg_{0.20}Ni_{2.95}Co_{0.55}Al_{0.15}$ alloys after various cycles (a), topology of $La_{0.80}Mg_{0.20}Ni_{2.95}Co_{0.70}$ alloy (b) and $La_{0.80}Mg_{0.20}Ni_{2.95}Co_{0.55}Al_{0.15}$ alloy (c) alloys after immersing in 6 M alkaline solution for 24 h.

of the Al-contained alloy is still a little higher than those of the corresponding Al-free ones, but the gap becomes smaller. After 20 cycles the oxygen content of the Al-containing alloy becomes lower than that of the corresponding Al-free alloy and the effect becomes more obvious as cycling. The lower oxygen content of the Al containing alloys after many cycles is due to its thin but effective oxide layer that prevents further oxidation of the alloys [22, 54]. Early X-ray photoelectron spectra study of the $La_{0.7}Mg_{0.3}Ni_{2.75}Co_{0.75-x}Al_x$ (x = 0, 0.3) alloys after 20 charge/discharge cycles showed that the formation of the Al_2O_3 layer played an effective role in protecting the elements of the alloys from being oxidized [22]. For example, both Co and CoO existed for the sample without Al addition after sputtering the samples for 540 s, whereas only metallic Co was detected for the Al-containing alloys [22]. Furthermore, the pulverization of $La_{0.80}Mg_{0.20}Ni_{2.95}Co_{0.55}Al_{0.15}$ alloy after 150 cycles is far less severe than that of the $La_{0.80}Mg_{0.20}Ni_{2.95}Co_{0.70}$ alloy sample, which exposes more surfaces to the alkaline solution and intensifies the flushing action of the alkaline electrolyte on the cracks of the alloy particles during cycling, resulting in higher oxygen content [53].

Al incorporation in the La–Mg–Ni-based alloy retards the severe and constant corrosion/oxidation of the alloys during discharge/discharge in alkaline electrolyte so that the consumption of the active material is lowered and the degradation of the discharge capacity is relived. Moreover, Al addition reduced the corrosion/oxidation products induced by cycling and keeps the metal elements of the alloy electrodes from oxidation in alkaline solution, so as to reduce both the degradation in the conductivity for electron transfer and the block for hydrogen diffusion, reliving the degradation in the electrochemical kinetics performance of the alloy electrodes.

4.3.2 Phase Structure and Electrochemical Performances of $La_{0.74}Mg_{0.26}Ni_{2.55}Co_{0.65-x}Fe_x$ (x = 0, 0.10, 0.20, 0.30) Alloys

4.3.2.1 Microstructure

XRD patterns of $La_{0.74}Mg_{0.26}Ni_{2.55}Co_{0.65-x}Fe_x$ (x = 0, 0.10, 0.20, 0.30) alloys and Rietveld refinement are shown in Figure 4.48. The phase abundance, lattice parameters and cell volume obtained from Rietveld refinements are listed in Table 4.16. The results show that the alloys consist mainly of $(La,Mg)_2Ni_7$ phase, $(La,Mg)_5Ni_{19}$ phase and $LaNi_5$ phase, except for absence of $LaNi_5$ phase in the non-substituted alloy. The Fe substitution facilitates formation of $(La,Mg)_5Ni_{19}$ phase and $LaNi_5$ phase, increasing their abundance from 25.87 wt.% (x = 0) to 36.13 wt.% (x = 0.30), and from 25.58 wt.% (x = 0.10) to 31.68 wt.% (x = 0.30), respectively. The abundance of $(La,Mg)_2Ni_7$ phase decreases from 74.14 to 32.19 wt.% accordingly. The lattice parameters (a and c) and cell volume of each phase all increase with increasing partial Fe substitution for Co, which may be ascribed to larger atomic radius of Fe atoms (1.72 Å) than Co atoms (1.67 Å).

The SEM images of $La_{0.74}Mg_{0.26}Ni_{2.55}Co_{0.65-x}Fe_x$ (x = 0, 0.10) alloys are shown in Figure 4.49. There two distinct areas, the white area and the grey area in image of non-substituted alloy, and they are corresponding to $(La,Mg)_2Ni_7$ phase and $(La,Mg)_5Ni_{19}$ phase by EDS analysis, respectively.

4.3.2.2 Electrochemical Properties

Figure 4.50 shows discharge capacity of $La_{0.74}Mg_{0.26}Ni_{2.55}Co_{0.65-x}Fe_x$ (x = 0, 0.10, 0.20, 0.30) alloy electrodes versus cycle number. When x equals to 0 or 0.10, the electrodes can be activated within two charge/discharge cycles and reach their maximum discharge capacity 410 and 401 mAh g^{-1}, respectively. While x is 0.20 or 0.30, the electrodes are activated after three cycles, and the maximum discharge capacity are both

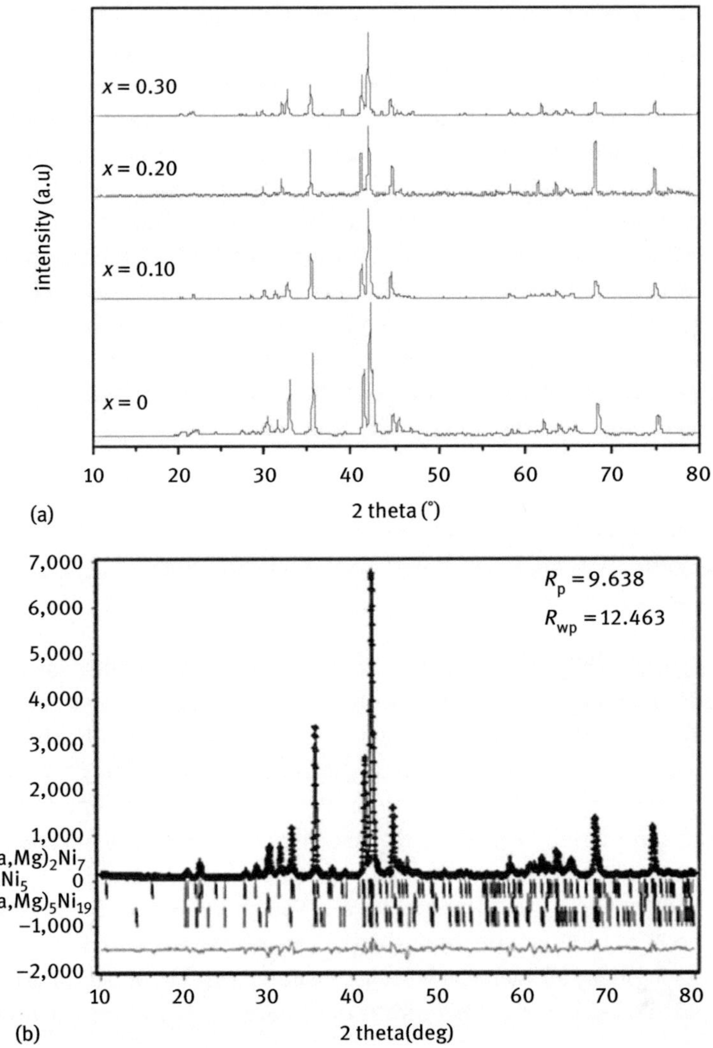

Figure 4.48: XRD patterns of $La_{0.74}Mg_{0.26}Ni_{2.55}Co_{0.65-x}Fe_x$ alloys and Rietveld refinements. (a) XRD patterns of $La_{0.74}Mg_{0.26}Ni_{2.55}Co_{0.65-x}Fe_x$ alloys; (b) Rietveld refinement for $x = 0.10$ sample.

368 mAh g^{-1}. The decrease in maximum discharge capacity can be ascribed to phase abundance and composition in alloys. The discharge capacity decreases with increasing cycle number for all the four electrodes; however, cycling stability can be improved after Co is substituted by Fe. When x equals to 0.10, the discharge capacity retention at the 200th cycle (S_{200}) increases from 68.5 % ($x = 0$) to 82.3 % ($x = 0.10$), 79.3 % ($x = 0.20$) and 82.3 % ($x = 0.30$). With partial Fe substitution for Co, the cell volume increases, which contributes to a smaller cell volume expanse rate and consequently a

Table 4.16: Phase characteristics of $La_{0.74}Mg_{0.26}Ni_{2.55}Co_{0.65-x}Fe_x$ alloys.

Alloys	Phases	Abundance (wt.%)	*a* (Å)	*c* (Å)	*V* (Å^3)
$x = 0$	$(La,Mg)_2Ni_7$	74.14	5.0460	24.3515	536.97
	$(La,Mg)_5Ni_{19}$	25.87	5.0508	48.5358	1072.29
$x = 0.10$	$(La,Mg)_2Ni_7$	45.68	5.0528	24.3589	538.58
	$(La,Mg)_5Ni_{19}$	28.74	5.0565	48.5369	1075.72
	$LaNi_5$	25.58	5.0432	3.9896	87.87
$x = 0.20$	$(La,Mg)_2Ni_7$	40.81	5.0576	24.4727	542.14
	$(La,Mg)_5Ni_{19}$	30.35	5.0588	48.7268	1078.94
	$LaNi_5$	28.84	5.0447	4.0323	88.87
$x = 0.30$	$(La,Mg)_2Ni_7$	32.19	5.0628	24.6003	546.08
	$(La,Mg)_5Ni_{19}$	36.13	5.0627	48.6995	1080.98
	$LaNi_5$	31.68	5.0449	4.0332	88.90

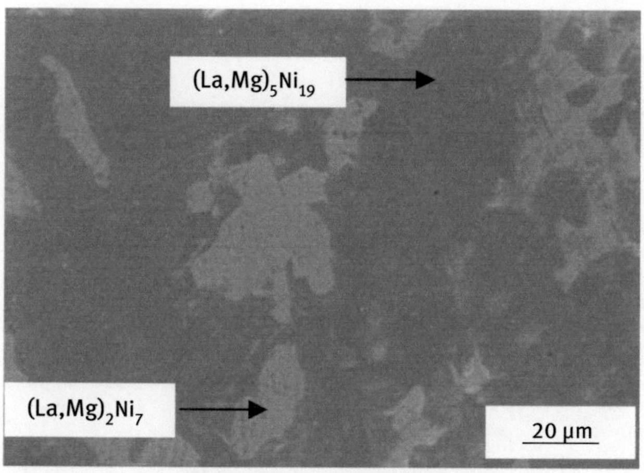

Figure 4.49: SEM images of $La_{0.74}Mg_{0.26}Ni_{2.55}Co_{0.65}$ alloy.

reduced pulverization. Therefore, the cycling stability is improved by Fe substitution. However, the discharge capacity upon cycling is not only related to phase structure, but also to the electrocatalytic activity on the surface, and hence the cycling stability does not change monotonously in one direction with increasing Fe content.

Self-dischargeability of electrodes is usually characterized by charge retention. Table 4.17 shows 24 h charge retention at 318 K for the $La_{0.74}Mg_{0.26}Ni_{2.55}Co_{0.65-x}Fe_x$ ($x = 0, 0.10, 0.20, 0.30$) alloy electrodes. The partial Fe substitution for Co greatly suppress self-discharge of the electrodes and the charge retention increases from 82.7 to 98.5 %. It is generally believed that self-discharge capacity consists of reversible loss and irreversible loss [19]. When x increases from 0 to 0.20, the irreversible loss between two neighbouring cycle can be calculated as $(C_a - C_c)/2$, and the reversible loss can

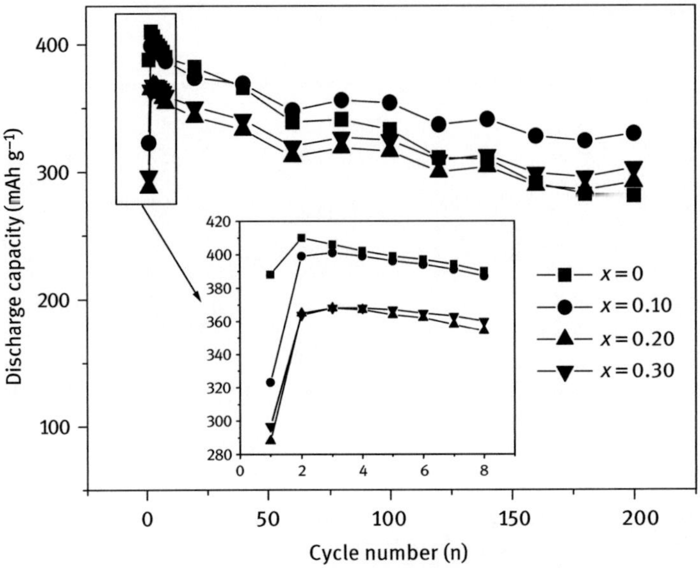

Figure 4.50: Discharge capacities of $La_{0.74}Mg_{0.26}Ni_{2.55}Co_{0.65-x}Fe_x$ alloy electrodes versus cycle number.

Table 4.17: Electrochemical characteristics of $La_{0.74}Mg_{0.26}Ni_{2.55}Co_{0.65-x}Fe_x$ alloy electrodes.

Samples	C_a (mAh g^{-1})	C_b (mAh g^{-1})	C_c (mAh g^{-1})	Reversible loss (mAh g^{-1})	Irreversible loss (mAh g^{-1})	CR (%)
$x = 0$	340	278	332	58	4	82.7
$x = 0.10$	296	272	291	21.5	2.5	92.7
$x = 0.20$	237	217	231	17	3.0	92.9
$x = 0.30$	228	233	245	–	–	98.5

be calculated as $(C_a + C_c)/2–C_b$. The calculating results are listed in Table 4.17. It can be seen that the irreversible capacity loss ranges from 2.5 to 4.0 mAh g^{-1}, no great difference existing. So the variation in self discharge results mainly from reversible loss. In the case of Fe substitution for Co, the cell volume increases with increasing Fe content for each phase, $(La,Mg)_2Ni_7$ phase, $(La,Mg)_5Ni_{19}$ phase and $LaNi_5$ phase. The increase in cell volume results in a larger tetrahedral or octahedral interstice for hydrogen storage, and thus a more stable metal hydride. So in the higher Fe-content alloy, less hydrogen atoms leave from the fully charged electrodes during laying aside under open circuit, and therefore the charge retention increases with increasing x from 0 to 0.30. Instead of decreasing after laying aside for 24 h, when x is 0.30, the discharge capacity increases. This is a different phenomenon from what we usually obtain;

several repetitions confirm the experimental results and the reason is under-revealed in our further work.

The *HRD* at 1,080 mA g^{-1} for the $La_{0.74}Mg_{0.26}Ni_{2.55}Co_{0.65-x}Fe_x$ (x = 0, 0.10, 0.20, 0.30) alloy electrodes has been shown in Table 4.17. With x increases from 0 to 0.30, *HRD* at 1,080 mA g^{-1} decreases from 51.4 to 15.8 %. *HRD* of electrodes is mainly controlled by kinetics, including charge transfer at the surface and hydrogen diffusion in the bulk. In order to illustrate the reason for HRD decrease, some kinetics parameters are measured.

Figure 4.51 shows cyclic voltammograms of $La_{0.74}Mg_{0.26}Ni_{2.55}Co_{0.65-x}Fe_x$ (x = 0, 0.10, 0.20, 0.30) alloy electrodes. In the cathodic branch of the voltammogram, we observe a plateau attributed to the adsorption of hydrogen atom on electrode surface and a peak in hydrogen evolution. In the anodic branch of the voltammogram we observe a peak attributed to the oxidation of hydrogen on electrode surface. This anodic current peak increases and its potential slightly shifts to the positive direction with increasing potential scan rate for each alloy. The difference between the anodic and the cathodic current peak positions has been taken as an estimate for reversibility of the redox reaction, which is an important parameter to assess the electrochemical properties. It

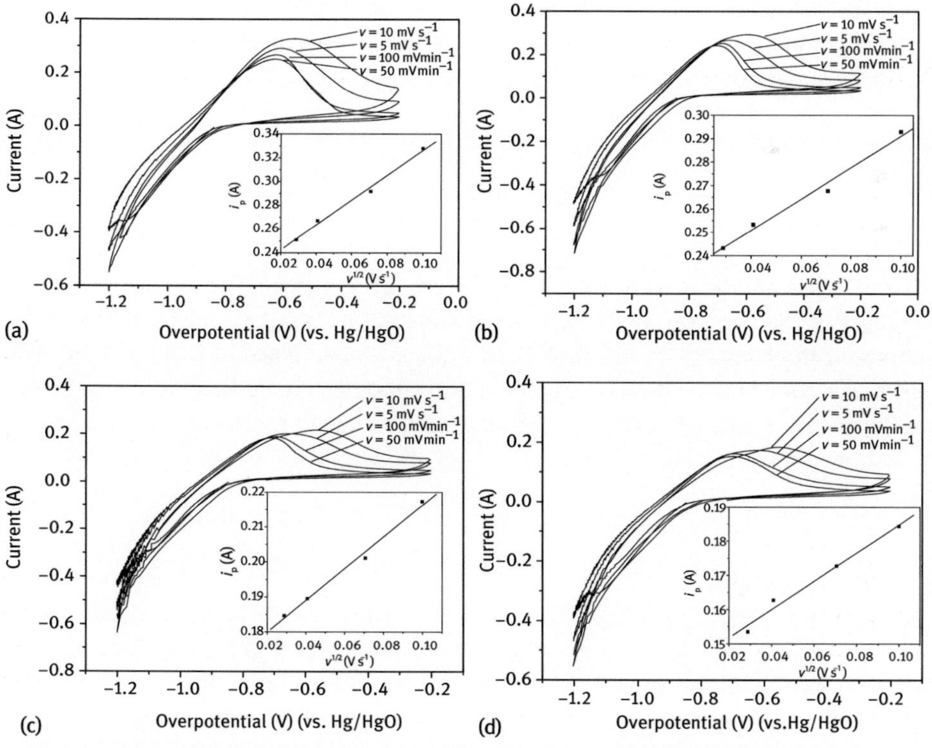

Figure 4.51: Cyclic voltammograms of $La_{0.74}Mg_{0.26}Ni_{2.55}Co_{0.65-x}Fe_x$ alloy electrodes.

Table 4.18: Electrochemical kinetics parameters of $La_{0.74}Mg_{0.26}Ni_{2.55}Co_{0.65-x}Fe_x$ alloy electrodes.

Samples	HRD_{1080} (%)	α	C_0 (mol cm^{-3})	D_H($\times 10^{-9}$cm^2 s^{-1})
$x = 0$	51.4	0.473	0.0453	3.57
$x = 0.10$	24.1	0.298	0.0388	3.19
$x = 0.20$	16.0	0.242	0.0311	2.79
$x = 0.30$	15.8	0.213	0.0311	2.61

can be also seen that the anodic peak shifts to negative position when Co is partially replaced by Fe. This indicates that the reversibility for the electrochemical redox reaction is improved by the partial Co substitution. The oxidation peak current (i_p) of the voltammogram can be written for a semi-infinite diffusion and an irreversible transfer under the following form [55]:

$$i_p = 2.98 \times 10^5 S C_0 \alpha^{1/2} D_H^{1/2} v^{1/2} \tag{4.1}$$

where α is the coefficient of transfer, F is the constant of Faraday, S is the geometrical surface of the work electrode (cm^2), C_0 is the concentration of the species diffuse (mol cm^{-3}), D is the diffusion coefficient of the electroactive species (cm^2 s^{-1}) and v is the scanning rate of the potential (V s^{-1}). α and C_0 are calculated according to the following equations [55, 56]:

$$\mathrm{d}E_p/\mathrm{dlog}v = 2.3RT/\alpha nF \tag{4.2}$$

$$C_0 = \frac{M}{mFV_m}\int_{t_1}^{t_2} i\mathrm{d}t \tag{4.3}$$

where i is the anodic current of voltammogram, M and m are the molar weight and the mass of the alloy respectively, F is the constant of Faraday and V_m is the molar volume of the alloys (~54 cm^3 mol^{-1}). The linear dependence of i_p on $v^{1/2}$ for each tested electrode is also shown in Figure 4.51(a–d). D_H calculated according to the linear dependence is listed in Table 4.18, and it decreases with increasing x. The decrease in hydrogen diffusion coefficient is related to stronger stability of hydride after Fe is added.

4.4 Conclusions

For $La_{2-x}Mg_xNi_7$ (x = 0.40–0.60) alloys, within the range of Mg concentration from 0.48 to 0.50, the alloys consist of allotropic structures 2H- and 3R-type phases, and Mg atoms tend to access into 3R-type phase rather than 2H-type phase. Increasing and decreasing content of Mg are benefit formations of $PuNi_3$-type and $CaCu_5$-type phases, respectively. Alloys with 2H- and 3R-type phases possess excellent properties in discharge capacity and cycling stability, and the capacity degradation is due to the oxidation of La and Mg elements, the pulverization of the alloys and the phase decomposition into $LaNi_5$ phase of $(La,Mg)_2Ni_7$ phase. *P–C* isotherms suggest that the discharge pressure of the 3R-type phase is higher than that of the 2H-type phase considering elevation of the plateau pressure with increasing of 3R- to 2H-type ratio. The appearance of $PuNi_3$-phase or $CaCu_5$-type phase reduces the cycling stability of an alloy, considering the boosted pulverization and oxidation of the alloys because of the difference in the expansion rates during hydrogen absorption/desorption and the phase boundaries between the phase structures.

For $Pr_{3-x}Mg_xNi_9$ (x = 0.40–0.60) alloys, increasing Mg content yielded the phase transformation from Gd_2Co_7-type to $PuNi_3$-type then to $MgCu_4Sn$-type phases and the maximum solid solubility of Mg in $PuNi_3$-type phase for $Pr_{3-x}Mg_xNi_9$ alloys was 1.0. Electrochemical studies showed that the impact of Mg on discharge capacity of Pr–Mg–Ni alloys was not obvious; however, the *HRD* decreased from 62.2 to 56.7 % for single-phase alloy and then increased by 2 %, and the cycling stability enhanced from 82.8 to 86.3 % for single-phase alloy and then reduced to 62.1 % due to appearance of $MgCu_4Sn$-type phase. Both of the *HRD* and the cycling stability prosperities of Pr_2MgNi_9 alloy were enhanced by 20 %. For $Nd_{1-x}Mg_xNi_{2.8}$ (x = 0.10–0.50) alloys, it is found that when Mg increases from 0.10 to 0.36, the Ce_2Ni_7-, $PuNi_3$- and $MgCu_4Sn$-type multiphase (x = 0.10 and 0.20) of the alloy changes to $PuNi_3$-type single phase (x = 0.36) with a gradual decreasing unit cell volume of $PuNi_3$-type phase. Further increasing the Mg to 0.5, the excess Mg cannot further enter into the $PuNi_3$-type phase and retains in an $MgNi_2$ secondary phase. Determined from the phase composition changes of the alloy and unit cell volume changes of the $PuNi_3$-type phase as increasing content of Mg, it can be found that the maximum solid solubility of Mg in the $PuNi_3$-type $(Nd,Mg)Ni_3$ phase is 0.36. The formation of $MgCu_4Sn$-type phase or $MgNi_2$ phase both is adverse to the electrochemical properties of the alloy. Effectively controlling the content of Mg is a key factor to optimize phase structure and improve the electrochemical properties of the Nd–Mg–Ni-based alloys.

For $La_{0.8-x}Pr_xMg_{0.2}Ni_{3.4}Al_{0.1}$ (x = 0–0.3) alloys, the alloys are mainly composed of $(La,Mg)_2Ni_7$ and $(La,Mg)_5Ni_{19}$ phases as well as a small amount of $LaNi_5$ phase. It is found that Pr tends to occupy La site in $[AB_2]$ subunit rather than $[AB_5]$ subunit. This occupancy tendency leads to the increase in the $(LaMg)_5Ni_{19}$ phase with higher ratio of $[AB_5]/[AB_2]$ subunits instead of $(LaMg)_2Ni_7$ phase. The $(La,Mg)_5Ni_{19}$ phase distributes among the $(La,Mg)_2Ni_7$ phase in the shape of continuous net, which enhances the ability of the alloys against pulverization and amorphization during repeated charge/discharge cycles. Consequently, the $La_{0.6}Pr_{0.2}Mg_{0.2}Ni_{3.4}Al_{0.1}$ alloy electrode exhibits a discharge capacity of 329 mAh g^{-1} and a cycling stability of 90.7 % after 100 cycles. For $La_{0.75-x}Nd_xMg_{0.25}Ni_{3.3}$ (x = 0, 0.15) alloys, XRD analysis shows that the main phases of the $La_{0.75}Mg_{0.25}Ni_{3.3}$ alloy are A_2B_7-type 2H- and 3R- phases along with minor $CaCu_5$-type phase, which disappears in the $La_{0.60}Nd_{0.15}Mg_{0.25}Ni_{3.3}$ alloy. Rietveld result presents that Nd atom prefers to enter La sites rather than Mg sites in the A_2B_4 subunits. In the sintering process, the molten $LaMgNi_4$ parcels around the $LaNi_5$, so that the replaced La atoms in the $LaMgNi_4$ alloy and the Ni atoms transform into more A_2B_4 subunits. Then the A_2B_4 subunits crystallize with the AB_5 subunits sufficiently, which consumes the $CaCu_5$-type phase. Therefore, Nd partial substitution for La is beneficial for the formation of the 2H- and 3R-A_2B_7 phases in expense of $CaCu_5$-type phase. The equilibrium plateau is elevated and broadened with Nd partial substitution for La, which coincide with linear shrinkage in the cell volume and the increase of 2H- and 3R-A_2B_7 phases. The maximum discharge capacity and the HRD_{1440} are dramatically increased due to the phase formation of $CaCu_5$-type phase into 2H- and 3R-A_2B_7 phases by Nd substitution.

For $La_{0.8}Mg_{0.2}Ni_{2.95}Co_{0.7-x}Al_x$ (x = 0, 0.05, 0.10, 0.15) alloys, the alloys mainly contain $(La,Mg)_2Ni_7$ and $LaNi_5$ phases. Al is more preferable to enter the $LaNi_5$ phase rather than the $(La,Mg)_2Ni_7$ phase so that the $LaNi_5$ phase abundance is increased while that of the $(La,Mg)_2Ni_7$ phase is decreased as x increases from 0 to 0.15. Moreover, the ratio of 2H- to 3R-type $(La,Mg)_2Ni_7$ phase increases and the cell volumes of the corresponding phases augment as Al content increases. Al incorporation renders tremendous influence on the alloy pulverization and oxidation/corrosion during charge/discharge cycles. Firstly, the cell volume expansion rate of $(La,Mg)_2Ni_7$ and $LaNi_5$ phases after hydrogenation is decreased so that the alloy pulverization is relieved. Moreover, Al oxide layer formed on the alloy surface when the alloy was immersed in 6 M alkaline electrolyte protects the alloy from further oxidation. Both discharge capacity retention and HRD_{1500} cycling retention was promoted by Al incorporation. For $La_{0.74}Mg_{0.26}Ni_{2.55}Co_{0.65-x}Fe_x$ (x = 0, 0.10, 0.20, 0.30) alloys, it is revealed by XRD and SEM that the alloys consist mainly of $(La,Mg)_2Ni_7$ phase, $(La,Mg)_5Ni_{19}$ phase and $LaNi_5$ phase. After Co is partially substituted by Fe, the abundance of $(La,Mg)_5Ni_{19}$ phase and $LaNi_5$ phase increases and that of $(La,Mg)_2Ni_7$ phase decreases with increasing x. The maximum discharge capacity decreases with increasing Fe content, while the cycling stability is greatly improved, and the discharge retention rate at the 200th cycle increases from 68.5 % (x = 0) to 82.3 % (x = 0.10). The self-discharge of electrodes is excellently suppressed by Fe-substitution, and the 24 h charge retention at 318 K increases from 82.7 % (x = 0) to 98.5 % (x = 0.30). The electrochemical kinetics is deteriorated with increasing Fe content, and therefore the high rate dischargeability at 1,080 mA g^{-1} decreases from 51.4 % (x = 0) to 15.8 % (x = 0.30).

References

[1] Kadir K, Sakai T, Uehara I. Synthesis and structure determination of a new series of hydrogen storage alloys; RMg_2Ni_9 (R=La, Ce, Pr, Nd, Sm and Gd) built from $MgNi_2$ Laves-type layers alternating with AB_5 layers. J Alloys Compd. 1997;257:115–21.

[2] Férey A, Cuevas F, Latroche M, et al. Elaboration and characterization of magnesium-substituted La_5Ni_{19} hydride forming alloys as active materials for negative electrode in Ni-MH battery. Electrochim Acta 2009;54:1710–14.

[3] Gal L, Charbonnier V, Zhang J, et al. Optimization of the La substitution by Mg in the La_2Ni_7 hydride-forming system for use as negative electrode in Ni-MH battery. Int J Hydrogen Energy 2015;40:17017–20.

[4] Denys RV, Riabov AB, Yartys VA, et al. Mg substitution effect on the hydrogenation behaviour, thermodynamic and structural properties of the La_2Ni_7–$H(D)_2$ system. J Solid State Chem 2008;181:812–21.

[5] Zhang L, Han SM, Li Y, et al. Effect of magnesium on the crystal transformation and electrochemical properties of A_2B_7-type metal hydride alloys. J Electrochem Soc 2014;161:A1844–50.

[6] Nwakwuo C, Holm T, Denys RV, Hu WK, Maehlen JP, Solberg JK, Yartys VA. Effect of magnesium content and quenching rate on the phase structure and composition of rapidly solidified La_2MgNi_9 metal hydride battery electrode alloy. J Alloys Compd 2013;555:201–8.

[7] Zhai TT, Yang T, Yuan ZM, Xu S, Bu WG, Qi Y, Zhang YH. Influences of hydrogen-induced amorphization and annealing treatment on gaseous hydrogen storage properties of $La_{1-x}Pr_xMgNi_{3.6}Co_{0.4}$ (x = 0–0.4) alloys. J Alloys Compd 2015;639:15–20.

[8] Iwase K, Terashita N, Mori K, Yokota H, Suzuki T. Crystal structure and cyclic hydrogenation property of Pr_4MgNi_{19}. Inorg Chem 2013;52:14270–4.

[9] Zhang YH, Li BW, Ren HP, Guo SH, Yan Q, Wang XL. Structures and electrochemical cycle stability of $La_{0.75-x}Pr_xMg_{0.25}Ni_{3.2}Co_{0.2}Al_{0.1}$ (x = 0–0.4) alloys prepared by melt spinning. Mater Chem Phys 2009;118:129–34.

[10] Pan HG, Ma S, Shen J, Tan JJ, Deng JL, Gao MX. Effect of the substitution of Pr for La on the microstructure and electrochemical properties of $La_{0.7-x}Pr_xMg_{0.3}Ni_{2.45}Co_{0.75}Mn_{0.1}Al_{0.2}$ (x = 0.0–0.3) hydrogen storage electrode alloys. Int J Hydrogen Energy 2007;32:2949–56.

[11] Zhang F, Luo Y, Deng A, Tang Z, Kang L, Chen J. A study on structure and electrochemical properties of $(La, Ce, Pr, Nd)_2MgNi_9$ hydrogen storage electrode alloys. Electrochim Acta 2006;52:24–32.

[12] Li Y, Han SM, Li JH, Zhu XL, Hu L. The effect of Nd content on the electro- chemical properties of low-Co La–Mg–Ni-based hydrogen storage alloys. J Alloys Compd 2008;458:357–62.

[13] Zhu J, Zhang J, Fang F, Zheng SY, Chen GR, Sun DL. Effect of A-side element on crystal structure of AB_3-type hydrogen storage alloy. Rare Met Mater Eng 2008;37:1377–9.

[14] Serin V, Zhang J, Magén C, Serra R, Hÿtch MJ, Lemort L, Latroche M, Ibarra MR, Knosp B, Bernard P. Identification of the atomic scale structure of the $La_{0.65}Nd_{0.15}Mg_{0.20}Ni_{3.5}$ alloy synthesized by spark plasma sintering. Intermetallics 2013;32:103–8.

[15] Li Y, Han D, Han SM, Zhu X, Hu L, Zhang Z, Liu Y. Effect of rare earth elements on electrochemical properties of La–Mg–Ni-based hydrogen storage alloys. Int J Hydrogen Energy 2009;34:1399–404.

[16] Tang R, Wei XD, Liu YN, Zhu CC, Zhu JW, Yu G. Effect of the Sm content on the structure and electrochemical properties of $La_{1.3-x}Sm_xCaMg_{0.7}Ni_9$ (x = 0–0.3) hydrogen storage alloys. J Power Sources 2006;155:456–60.

[17] Zhang YH, Hou ZH, Li BW, Ren HP, Zhang GF, Zhao DL. A study on structure and electrochemical properties of the as-cast and –annealed $La_{0.8-x}Sm_xMg_{0.2}Ni_{3.35}Al_{0.1}Si_{0.05}$ (x = 0–0.4) alloys. J Alloys Compd 2012;537:175–82.

[18] Li P, Hou ZH, Yang T, Shang HW, Qu XH, Zhang YH. Structure and electrochemical hydrogen storage characteristics of the as-cast and annealed $La_{0.8-x}Sm_xMg_{0.2}Ni_{3.15}Co_{0.2}Al_{0.1}Si_{0.05}$ (x = 0–0.4) alloys. J Rare Earth 2012;30:696–704.

[19] Zhang Y, Yang T, Cai Y, et al. Electrochemical hydrogen storage characteristics of $La_{0.75-x}M_xMg_{0.25}Ni_{3.2}Co_{0.2}Al_{0.1}$ (M = Zr, Pr, x = 0, 0.1) alloys prepared by melt spinning. Rare Metals 2012;31:457–65.

[20] Zhang YH, Li BW, Ren HP, et al. Structures and electrochemical hydrogen storage characteristics of $La_{0.75-x}Pr_xMg_{0.25}Ni_{3.2}Co_{0.2}Al_{0.1}$ (x = 0–0.4) alloys prepared by melt spinning. J Alloys Compd 2009;485:333–9.

[21] Zhang FL, Luo YC, Chen JP, Yan RX, Kang L, Chen JH. Effect of annealing treatment on structure and electrochemical properties of $La_{0.67}Mg_{0.33}Ni_{2.5}Co_{0.5}$ alloy electrodes. J Power Sources 2005;150:247–54.

[22] Zhao Y, Gao MX, Liu YF, Huang L, Pan HG. The correlative effects of Al and Co on the structure and electrochemical properties of a La–Mg–Ni-based hydrogen storage electrode alloy. J Alloys Compd 2010;496:454–61.

[23] Liu YX, Xu LQ, Jiang WQ, Li GX, Wei WL, Guo J. Effect of substituting Al for Co on the hydrogen-storage performance of $La_{0.7}Mg_{0.3}Ni_{2.6}Al_xCo_{0.5-x}$ (x = 0.0–0.3) alloys. Int J Hydrogen Energy 2009;34:2986–91.

[24] Moussa MB, Abdellaoui M, Mathlouthi H, Lamloumi J, Guégan AP. Electrochemical properties of the $MmNi_{3.55}Mn_{0.4}Al_{0.3}Co_{0.75-x}Fe_x$ (x = 0.55 and 0.75) compounds. J Alloys Compd 2008;458:410–14.

[25] Zhao YP, Zhang YH, Wang GQ, Dong XP, Guo SH, Wang XL. Effects of substituting Co with Fe on the microstructures and electrochemical characteristics of the as-cast and quenched $Mm(NiMnSiAl)_{4.3}Co_{0.6-x}Fe_x$ (x = 0–0.6) electrode alloys. J Alloys Compd 2005;388:284–92.

[26] Buschow KH, Van Der Goot AS. The crystal structure of rare-earth nickel compounds of the type R_2Ni_7. J Less-Common Met 1970;22:419–28.

[27] Zhang QA, Zhao B, Fang MH, Liu CR, Hu QM, Fang F, Sun DL, Ouyang LZ, Zhu M. Comparison of the effect of functional groups on gas-uptake capacities by fixing the volumes of cages A and B and modifying the inner wall of cage C in rht-type MOFs. Inorg Chem 2012;51:10350–5.

[28] Young K, Ouchi T, Huang B, Chao B, Fetcenko MA, Bendersky LA, Wang K, Chiu C. Fabrication, properties and laser performance of Ho:YAG transparent ceramic. J Alloys Compd 2010;506:745–8.

[29] Chai YJ, Asano K, Sakaki K, Enoki H, Akiba E. $Li_4Ti_5O_{12}$/C composite electrode material synthesized involving conductive carbon precursor for Li-ion battery. J Alloys Compd 2009;485:93–7.

[30] Denys RV, Yartys VA. Effect of magnesium on the crystal structure and thermodynamics of the $La_{3-x}Mg_xNi_9$ hydrides. J Alloys Compd 2011;509 S540–8.

[31] An XH, Gu QF, Zhang JY, Chen SL, Yu XB, Li Q. Experimental investigation and thermodynamic reassessment of La–Ni and $LaNi_5$–H systems. Calphad 2013;40:48–55.

[32] Srivastava S, Srivastava ON. Hydrogenation behaviour with regard to storage capacity, kinetics, stability and thermodynamic behaviour of hydrogen storage composite alloys, $LaNi_5/La_2Ni_7$, $LaNi_3$. J Alloys Compd 1999;290:250–6.

[33] Hirano K, Kadono J, Yamamoto S, Tanabe T, Miyake H. Hydrogen separation using proton-conducting perovskites. J Alloys Compd 2006;408–412:456–62.

[34] Zheng G, Popov BN, White RE. Electrochemical determination of the diffusion coefficient of hydrogen through an $LaNi_{4.25}Al_{0.75}$ electrode in alkaline aqueous solution. J Electrochem. Soc. 1995;142:2695–2698.

[35] Zhang JL, Han SM, Li Y, Liu JJ, Yang SQ, Zhang L, Wang JD. Effects of $PuNi_3$- and Ce_2Ni_7-type phase abundance on electrochemical characteristics of La–Mg–Ni-based alloys. J Alloys Compd 2013;581:693–8.

[36] Miao H, Gao MX, Liu YF, Zhu D, Pan HG. An improvement on cycling stability of Ti–V–Fe-based hydrogen storage alloys with Co substitution for Ni. J Power Sources 2008;184:627–32.

[37] Liu JJ, Han SM, Li Y, Zhang JL, Zhao YM, Che L. Effect of crystal transformation on electrochemical characteristics of La–Mg–Ni-based alloys with A_2B_7-type super-stacking structures. Int J Hydrogen Energy 2013;38:14903–11.

[38] Srivastava S, Srivastava ON. Synthesis, characterization and hydrogenation behavior of composite hydrogen storage alloys, $LaNi_5/La_2Ni_7$, $LaNi_3$. J Alloys Compd 1999;282:197–205.

[39] Wang ZM, Zhou HY, Zou RP, Yao QR. Effect of sintering conditions on the formation of single-phase $NdMgNi_4$ compound and its hydrogen storage properties. J Alloys Compd 2007;429:260–3.

[40] Zhou HY, Zhang SL, Q.R.Yao, Li WJ. The isothermal sections of the phase diagram of the Nd–Mg–Ni ternary system at 1123 and 673 K (Ni-rich part). J Alloys Compd 2007;429:116–18.

[41] Wang JD, Han SM, Li Y, Liu JJ, Che L, Zhang L, Zhang JL. Study on phase formation mechanism and electrochemical properties of $La_{0.75-x}Nd_xMg_{0.25}Ni_{3.3}$ ($x = 0, 0.15$) alloys prepared by powder sintering. J Alloys Compd 2014;582:552–557.

[42] Oesterreicher H, Clinton J, Bittner H, Hydrides of La–Ni compounds. Mater Res Bull 1976;11:1241–7.

[43] Liu JJ, Li Y, Han SM, Yang SQ, Shen WZ, Chen XC, Zhao YM. Microstructure and electrochemical characteristics of step-wise annealed $La_{0.75}Mg_{0.25}Ni_{3.5}$ alloy with A_2B_7- and A_5B_{19}-type super-stacking structure. J Electrochem Soc 2013;160:A1139–45.

[44] Liu JJ, Han SM, Han D, Li Y, Yang SQ, Zhang L, Zhao YM. Enhanced cycling stability and high rate dischargeability of $(La,Mg)_2Ni_7$-type hydrogen storage alloys with $(La,Mg)_5Ni_{19}$ minor phase. J Power Sources 2015;287:237–46.

[45] Zhang L, Han SM, Han D, Li Y, Zhao X, Liu JJ. Phase decomposition and electrochemical properties of single phase $La_{1.6}Mg_{0.4}Ni_7$ alloy. J Power Sources 2014;268:575–83.

[46] Yang SQ, Han SH, Li Y, Liu JJ. Study on the microstructure and electrochemical kinetic properties of $MmNi_{4.50-x}Mn_xCo_{0.45}Al_{0.30}$ ($0.25 \leq x \leq 0.45$) hydrogen storage alloys. Mat Sci Eng B-Solid 2013;178:39–44.

[47] Shen XQ, Chen YG, Tao MD, Wu CL, deng G, Kang ZZ. Properties of large-scale methane/hydrogen jet fires. Int J Hydrogen Energy 2009;34:9611–19.

[48] Mendelsohn MH, Gruen DM, Dwight AE. The effect of aluminum additions on the structural and hydrogen absorption properties of AB_5 alloys with particular reference to the $LaNi_{5-x}Al_x$ ternary alloy system. J Less Comm Met 1979;63:193–207.

[49] Notten PH, Hokkeling P. Double-phase hydride forming compounds: a new class of highly electrocatalytic materials. J Electrochem Soc 1991;138:1877–85.

[50] Akiba E, Hayakawa H, Kohno T. Crystal structures of novel La–Mg–Ni hydrogen absorbing alloys. J Alloys Compd 2006;408:280–3.

[51] Zhang P, Liu YN, Zhu JW, Wei XD, Yu G. Effect of Al and W substitution for Ni on the microstructure and electrochemical properties of $La_{1.3}CaMg_{0.7}Ni_{9-x}(Al_{0.5}W_{0.5})_x$ hydrogen storage alloys. Int J Hydrogen Energy 2007;32:2488–93.

[52] Li SL, Chen W, Luo G, Han XB, Chen DM, Yang K, Chen WP. Effect of hydrogen absorption/desorption cycling on hydrogen storage properties of a $LaNi_{3.8}Al_{1.0}Mn_{0.2}$ alloy. Int J Hydrogen Energy 2012;37:3268–75.

[53] Liu YF, Pan HG, Gao MX, Lei Y, Wang QD. Degradation mechanism of the La-Mg-Ni-based metal hydride electrode $La_{0.7}Mg_{0.3}Ni_{3.4}Mn_{0.1}$. J Electrochem Soc 2005;152:A1089–95.

[54] Balogun MS, Wang ZM, Chen HX, Deng JQ, Yao QR, Zhou HY. Effect of Al content on structure and electrochemical properties of $LaNi_{4.4-x}Co_{0.3}Mn_{0.3}Al_x$ hydrogen storage alloys. Int J Hydrogen Energy 2013;38:10926–31.

[55] Khaldi C, Mathlouthi H, Lamloumi J, Percheron-Guégan A. Electrochemical study of cobalt-free AB_5-type hydrogen storage alloys. Int J Hydrogen Energy 2004;29:307–11.

[56] Chen W. Cyclic voltammetry and electrochemical impedance of $MmNi_{3.6}Co_{0.7}Mn_{0.4}Al_{0.3}$ alloy electrode before and after treatment with a hot alkaline solution containing reducing agent. J Power Sources 2000;90:201–5.

5 Effect of Surface Treatment on Electrochemical Characteristics of RE–Mg–Ni-Based Hydrogen Storage Alloys

Rare earth–Mg–Ni-based hydrogen storage alloys were recently considered as alternative negative electrode materials for Ni/MH batteries owing to their high capacity [1]. Since the last decade, lots of reports have investigated microstructure, gas–solid hydrogen storage properties and electrochemical properties [2–4]. Rare earth–Ni-based hydrogen storage alloys were reported to have different super-stacking structure according to the stoichiometry [5, 6]. Among the super-stacking phases, $PuNi_3$-type phase was reported to show high capacity, Ce_2Ni_7-type phase was reported to have high stability and contribute to excellent cycling stability and Pr_5Co_{19}-type phase was reported to improve the HRD [7–9]. Partial or complete element substitution was believed to be an effective method for improving electrochemical properties. Mg substitution for rare earth elements in super-stacking structure contributed to a stable structure during hydrogenation/dehydrogenation. For A-side elements, La was usually substituted by Ce, Pr, Nd and Ca; otherwise, for B-side elements Mn, Co and Al substitution for Ni was widely reported, and recently some co-substitution by two or more metals has been reported [10].

However, rare earth elements and magnesium element are unstable and easy to be oxidized into oxides or hydroxides in concentrated KOH (6–7 mol L^{-1}) solution, which causes a fast degradation in discharge capacity of rare earth-magnesium–nickel hydrogen storage alloys. The electrochemical properties are not only influenced by bulk structure but also by surface state of the alloy electrodes, for the electrochemical properties are determined by the interface reaction between the alloy solid surface and the electrolyte [11, 12]. For the melted or annealed alloys, unexpected oxide layer usually forms in the air atmosphere and hinders electrochemical reaction at the alloy surface. Thereby some reports used acid, base, heavy ion irradiation and reductive atmosphere to remove the passive layer and to improve the surface catalytic activity of the alloys [13–16]. Consequently, electrochemical kinetics and HRD are remarkably enhanced [15, 17]. Fluorination treatment helps to get MgF_2, preventing MgO formation at the alloy surface and makes the hydrogen absorption become faster [18, 19]. Surface coating is another effective pathway to electrochemical property improvement. As is reported, metallic layers, such as Ni, Co and Pd, reduced graphite oxide, polymer have been reported to coat at the alloy surface [20–23] and the authors get improvement to the hydrogen storage or electrochemical properties.

In order to improve the poor cycling stability of the alloys, several methods, such as additive in negative electrodes, electrodeposition and electroless plating on negative electrode materials, have been tried to form protective species at the alloy surface. Electrochemical properties and functional mechanism have been investigated.

DOI 10.1515/9783110501483-005

5.1 Additive in Negative Electrode

5.1.1 Reactive Additive: CuO

5.1.1.1 Preparation and Testing Methods

The CuO compound was synthesized by solid-state reaction. 2.00 g NaOH and 4.28 g $CuCl_2 \cdot 2H_2O$ were separately milled in the mortar for 10 min and then they were mixed and the mixture was milled for another 30 min until they turned completely black. Then the paste was washed with diluted water and ethanol for several times. The precipitation was finally dried under vacuum at 323 K.

CuO additives (0.5, 1.0, 1.5, 2.0, 2.5 and 3.0 wt.% of the alloys) and 0.06 g carbonyl nickel powders were mixed with two drops of 3 wt.% poly vinyl alcohol (PVA) solution. The slurry was pasted onto both side of a foamed nickel (1.8 cm × 1.8 cm), dried completely in vacuum at 333 K and finally cold pressed under 10 MPa. Electrochemical tests were performed in a half-cell consisting of metal hydride as working electrode, $Ni(OH)_2$/NiOOH as counter electrode and 6 mol L^{-1} KOH solution as electrolyte on DC-5 battery testing instrument at 298 K. The electrodes were fully charged (overcharged ratio was approximately 50 %) at current density of 72 mA g^{-1}, and then discharged at the same current density to cut-off voltage of 1.0 V during charge/discharge test. Discharge capacities were measured at different discharge current densities when HRD was measured. The interval between charge and discharge is 10 min.

XRD patterns were obtained on a D/Max-2500/PC X-ray diffractometer (Cu Kα radiation). Phase identification was carried out using Jade 5.0 software. Morphology of the alloys was observed by KYKY-2800 SEM and KEVEX-SIGMA LEVEL4 EDS. Cyclic voltammograms were obtained on a ZF-9 potentiostat by scanning electrode potential at the rate of 50 mV min^{-1} from –1,200 to 400 mV (vs. Hg/HgO reference electrode).

5.1.1.2 Structure and Electrochemical Properties

Figure 5.1 shows XRD pattern of the as-prepared $La_{0.88}Mg_{0.12}Ni_{2.95}Mn_{0.10}Co_{0.55}Al_{0.10}$ alloy and CuO compound. It can be seen that the alloy consists of $CaCu_5$-type $LaNi_5$ phase, $PuNi_3$-type $LaNi_3$ phase and some minor LaNi impurity. As for the patterns of CuO compound, XRD analysis shows that it is identical with the CuO patterns, no diffraction peaks for $CaCl_2$ or Cu_2O can be detected.

Capacities of the electrodes with and without CuO during activation are shown in Figure 5.2. It can be noticed that the electrode without additive need eight cycles to be activated and the maximum discharge capacity is 314 mAh g^{-1}. The CuO additive not only reduces the cycle number for activation, but also increases the discharge capacity. When 0.5 wt.% CuO is added the activation cycle number is 7 and the maximum discharge capacity is 326 mAh g^{-1}. When 1.0 wt.% or more CuO is added the activation number is 4 or 5 and the maximum discharge capacity is more than 335 mAh g^{-1}; when 3.0 wt.% CuO is added the discharge capacity increases to 341 mAh g^{-1}.

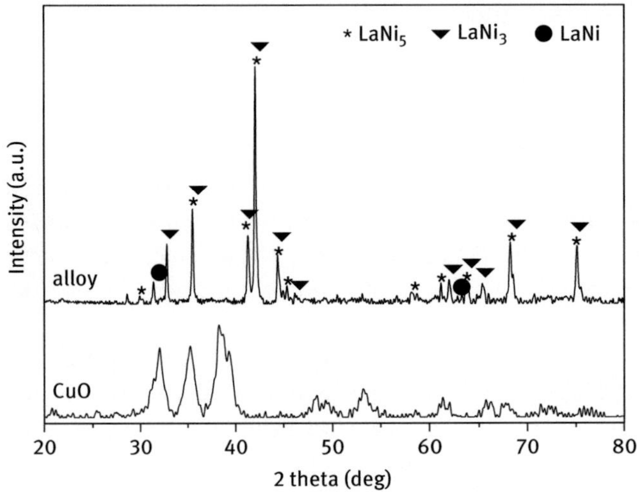

Figure 5.1: XRD pattern of $La_{0.88}Mg_{0.12}Ni_{2.95}Mn_{0.10}Co_{0.55}Al_{0.10}$ alloy and as-prepared CuO.

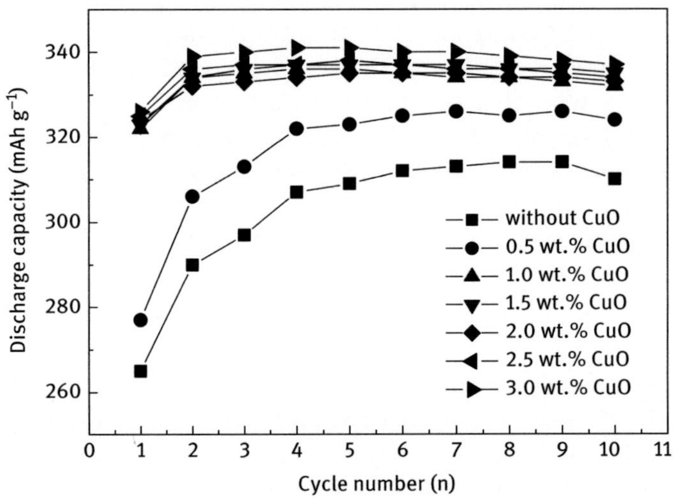

Figure 5.2: Activation curves of electrodes with and without CuO additive.

Figure 5.3 shows SEM image and EDX analysis of the electrode with 2.0 wt.% CuO addition after 100 cycles. It can be seen some fine spherical particles appear at surface of the alloy particles. Micro-area A corresponds to the alloy basis, B to the spherical particles and C to the particles around the alloy particles. The EDX analysis results are also listed. It is shown that the element composition of micro-area A is close to that of the alloy except for Mn and Mg element. Mn and Mg element cannot be detected by EDX for they easily enter into the electrolyte during charging/discharging. However, in this micro-area, 13.12 wt.% elemental O can be detected, indicating the

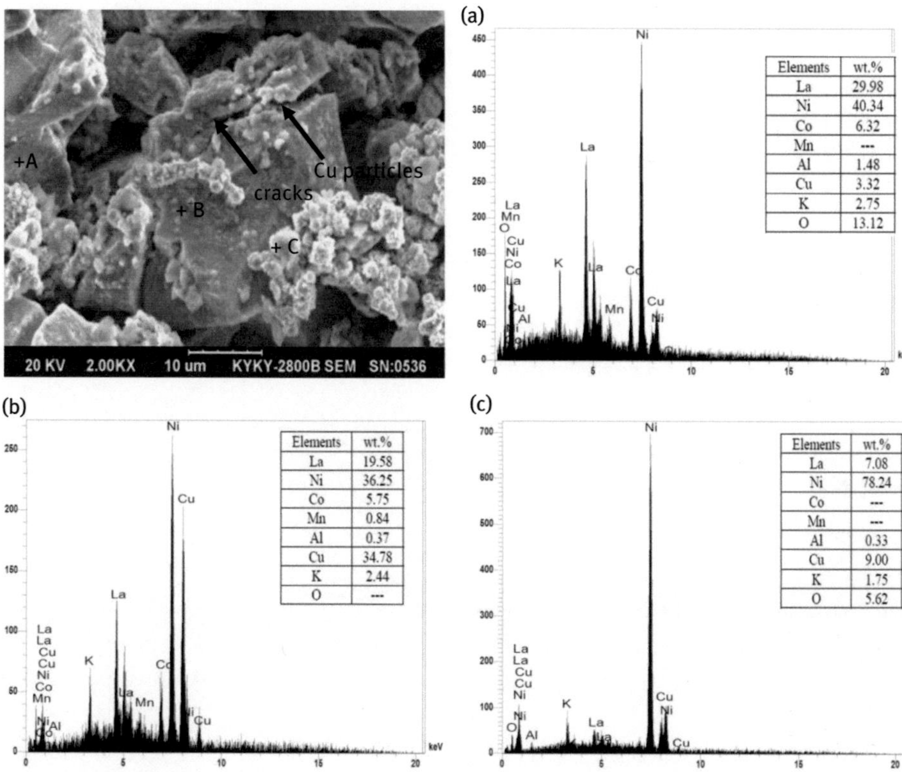

Figure 5.3: SEM image and EDX analysis of electrodes with 2.0 wt.% CuO addition.

alloy is seriously oxidized during the 100 cycles. In micro-area B, a spherical particle can be seen and the Cu content is more obvious, up to 34.78 wt.%. The particle can be considered as Cu reduced from CuO. In micro-area B a little of Mn can also be detected than the alloy basis, which may result from protection from Cu deposition to the alloy dissolution and in this area no elemental O is detected. Many similar particles as in micro-area B can be found in the SEM image, even between particles and in the cracks of the alloy particles. These particles may be helpful to enhance the electric and heat conductivity between alloy particles and are beneficial to protecting the electrodes from fast capacity degradation. Micro-area C is mainly carbonyl nickel added into the electrode as conductor and linker between alloy particles. As for the elemental K, it can be considered residual electrolyte solution at the alloy surface.

Figure 5.4 shows the cyclic voltammogram of CuO. It can be seen that cyclic voltammogram of CuO shows five noticeable current peaks which correspond to redox reactions in alkaline solution. Two anodic current peaks of E_{O1} (−0.36 V) and E_{O2} (−0.02 V) and three cathodic current peaks of E_{R1} (−0.70 V), E_{R2} (−0.95 V) and E_{R3} (−1.19 V) are observed in the figure. The reduction process of cupric oxide in alkaline

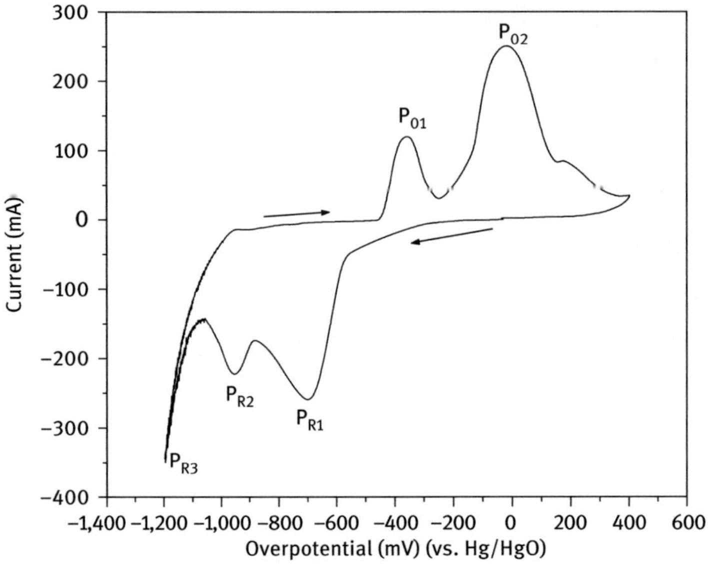

Figure 5.4: Cyclic voltammograms of as-prepared CuO.

solution consisted of two steps [24, 25], which can be described by the following reactions.

$$2CuO + H_2O + 2e^- \rightarrow Cu_2O + 2OH^- \quad (5.1)$$

$$Cu_2O + H_2O + 2e^- \rightarrow 2Cu + 2OH^- \quad (5.2)$$

Therefore the cathodic current peak of E_{R1} is attributed to the reduction reaction of CuO to Cu_2O eq. (5.1) and the cathodic current peak of E_{R2} is attributed to the reduction reaction of Cu_2O to Cu eq. (5.2). Correspondingly, the two anodic current peaks are attributed to the oxidation reaction of Cu to Cu_2O (E_{O1}) and Cu_2O to CuO (E_{O2}). In the potential range −1.2 to 0.6 V, only reduction reaction of CuO can occur, so in the first charging, Cu particles form in the hydride electrode, while the as-deposited Cu particles will not transfer into CuO again. It is well known that Cu has good heat and electronic conductivity, which can reduce polarization of the anodic electrode reaction, and therefore the maximum discharge capacity is increased.

Figure 5.5(a) shows HRD of electrodes with and without CuO additives. It is noticed that CuO additives improve the HRD when the discharge current density is larger than 720 mA g^{-1}. At discharge current density of 1,440 mA g^{-1}, the HRD is noticeably higher than that of electrode without CuO additives. As the case of CuO additive being 0.5 wt.%, HRD of the electrode reaches up to 59.9 %. The increase in HRD of CuO-added electrode may be due to good electronic and heat conductivity of as-deposited Cu particles during the charging process.

Figure 5.5(b) shows decrease in discharge capacity of electrodes varying with cycle number. The ratio of discharge capacity at the 40th cycle to maximum discharge capacity (C_{40}/C_{max}) for the CuO-added electrodes is smaller than that of the electrode

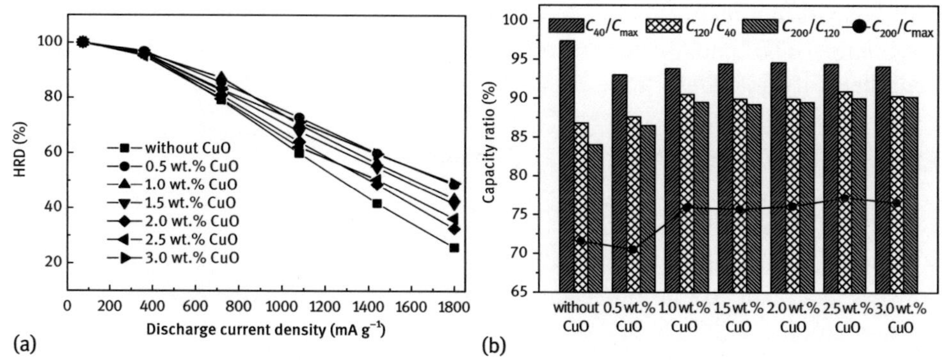

Figure 5.5: (a) HRD of electrodes with and without CuO; (b) cycling stability of electrodes with and without CuO.

without CuO addition, indicating faster capacity degradation at the first 40 cycles. However, C_{120}/C_{40} and C_{200}/C_{120} for the CuO-added electrodes is higher than that of the electrode without addition, suggesting that capacity degradation in CuO-added electrodes becomes slower after 40 charge/discharge cycles. This leads directly to improvement in cycling stability, and the discharge capacity retention at the 200th cycle increases from 71.6 % (blank electrode) to 77.2 % (2.5 wt.% CuO electrode).

As can be seen from SEM and EDX analysis in Figure 5.3 CuO transfers to fine Cu particles and deposit at surface of the alloy particles, protecting constituent elements of the alloy, such as La, Mg and Mn from corrosion by the strong alkaline electrolyte, and thus reduces discharge capacity reduction of the electrodes.

Figure 5.6 shows high-temperature dischargeability (HTD) of the electrodes with and without CuO additives. CuO additives improve HTD of the electrodes, especially

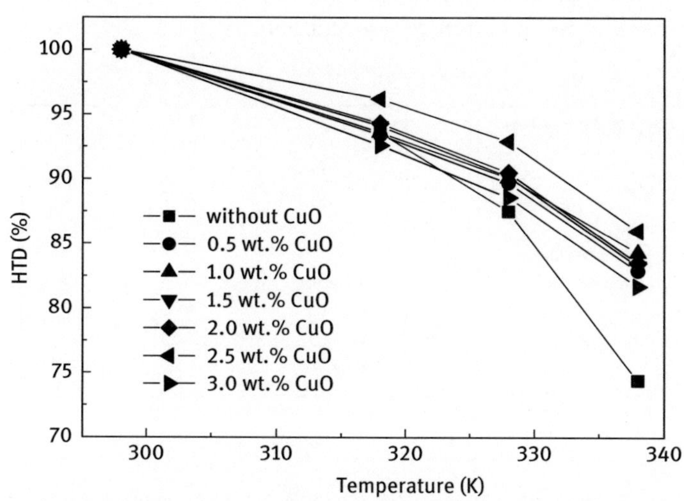

Figure 5.6: HTD of electrodes with and without CuO.

with temperature higher than 328 K. The electrode with 2.5 wt.% CuO additive has much better HTD than the others and its HTD at 333 K reaches 86.0 %, 11.6 % higher than that of the blank electrode (74.4 %).

5.1.2 Non-reactive Additive: TiO_2, Er_2O_3 and ZnO

5.1.2.1 Experimental

2 wt.% additives (TiO_2, Er_2O_3 and ZnO) (compared with hydrogen storage alloy powders) were added into the electrodes. The testing and characterization were the same as 5.1.1.

5.1.2.2 Electrochemical Properties

Capacities of the electrodes with and without additives during activation are shown in Figure 5.7(a). It can be noticed that the electrode without additive needs eight cycles to achieve the maximum discharge capacity, 334 mAh g^{-1}. ZnO additive reduces the activation cycle number to 4 and increases the discharge capacity slightly to 338 mAh g^{-1}. TiO_2 and Er_2O_3 decrease the discharge capacity slightly from 334 mAh g^{-1} to 326 mAh g^{-1} and 333 mAh g^{-1}, respectively.

The Ni/MH batteries will be more widely used if they can be rapidly charged, and therefore the charge acceptance capability, especially for the negative electrodes, should be improved [26]. The high-rate charge capability (HRC) at different current density is shown in Figure 5.7(b). It can be seen that the electrodes with metal oxides have higher HRC values, and the difference is obviously seen when the current density increases to more than 360 mA g^{-1}. When the charge current density increases to 1,440 mA g^{-1}, the HRC increases from 85.1 % (blank) to 94.1 % (TiO_2), 93.3 % (Er_2O_3) and 90.5 % (ZnO).

The effect of temperature on the electrochemical characteristics is investigated by charging/discharging the electrodes at 60 mA g^{-1} current density in the temperature

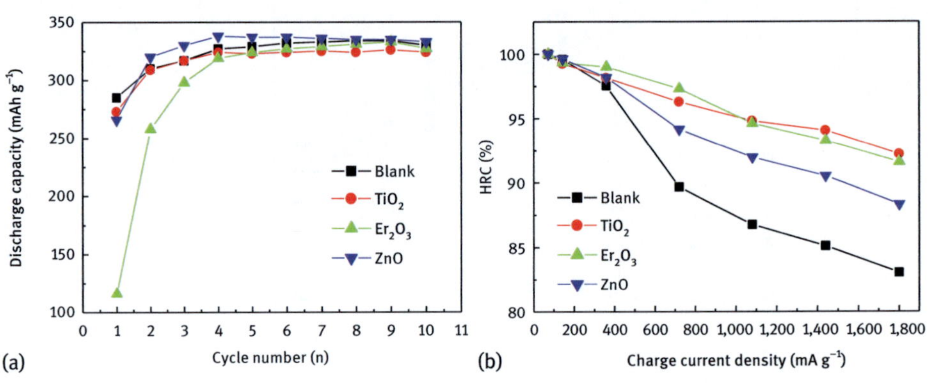

Figure 5.7: (a) Activation curves of electrodes with and without additives; (b) HRC of electrodes with and without metal oxides.

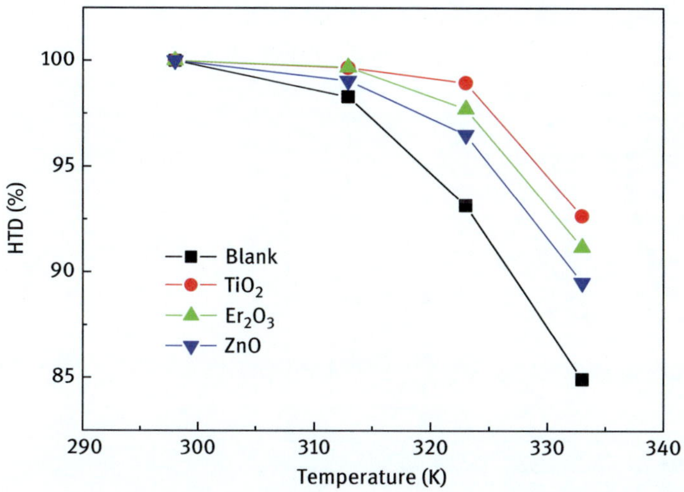

Figure 5.8: High-temperature discharge characteristics of electrodes with and without additives.

range of 298–333 K. The HTD at different temperature is calculated as a ratio of the discharge capacity at a certain temperature to the discharge capacity at 298 K and is shown in Figure 5.8.

It can be seen that the discharge capacity of blank electrode is very sensitive to temperature and decreases sharply with increasing temperature when the temperature is higher than 323 K. When metal oxides, TiO_2, Er_2O_3 and ZnO are added, the decrease in discharge capacity is slow down. It should be pointed out that TiO_2 addition help to keep capacity when the temperature is below 323 K, and the electrode with TiO_2 addition keeps 99.0 % of its capacity at room temperature. When the temperature increases to 333 K, the discharge capacity of all the four electrodes decreases sharply, but the capacity of electrode with additives is higher than that of the blank one. The HTD increases from 84.9 % (blank) to 92.7 % (TiO_2), 91.2 % (Er_2O_3) and 89.5 % (ZnO).

The $i-V$ polarization curves for hydrogen evolution reaction (HER) are provided in Figure 5.9. As shown, hydrogen evolution overpotential increases as TiO_2, Er_2O_3 and ZnO are added to the electrodes. The increase in hydrogen evolution overpotential suppress auxiliary hydrogen evolution at last charging stage, improving charge efficiency, even at high charge current density and at high temperature. Therefore, HRC and HTD are improved after metal oxides, TiO_2, Er_2O_3 and ZnO are added.

In order to analyse effect of additives on cycling stability of the electrodes further, capacity retention rate S_n is defined as a ratio of the capacity retention at Cycle n to the maximum discharge capacity. From Figure 5.10 it can be clearly seen that the discharge capacity retention rate of electrode with additives is higher than that of the blank electrode, indicating the cycling stability is improved by adding metal oxides, TiO_2, Er_2O_3 and ZnO. The discharge capacity retention rate at the 200th cycle

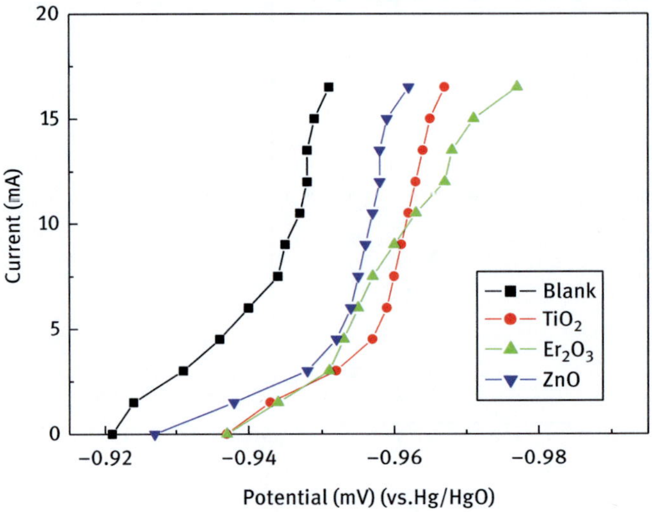

Figure 5.9: Galvanostatic polarization measurement for the electrodes with and without additives.

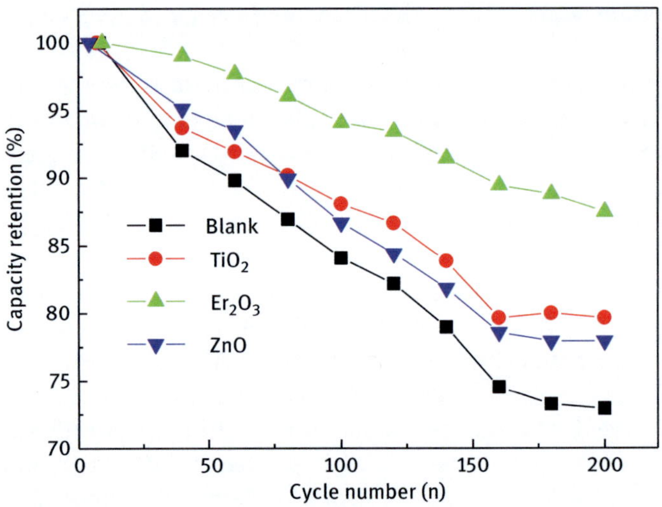

Figure 5.10: Cycling stability of electrodes with and without additives.

increases from 72.9 % (unmodified) to 79.6 % (TiO_2), 87.5 % (Er_2O_3) and 77.9 % (ZnO), respectively. Among the metal oxides, Er_2O_3 is the most effective additive to improve the cycling stability.

It is generally believed that oxidation and pulverization of alloy are the mainly reason for discharge capacity decay for metal hydride electrodes. In order to see changes in phase structure after cycling, the XRD patterns are measured and shown

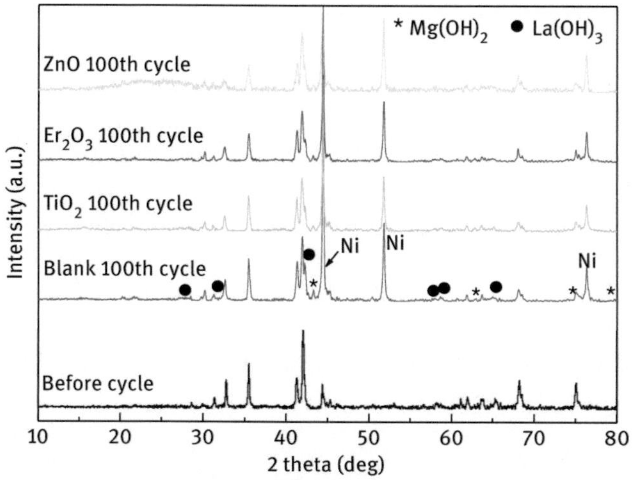

Figure 5.11: XRD patterns of electrodes before and after 100 charge/discharge cycles.

in Figure 5.11. It can be seen that the main crystal structure remained even after 100 charging/discharging cycles. Compared with the pattern for alloy before cycling, some new peaks appear corresponding to $Mg(OH)_2$ and $La(OH)_3$ which are formed during the charging/discharging process. In the four patterns for alloys after cycling, the intensity for peaks corresponding to $Mg(OH)_2$ and $La(OH)_3$ varies for alloy

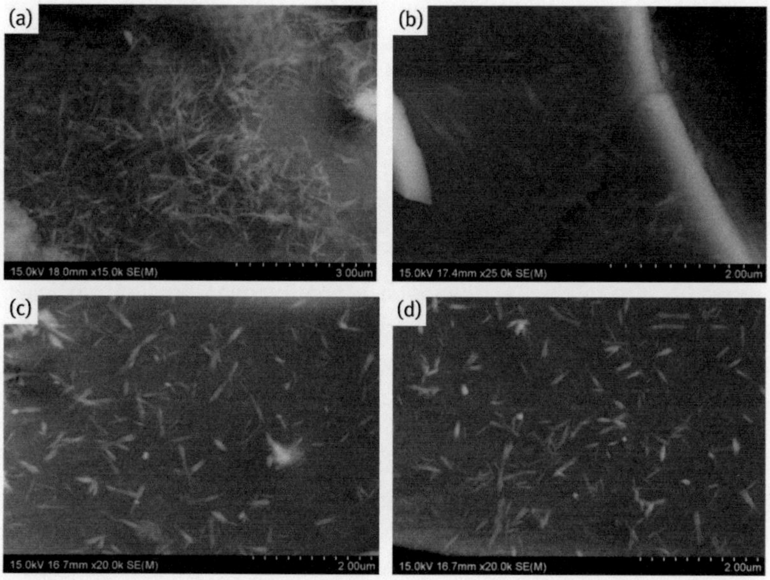

Figure 5.12: SEM images of electrodes after 100 charge/discharge cycles. (a) Blank electrode; (b) TiO_2; (c) Er_2O_3; (d) ZnO.

electrodes with and without additives. The XRD patterns have weaker peak intensity of $Mg(OH)_2$ and $La(OH)_3$ for electrodes with metal oxide additives. This indicates that the alloy electrode with additives suffers from less serious corrosion during charging/discharging process, which contributes to a better cycling stability.

In light of small difference in XRD patterns, FESEM images of electrodes with and without additives after cycling are shown in Figure 5.12. In Figure 5.12(a), surface of the electrode without additive is covered with tiny needle-like substance. EDS shows that elemental oxygen increases remarkably, that is, the substance covered on the surface is $Mg(OH)_2$ and $La(OH)_3$. In Figure 5.12(b–d) scattered needle-shaped hydroxide distribute on surface of the alloy particles, indicating that the oxide additives (TiO_2, Er_2O_3 and ZnO) are helpful to suppress oxidation of the alloy in aqueous alkali, contributing to less serious degradation in discharge capacity.

5.2 Electrodeposition on Negative Electrode Materials

5.2.1 Electrodeposition on Electrodes

5.2.1.1 Electrodeposition Treatment

Electrochemical deposition of Co–Ni alloy was performed onto the electrodes at room temperature. The electroplating solution consisted of 0.3 mol L^{-1} $NiSO_4$, 0.025 mol L^{-1} $CoSO_4$ and 0.15 mol L^{-1} H_3BO_3, the foamed nickel was immersed in plating solution for 15 min, different current densities were applied in the experiment. As soon as the treatment was finished, the nickel foam substrate was rinsed thoroughly with diluted water and dried completely in vacuum.

5.2.1.2 Effect of Electrodeposition on Morphology and Electrochemical Properties

The microstructure of the coatings on surface of alloy electrodes was characterized by SEM and by XRD, and the results are shown in Figure 5.13. Figure 5.13(a) shows that the surface of the bare alloy electrodes is smooth and some small particles attach to the surface. For the following electrodes electroplating Co–Ni, the surface becomes rather rough because of the nanometre-scale coating over the alloy particles, while more homogeneous acicular layer appears as shown in Figure 5.13(c). In order to get changes in chemical composition at the alloy surface, EDS was applied in the detection. Conspicuously, the results indicate that the emergence of cobalt and the increase of nickel contents clarify that Co–Ni alloy coating deposited on the alloy surface.

As a high catalytic activity element, Ni combined with other metal elements can demonstrate electrocatalytic performance by improving effective surface areas [27]. In addition, Co–Ni-based alloys can be reduced by electroplating method and show excellent adhesion strength on substrates as well. The surface modification plays a positive role in improving the activation properties of the MH electrode in alkaline

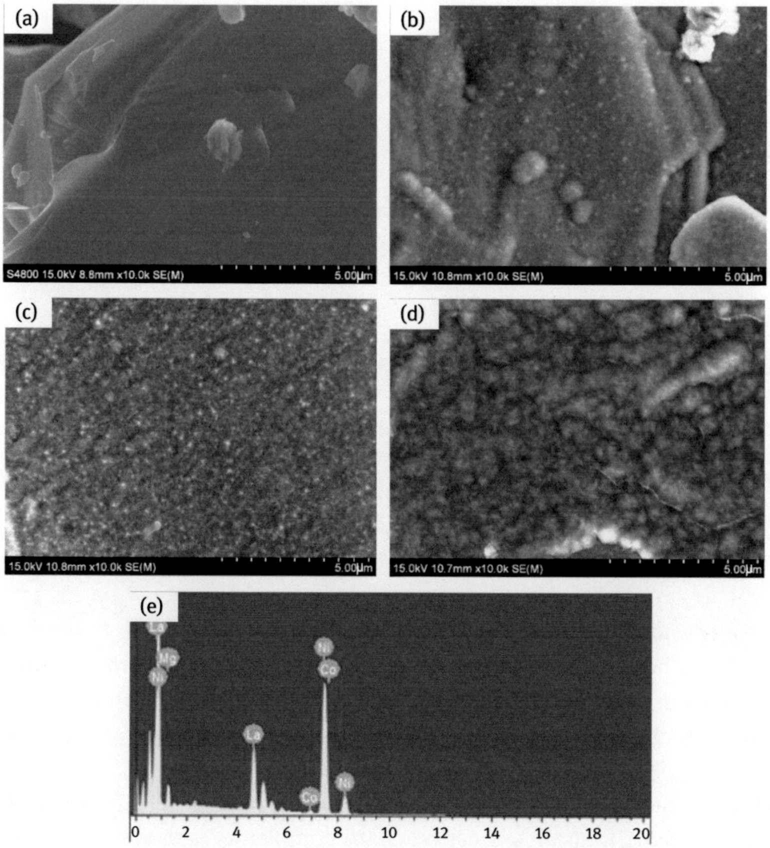

Figure 5.13: SEM images and EDS analysis of the bare and Co–Ni-coated alloy electrodes at different current densities: (a) bare alloy; (b) 7.5 mA cm^{-2}; (c) 10 mA cm^{-2}; (d) 12.5 mA cm^{-2}; (e) EDS after treatment (plating current density = 10 mA cm^{-2}).

solution. The results can be ascribed to the modified layer, which has selective penetrability for hydrogen atoms, thereby improving anti-corrosion of the alloy surface and promoting hydriding/dehydriding reaction.

The electrochemical kinetic performances of alloy electrodes are improved by the coating layer formed close to the surface after electroplating. The condition of alloy coating changes along with the different current densities, varying morphology and even characteristics of alloy electrodes. During the low current density electroplating, applied external energy is relatively small, and the particles generated on alloy surface become increasingly large. As electroplating current density increases, the alloy coating gets homogenous and more fine particles appear owing to the enough external energy which overcomes the greater Gibbs free energy. When the plating current density increases to 12.5 mA cm^{-2}, the layer cannot afford so large current and cracks appear on the electrode surface, hindering the charge transfer process, which is

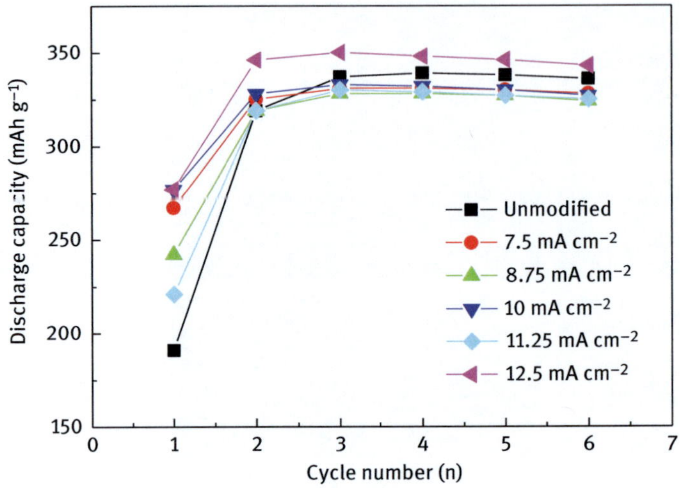

Figure 5.14: Activation curves of La–Mg–Ni-based bare and Co–Ni-coated alloy electrodes.

harmful to alloy electrodes conductivity. In general, the alloy electrode treated at 10 mA cm^{-2} possesses the most continuous coating, and the superior surface coverage guarantees the occurrence of charge transfer reaction at high discharge current density. Therefore, the alloy electrode exhibits the best electrochemical kinetics in this case.

Figure 5.14 shows the discharge capacities of the bare and the Co–Ni-coated alloy electrodes. It is obvious that alloy electrodes are completely activated after two cycles, and the initial discharge capacity of alloy increases from 191 mAh g^{-1} (bare) to 277 mAh g^{-1} (plating current density = 10 mA cm^{-2}) with electroplating treatment. This result can be ascribed to the role of the Co–Ni alloys which formed on the alloy surface facilitating the catalytic activity of the electrodes.

The discharge curve is an important measurement to describe discharge characteristics, which is crucial to hydrogen storage alloy evaluation. Figure 5.15 represents the discharge curves of the bare and Co–Ni-coated alloy electrodes.

Although the alloy is coated with a thin film containing Co, the curves of treated alloy only have one plateau owing to small Co quantity [28]. It can be seen when the discharge current density is 75 mA g^{-1} the discharge voltage plateau of alloy electrode after treatment (plating current density = 10 mA cm^{-2}) becomes longer and flatter, and the middle discharge voltage also rises by 0.029 V, which confirm that the electrochemical polarization is suppressed by surface modification. As the discharge current density increases to 1,875 mA g^{-1}, the discharge plateau gets narrow, and the discharge plateau voltage is lower throughout the discharge process compared to that at a discharge current density of 75 mA g^{-1}. The results can be ascribed to the aggravation of polarization. Moreover, the alloy electrode coated at 10 mA cm^{-2} shows

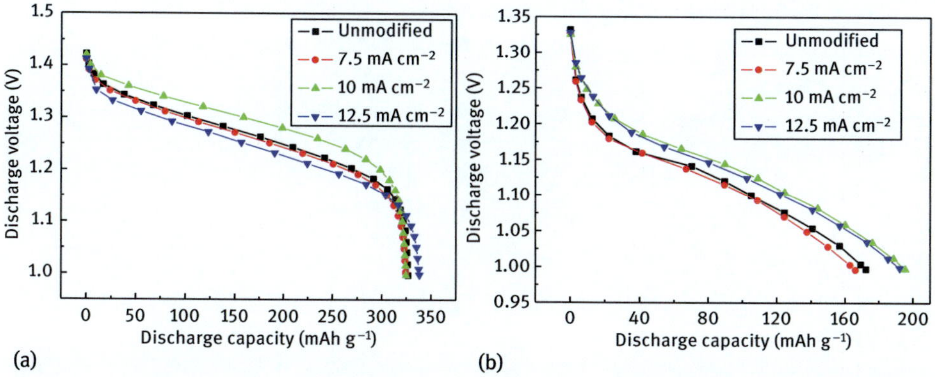

Figure 5.15: Discharge curves of the bare and Co–Ni-coated alloy electrodes at different discharge current densities: (a) 75 mA g^{-1}; (b) 1,875 mA g^{-1}.

a higher discharge voltage and a larger discharge capacity attributed to the excellent conductive layer on the alloy surface.

The HRD of the alloy as a function of discharge current density is shown in Figure 5.16. The trend of the graph clearly indicates that the alloy electrode electroplated at 10 mA cm^{-2} shows better discharge performance compared to the bare one. Moreover, the HRD of alloy electrode electroplated at 10 mA cm^{-2} can reach 57.5 % at 1,875 mA g^{-1}. It is well known that the HRD is controlled by charge–transfer reaction at the surface and the hydrogen diffusion ability in the bulk. The relatively slower step is the main controlling factor [29]. After the electroplating, the metallic layer accelerates the charge transfer reaction on the surface of alloy, reducing the polarization of

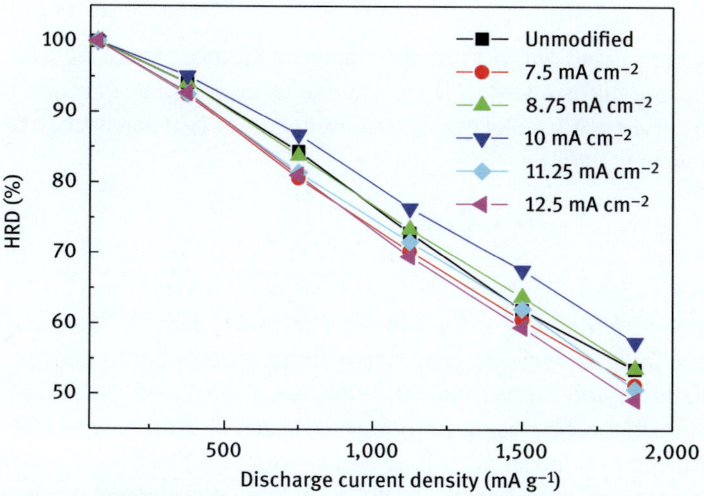

Figure 5.16: HRD of the bare and Co–Ni-coated alloy electrodes.

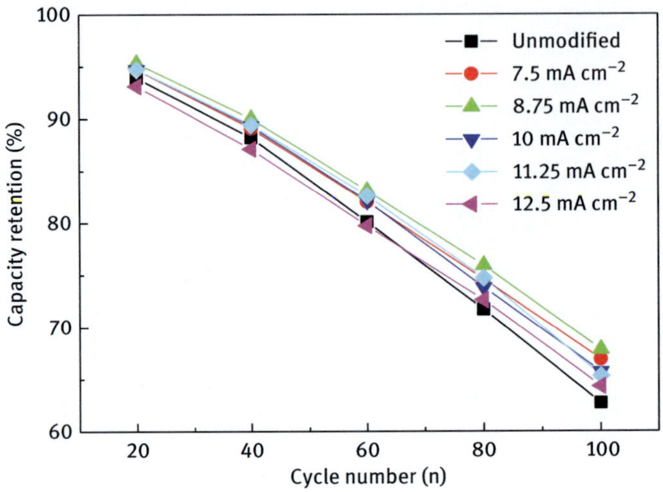

Figure 5.17: Discharge capacity of the bare and Co–Ni-coated alloy electrodes versus cycle number.

alloy electrode during the discharge process, which guarantees the HER smoothly. It is obvious that the treated alloy reveals better discharge characteristics.

The relationship between capacity retention and cycle number of the alloy electrodes is shown in Figure 5.17. When the cycling number is small, the capacity retention of coated alloy electrode is slightly higher than the bare alloy electrode. Moreover, with the increasing cycling number, the differences gradually become larger. Clearly, the bare alloy suffers from severe degradation through charge/discharge cycles, which can be attributed to the pulverization and oxidation phenomenon of the alloy. The protective layer keeps the alloy from being oxidized, and thereby enhances the cycling stability [30].

The kinetic tests were carried out to study the change in surface charge transferring rate and bulk hydrogen diffusion rate. Figure 5.18 presents a complete contrast of the linear polarization curves of alloy electrodes. The exchange current density can be calculated by the following equation [31]:

$$I_0 = RTI/F\eta \tag{5.3}$$

where I_0 is the exchange current density (mA g^{-1}), R is the gas constant (J mol^{-1} K^{-1}), T is the absolute temperature (K) and F is the Faraday constant (C mol^{-1}). The calculated results are listed in Table 5.1. It clearly proves that the alloy treated at 10 mA cm^{-2} is superior to the bare alloy, and further indicates that the presence of Co–Ni layer enhances the charge transfer process between the electrode and electrolyte interface.

The hydrogen diffusion rate in the alloy bulk can be measured by anode polarization curve. The curves of current density versus overpotential response of alloy electrode are shown in Figure 5.19. It shows that the current density increases along

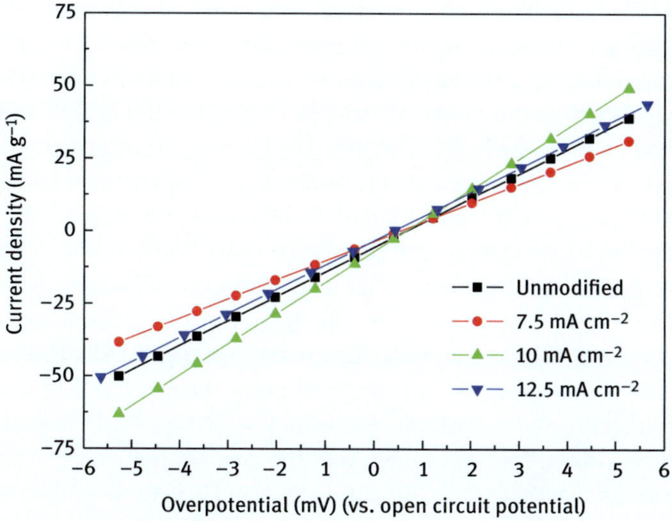

Figure 5.18: Linear polarization curves of the bare and Co–Ni-coated alloy electrodes at 50 % DOD.

Table 5.1: Kinetic parameters of bare and the Co–Ni-coated alloy electrodes at different electroplating current densities.

Samples	Bare	7.5 mA cm^{-2}	10 mA cm^{-2}	12.5 mA cm^{-2}
I_0 (mA g^{-1})	213.8	166.9	269.2	211.3

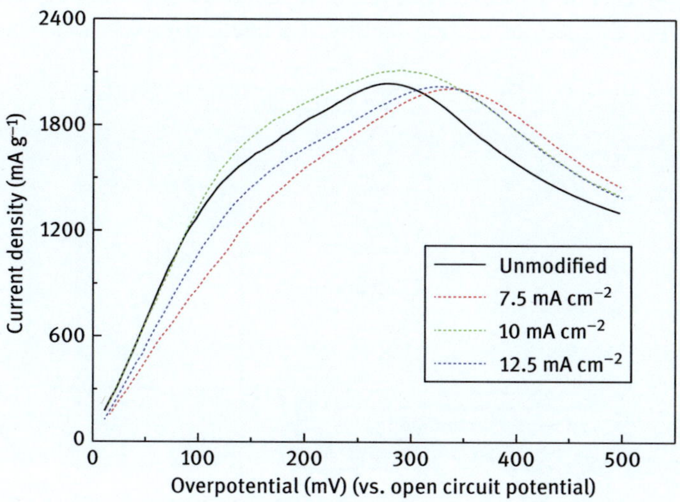

Figure 5.19: Anodic polarization curves of the bare and Co–Ni-coated alloy electrodes at 50 % DOD.

with the value of overpotential, reaching a maximum value defined as the limiting current density I_L which is mainly controlled by the hydrogen diffusion within the bulk alloy. As can be seen from the results, the limiting current density of alloy electrode treated at 10 mA cm^{-2} is higher than that of the other alloy electrodes. It is considered that the Co–Ni layer coated on the alloy leads to an increase in charge transfer rate on the surface, reducing hydrogen concentration at the surface, and thereby promoting the hydrogen diffusion in the alloy bulk under a larger concentration difference [32].

Figure 5.20 illustrates the cyclic voltammograms for the bare and Co–Ni-coated alloy electrodes at room temperature, carried out at a scan rate 50 mV min^{-1}, with the potential ranging from −1,200 to −200 mV (vs. Hg/HgO). In cathodic branch of cyclic voltammograms, there is a plateau dominated by the adsorption of the hydrogen atom on the electrode surface and a peak of hydrogen evolution. The anodic current is often used to evaluate the electrocatalytic activity, and after Co–Ni treatment the peak anode current is found to increase. It is speculated that the coated alloy exhibits an improvement in electrocatalytic activity for hydrogen oxidation [1]. The surrounding area of the curve is related to the capacity of the alloy electrode, and the curve area increases after treatment indicates that the capacity of electrode is increased.

5.2.2 Electrodeposition on Alloy Powders

5.2.2.1 Electrodeposition Treatment

The La–Mg–Ni-based $La_{0.88}Mg_{0.12}Ni_{2.95}Mn_{0.10}Co_{0.55}Al_{0.10}$ alloy powders with a particle size less than 74 μm were used for electroplating. The powders were first immersed into HCl (10^{-3} mol L^{-1}) for 1 min to remove the native oxides and then rinsed with diluted water to pH = 7. Cobalt or nickel electroplating was carried out using a plating facility. The schematic diagram of plating facility is shown in Figure 5.21. The

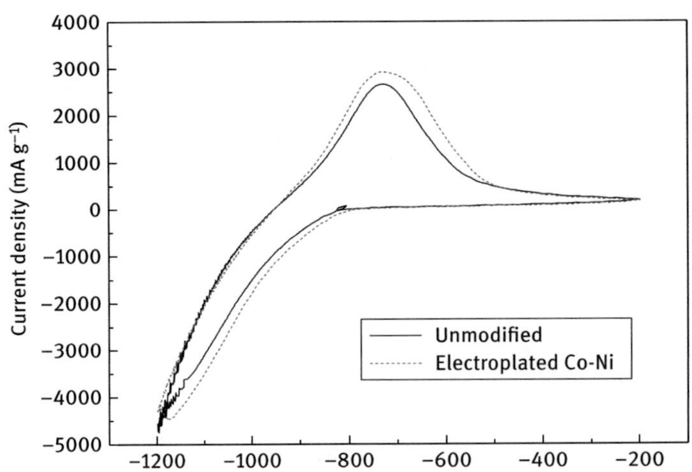

Figure 5.20: Cycle voltammograms of the La–Mg–Ni-based alloy electrodes.

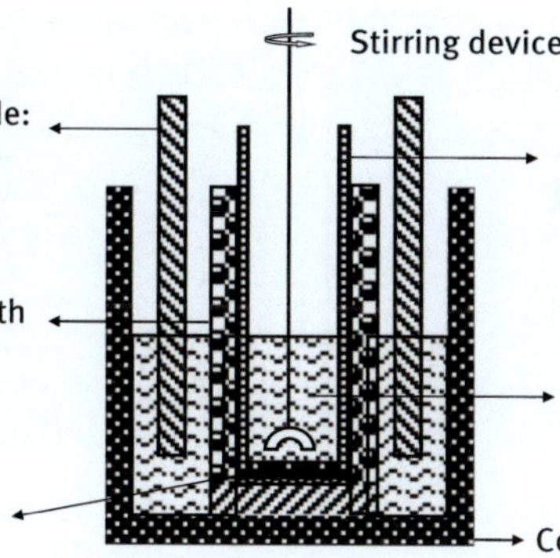

Figure 5.21: Schematic diagram of electroplating facility.

Table 5.2: Operating conditions and composition of the electroplating solutions.

Coating	Composition or operating conditions	
Co	$CoSO_4 \cdot 6H_2O$	330 g L^{-1}
	$CoCl_2 \cdot 6H_2O$	40 g L^{-1}
	H_3BO_3	30 g L^{-1}
	Current density	1.3 A dm^{-3}
	Temperature	35°C
	Time	15 min
	pH	5
Ni	$NiSO_4 \cdot 6H_2O$	100 g L^{-1}
	NaCl	10 g L^{-1}
	H_3BO_3	20 g L^{-1}
	$C_6H_5Na_3O_7 \cdot 2H_2O$	120 g L^{-1}
	Current density	0.8 A dm^{-3}
	Temperature	55°C
	Time	15 min
	pH	6

operating conditions and chemical composition of the plating solutions are given in Table 5.2. In the process, the electrolyte was continuously stirred in order to promote ion transport and the solution temperature was kept constant by a thermostatic bath. As soon as the process of the electroplating has been finished, filtered the alloy powders and washed the powders thoroughly with diluted water and anhydrous alcohol. Then the alloy powders were dried in vacuum at 60°C for 4 h.

5.2.2.2 Morphology and Electrochemical Properties

The FESEM images of the bare and coated alloy powders are shown in Figure 5.22. The bare alloy powders represent a smooth surface (Figure 5.22(a)). However, surface of the cobalt-coated specimens is rather rough (Figure 5.22(b)), and uniform spherical particles with a diameter of ~130 nm distribute on the surface. Lamellar nickel is

Figure 5.22: FESEM image of bare and coated alloy powders (a) bare; (b) Co-coated; (c) Ni-coated.

observed from the surface of the alloy particles in Figure 5.22(c). The cobalt or nickel metallic grains are softer and less brittle than the intermetallic alloy, and thus provide both the necessary contact between alloy powders, and good electrical and thermal conduction. In addition, the metallic coatings on the surface of the alloy powders are stable in alkaline solution and have selectivity for hydrogen atoms and oxygen atoms, that is, the hydrogen atoms could easily penetrate these coatings while the oxygen atoms (or OH^-) could not [33]. This indicates that metallic coatings can protect the alloy powders from oxidation whilst they do not block the hydriding/dehydriding reaction of the electrode in an alkaline solution.

The discharge capacities of the bare and coated alloy electrodes are plotted in Figure 5.23 as a function of cycle number. The maximum discharge capacity of bare alloy electrode is 316 mAh g^{-1} and that of cobalt and nickel encapsulated alloy electrodes are 336 and 335 mAh g^{-1}, respectively. These results show that the metallic coatings increase the discharge capacity effectively. This is due to the improvement in

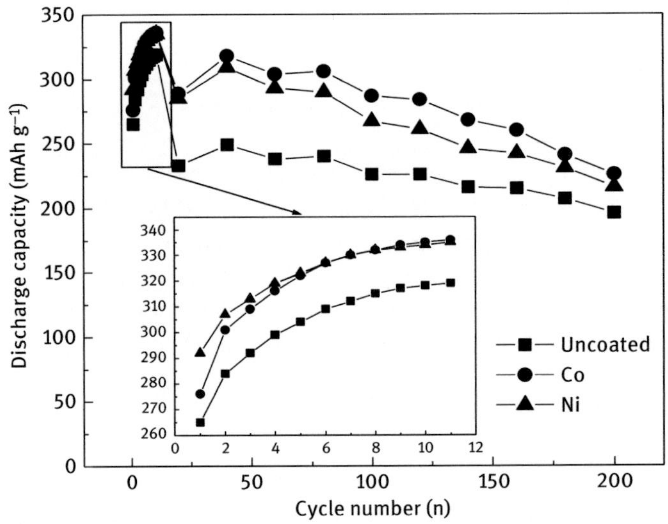

Figure 5.23: Capacity retention of bare and coated alloys versus cycle number.

discharge efficiency of the MH electrode by the faster charge transfer reaction arising from superior electrical conductivity of cobalt or nickel metallic layer.

Cycle life is a key factor for hydrogen storage alloy electrodes. From Figure 5.23, it can be seen that the capacity decay of the coated alloy electrodes proceeds more slowly than that of the bare alloy electrode. After 200 charge/discharge cycles, the capacity retention is approximately 67, 65 and 60 % for cobalt, nickel-coated and bare alloy electrodes, respectively. Alloy oxidation is believed to be one of the most important reasons for capacity decay. Studies have shown that the diffusion of fresh lanthanum and magnesium to the surface from the bulk, where they get oxidized. The oxidation results in the decrease in the dehydriding capacity. Cobalt or nickel encapsulation forms as a protective cover over the alloy surface and prevents the alloy from oxidation or corrosion. In addition, the metallic coatings work as catalyst for accelerating electrochemical reaction on the surface.

The self-dischargeability is an important property for the unloaded condition of the batteries and self-discharge rate must be as low as possible for practical applications. The rate at which a battery loses its powder in an open-circuit condition is called self-discharge rate. The self-discharge rate can be calculated according to the following equation:

$$\text{Self-discharge rate (\%)} = \left(1 - \frac{2C_b}{C_a + C_c}\right) \times 100 \tag{5.4}$$

C_b is the discharge capacity after laying open circuit for 24 h at 298 K and C_a and C_c are the discharge capacity before and after C_b is tested, respectively.

Figure 5.24 shows that the self-discharge rate of cobalt- or nickel-coated alloy electrode is lower than that of bare alloy electrode. The self-discharge rate decreases from

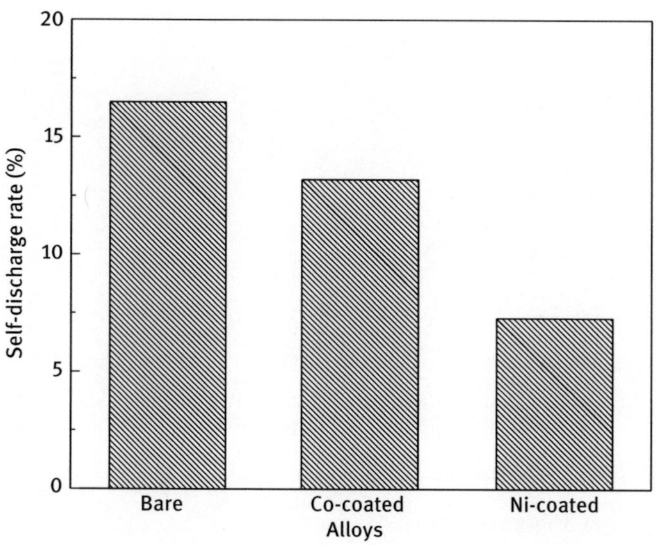

Figure 5.24: Self-discharge rate of bare and coated alloys at 298 K.

17 % (bare electrode) to 7 % (Ni-coated electrode). The loss of discharge capacity during the storage in open-circuit conditions can be divided into two parts, that is, reversible and irreversible loss. Reversible capacity is attributed to the desorption of hydrogen from the hydride electrode, while irreversible capacity loss is due to the degradation of the hydrogen-absorbing alloy The presence of cobalt or nickel is helpful to prevent corrosion or oxidation of the active material as mentioned above and further decrease the irreversible capacity loss. Thus, the self-discharge rate of the coated alloy electrodes is much lower than that of the bare alloy electrode.

The results given in Figure 5.25 clearly show that the cobalt or nickel coating greatly increases the HRD. As the discharge current I_d is 1,500 mA g^{-1}, the HRD is 6.1, 15.5 and 23 % for bare, cobalt-coated and nickel-coated alloy electrodes, respectively. The results indicate that the electrodeposits play an important role in improving the kinetic properties.

The main electrode reactions that occur at the surface of a metal hydride electrode are

$$M + H_2O + e^- \Leftrightarrow MH_{ads} + OH^-$$

$$MH_{ads} \Leftrightarrow MH_{abs}$$

where H_{ads}, H_{abs} represent hydrogen adsorbed on the surface, absorbed on the subsurface, respectively. Reactions above represent the charge transfer at the electrode/electrolyte interface and the transfer of hydrogen atoms from adsorbed to absorbed state at the surface, respectively.

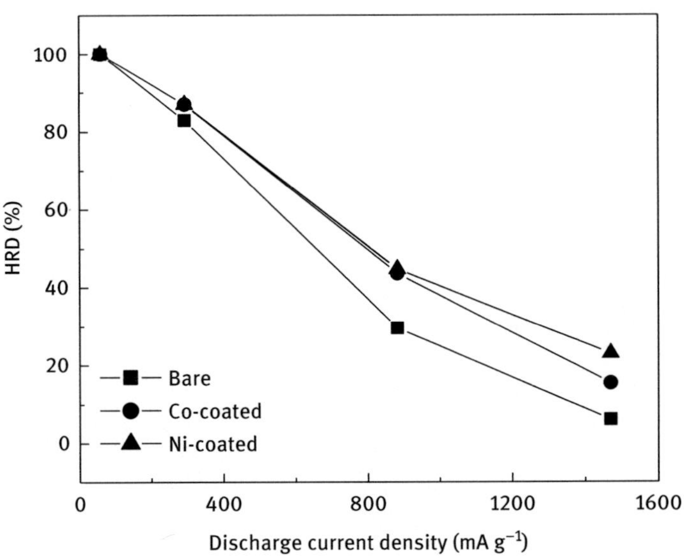

Figure 5.25: HRD of the bare and coated alloy electrodes.

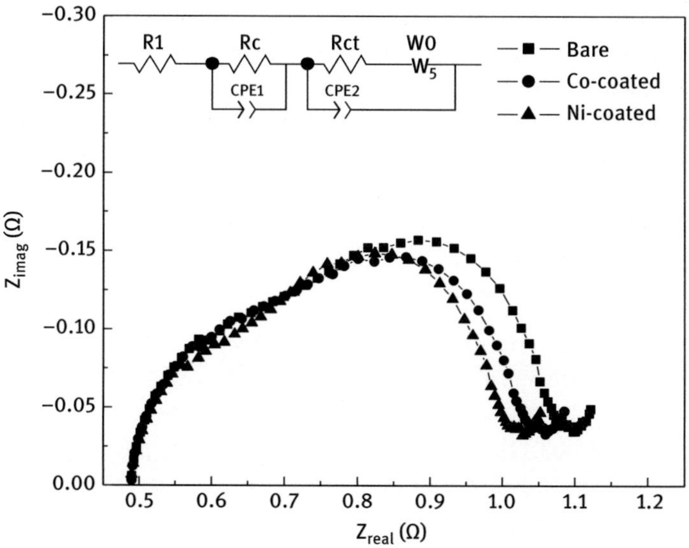

Figure 5.26: EIS of the bare and coated alloy electrodes.

In order to investigate the properties of charge–transfer reaction electrochemical impedance are performed on the alloy electrode at different temperatures. The selected typical EIS obtained at 298 K of the bare and coated $La_{0.88}Mg_{0.12}Ni_{2.95}Mn_{0.10}Co_{0.55}Al_{0.10}$ alloy electrodes as well as the equivalent circuit are shown in Figure 5.26. It can be seen that each EIS contains two semicircles followed by a straight line. Kuriyama et al. [34] have explained that the smaller semicircle in the higher frequency region is mainly related to the contact resistance and capacitance between the alloy particles and the current collectors, while the larger semicircle in the lower frequency region is attributed to the charge–transfer resistance (R_{ct}) for hydrogenation reaction on the surface, and the sloped straight line corresponds to Warburg impendence. On the basis of the equivalent circuit, the R_{ct} values are obtained by means of the fitting program Z-View.

Table 5.3 gives the values of R_{ct} at different temperatures. It is clear that the R_{ct} decreases with increasing temperature. This indicates increasing temperature is

Table 5.3: Electrochemical kinetic parameters of the bare and coated alloy electrodes.

Samples	**R_{ct} (Ω)**				**I_0 (mA g^{-1})**
	298 K	**308 K**	**318 K**	**328 K**	**298 K**
Bare	0.6779	0.3999	0.2674	0.1881	265.8
Co-coated	0.6275	0.4232	0.2883	0.2182	317.9
Ni-coated	0.5967	0.4430	0.3292	0.2374	319.3

Note: R_{ct}, charge–transfer resistance; I_0, exchange current density.

beneficial to the charge–transfer reaction at the electrode/electrolyte interface. A change in the R_{ct} is caused by both a change in reactivity of the alloy surface and a change in the reaction area due to disintegration of the alloy particles. Therefore, in order to clarify the electrochemical activity of the metal hydride surface, apparent activation enthalpy $\Delta_r H^*$ is applied and it can be determined by the following equation [35]:

$$\ln\left(\frac{T/R_{ct}}{K/\Omega}\right) = C_0 - \frac{\Delta_r H^*}{RT} \tag{5.5}$$

where R_{ct} is the charge transfer resistance for the metal hydride electrode, R is the gas constant, T is the temperature and C_0 is a constant in which the surface area is included. The R_{ct} data in Table 5.3 are used to plot the $1/T$ dependence of $\ln(T/R_{ct})$ as shown in Figure 5.27. The straight line of $\ln(T/R_{ct})$ versus $1/T$ in Figure 5.27 gives the apparent activation energy of 36.8, 31.4 and 27.5 kJ mol^{-1} for the bare, Co- and Ni-coated alloys, respectively. This indicates that microencapsulation by cobalt or nickel enhances the catalytic activity of the electrochemical reaction. In other words, the metallic coatings facilitate the electrochemical reaction for the MH electrode.

The exchange current density (I_0) is used to characterize electrocatalytic activity for charge–transfer reaction on surface of the alloy electrode. Figure 5.28 presents the linear polarization curves of bare and coated $La_{0.88}Mg_{0.12}Ni_{2.95}Mn_{0.10}Co_{0.55}Al_{0.10}$ alloy electrodes. Obviously, the polarization current density is almost linearly dependent on the overpotential within a small range of overpotential (±5 mV). As listed in Table 5.3, the I_0 increases from 265.8 mA g^{-1} (bare) to 317.1 mA g^{-1} (Co-coated) and 319.3 mA g^{-1} (Ni-coated). The I_0 represents the charge–transfer reaction rate on the surface of the

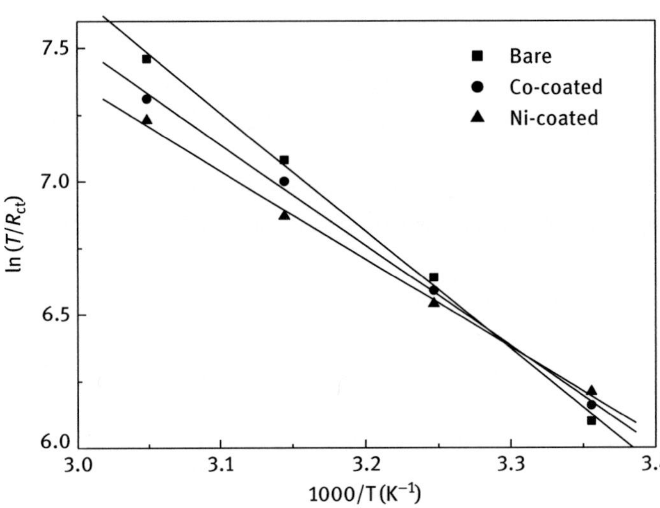

Figure 5.27: The relation between $\log(T/R_{ct})$ and $1/T$ of the bare and coated alloy electrodes.

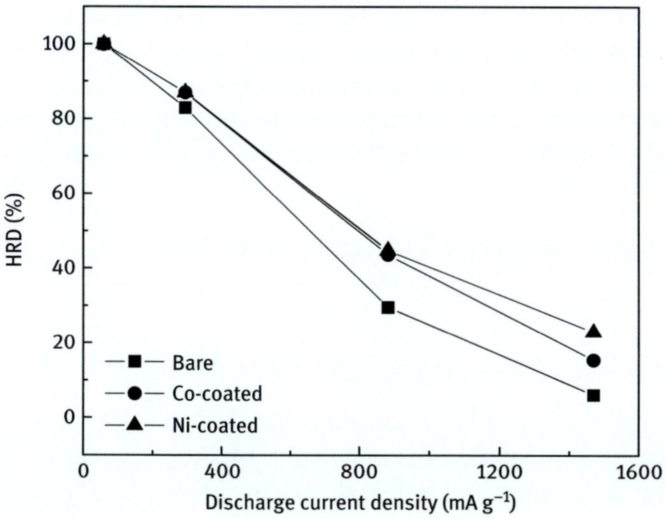

Figure 5.28: Linear polarization curves of bare and coated alloy electrodes.

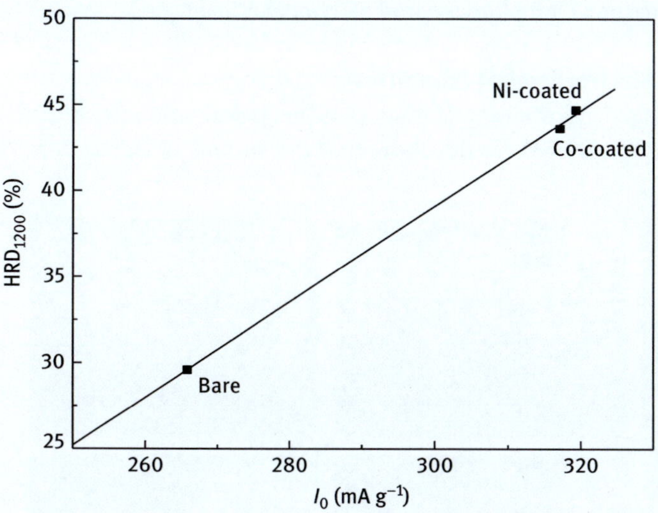

Figure 5.29: Dependence of HRD_{1200} on I_0 of the bare and coated alloy electrodes.

alloy electrode, so the increase in I_0 indicates that the cobalt or nickel coating increases the charge–transfer rate on the surface. This may be one of the reasons for the improvement of the kinetic performances of the alloy electrodes.

The relationship between the HRD_{1200} and exchange current density (I_0) of the bare and coated alloy electrodes is presented in Figure 5.29. It can be seen that the HRD_{1200} increases gradually with the increasing I_0 value, and there is a good linear relationship between the HRD_{1200} and the I_0. Iwakura et al. [36] have noticed that, if the electrochemical reaction (charge–transfer reaction) on the surface of the alloy electrodes is the rate-determining factor, a linear dependence of the *HRD* on the I_0 would be obtained. In the present study, the linear relationship between HRD_{1200} and I_0 implies that the *HRD* is essentially controlled by the charge–transfer reaction.

5.3 Electroless Plating on Negative Electrode Materials

5.3.1 Metallic Coating

5.3.1.1 Surface Treatment

The solution used for surface treatment was with a composition of 0.05 mol L^{-1} $NiSO_4$ ($CoSO_4$), 0.05 mol L^{-1} $N_2H_4{\cdot}H_2SO_4$ (solution I) and 0.64 g L^{-1} $NaBH_4$, and 0.2 g L^{-1} NaOH (solution II). During the treating processing, the alloy particles were merged into the solution I and stirred for 5 min, then solution II was added into solution I and the mixture was stirred slowly for another 20 min. After sufficient stirring, the alloy particles were filtered from the solution and dried at 60°C in vacuum. The surface treated alloy particles were then characterized and electrochemical tested.

5.3.1.2 Morphology and Electrochemical Properties

Figure 5.30 shows the surface morphology of alloy powders before and after electroless plating, In Figure 5.30(a–c) SEM results show that the surface of the untreated

Figure 5.30: Morphology of alloy powders before and after electroless plating (a) untreated alloy SEM; (b) Ni-coated alloy SEM; (c) Co-coated alloy SEM; (d) untreated alloy TEM; (e) Ni-coated alloy TEM; (f) Co-coated alloy TEM.

Table 5.4: EDS analysis of alloy surface before and after electroless plating.

	Nd[a]	Mg	Ni	Co
Untreated	0.171	0.070	0.759	–
Ni-coated	0.158	0.065	0.777	–
Co-coated	0.160	0.065	0.715	0.060

[a] The data here is atomic ratio.

alloy powders is smooth and some small particles formed during mechanical crashing are adhere to the surface. After Ni- and Co-plating, modified layers appear on the surface. For the Ni-coating alloy, some honeycomb-like depositions interlink and tightly adhere to the alloy surface. The interlinked species are nickel indicated by EDS results as nickel content increases compared with untreated alloy (Table 5.4). For the Co-coating alloy, some flake-like depositions covered at the alloy surface, indicated to be cobalt by EDS. In Figure 5.30(d–f) the TEM results show that the untreated alloy is with clear edge, while after Ni- and Co-coating a 10–20-nm-thick modifying layer appears at the surface. And the Co-coating layer is smoother than the Ni-coating layer. It is found that the plating layers adhere tightly to the alloy surface by morphology observation, and the tight adhering may be ascribed to the etching by acidic solution and induction by complex agent. Here we speculate the deposition layer formation as below. In solution I, $N_2H_4{\cdot}H_2SO_4$ dissociates according to eq. (5.3). As alloy particles are merged into the solution I, the surface is etched by acidic solution and becomes clean with metallic elements (Nd, Mg and Ni) left at the surface. The metallic elements display positive charges with electrons being lost into the near solution owing to solvation effect. Subsequently, one coordinating atom N complexes with metals at alloy surface, meanwhile the other atom N with depositing transition metal atoms in the solution. Finally transition metal (Ni or Co) ions were reduced and adhere tightly to the alloy surface. The formation process of modified layer can be described in Figure 5.31.

$$N_2H_4 \cdot H_2SO_4 \rightarrow N_2H_5^+ + HSO_4^- \tag{5.6}$$

The Ni- or Co deposition adheres tightly to the surface and protects the fresh surface from oxidation in the strong alkaline solution. In addition, the deposited layer has larger specific surface area and higher catalytic activity compared with originally smooth and rare earth-containing alloy surface, which may be effective to improve the electrochemical kinetic properties of the alloy electrodes.

Figure 5.32 shows activation profiles of alloy electrodes before and after electroless plating. The electroless treatment improves not only the initial discharge capacity but also the maximum discharge capacity. The initial discharge capacity increases from 237 mAh g^{-1} (untreated) to 305 mAh g^{-1} (Ni-coated) and 318 mAh g^{-1} (Co-coated), as well as the maximum discharge capacity increases from 327 (untreated) to 335 mAh g^{-1} (Ni-coated) and 337 mAh g^{-1} (Co-coated). The increase in discharge capacity is ascribed to the clean and catalytic alloy surface after surface treatment.

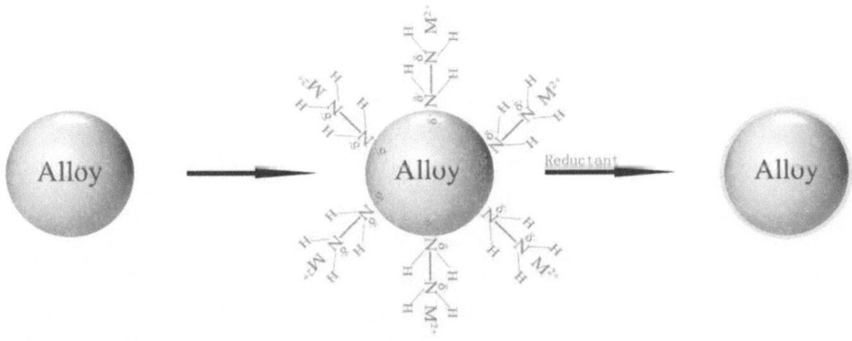

Figure 5.31: Formation of coating layer at alloy surface.

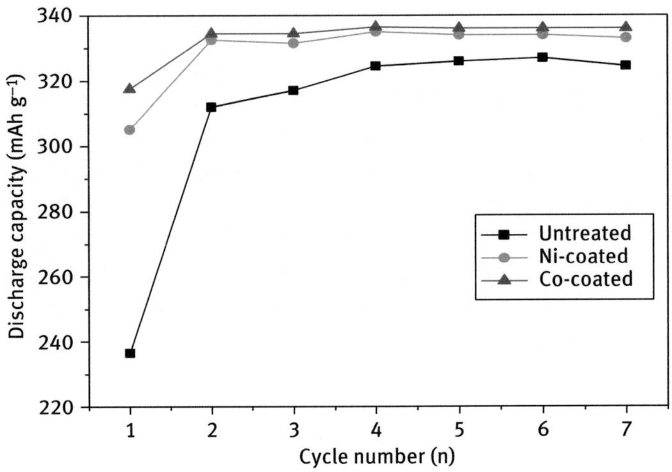

Figure 5.32: Discharge capacity versus cycle number for alloy electrodes before and after electroless plaiting.

In the treating process, the $N_2H_4{\cdot}H_2SO_4$ solved plating solution is with weak acidity, removing the passive layer at the surface, and then Ni or Co coatings deposit at the surface. The catalytic surface coating contributes to a lower electrochemical polarization, preventing the negative electrode potential from positive shift during discharge and makes the hydrogen absorption/desorption complete. Besides increase in discharge capacity, the activation processing decreases from five cycles for the untreated electrode to two cycles for both Ni- and Co-coated electrodes owing to surface treatment.

Figure 5.33 shows HRD of the alloy electrodes before and after electroless plating. When the discharging current density increases, the discharge capacity decreases. However, the decrease is suppressed by surface coating treatment, and the

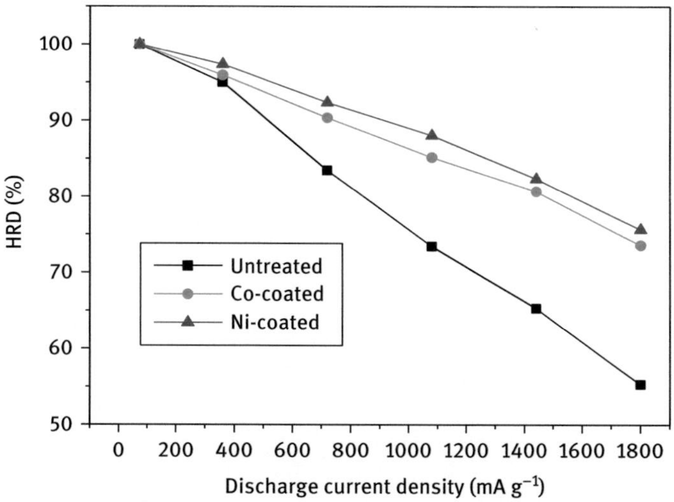

Figure 5.33: HRD of the alloy electrodes before and after electroless plating.

differences become more remarkable as the discharge current increases to 720 mA g^{-1} or larger. The coatings both bring great improvement for HRD. At the discharge current of 1,800 mA g^{-1}, the HRD increases from 55.3 % (untreated) to 75.7 % (Ni-coating) and 73.6 % (Co-coating), respectively. Ni- and Co-coating layer shows high electrocatalytic activity, suppressing electrochemical polarization during discharge process and delaying cut-off of the discharge process.

Figure 5.34 shows linear polarization curves of alloy electrodes before and after electroless plating. A high value of I_0 corresponds to good kinetics for hydriding/dehydriding. The I_0 values are listed in Table 5.5.

Figure 5.35 shows anodic polarization curves of alloy electrodes. In the over potential range from 0 to 400 mV, the polarization current density increases and then decreases, with a peak value appearing. The peak current density is known as limiting current density (I_L) and is controlled by hydrogen diffusion in the bulk. As metallic Ni and Co is coated on the alloy surface, the I_L increases remarkably from 1,775 mA g^{-1} (untreated) to 1,903 mA g^{-1} (Ni-coating) and 2,073 mA g^{-1} (Co-coating).

Figure 5.36 shows the response of discharge current density under +500 mV potential step. The current density decreases first sharply and then slowly. Each curve can be divided into two regions, before 2,500 s, the curves dope sharply and in a non-linear form, and after 2,500 s, the curves decreases slowly and in a close-to-linear form. D can be calculated according to the slope of log(i) versus t and is also listed in Table 5.5.

Figure 5.37 shows morphology images for alloys after 100 charge/discharge cycles. The alloy particles before and after electroless plating show great differences. In the SEM images, Figure 5.37(a–c), some tiny and needle-shaped species appear at the untreated alloy surface. The species distribute well on the surface, however, in a

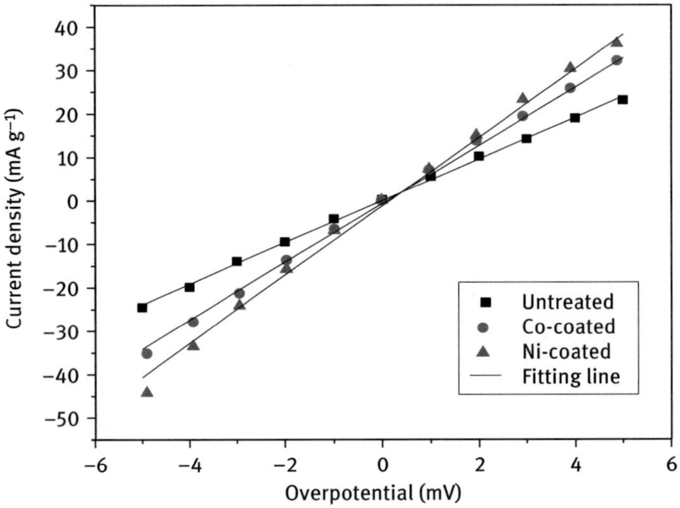

Figure 5.34: Linear polarization curves of alloy electrodes before and after electroless plating at 50 % DOD.

Table 5.5: Kinetic parameters of alloy electrodes before and after surface coating.

Alloys	Exchange current density I_0 (mA g^{-1})	Limiting current density I_L (mA g^{-1})	Hydrogen diffusion coefficient D($\times 10^{-10}$ cm^2 s^{-1})
Untreated	123.1	1,775	1.08
Ni-coated	202.6	2,073	1.40
Co-coated	163.6	1,903	1.34

loose form, and is not able to protect the alloy from corrosion in the solution. EDS results show that oxygen element content increases remarkably. These results suggest that the newly formed substance is $La(OH)_3$ according to its shape and its composition. On the surface of alloy with Ni-coating, the surface is still smooth with few hydroxides formed. Only some small particles can be found on the surface and EDS indicates much less oxygen content is detected on the Ni-coating alloy surface. For the Co-coating, hydroxides can be scarcely seen, and very little oxygen content can be detected, instead some flaky texture is found. It is the coated Co on the surface. However, the surface of the alloy particles becomes poorly conductivity owing to the formation of hydroxides and/or hydrides during repeat hydriding/dehydriding. Therefore, images with higher magnification are not clear. Luckily, it can be also seen from the low magnification images that Ni-coating and Co-coating protect the alloy in alkaline solution and keep them from corrosion during charge/discharge cycling. In Figure 5.37(d–f), TEM results show that a loose layer with needle-like species appears at the untreated alloy surface, while protective layer is found at the treated

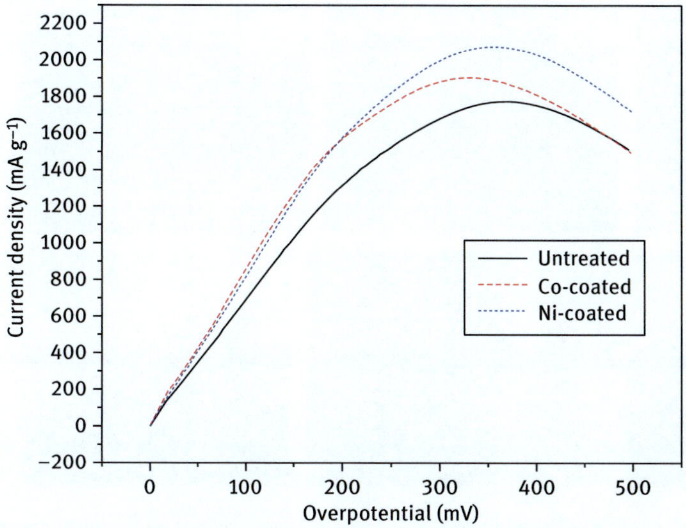

Figure 5.35: Anodic polarization of alloy electrodes before and after electroless plating.

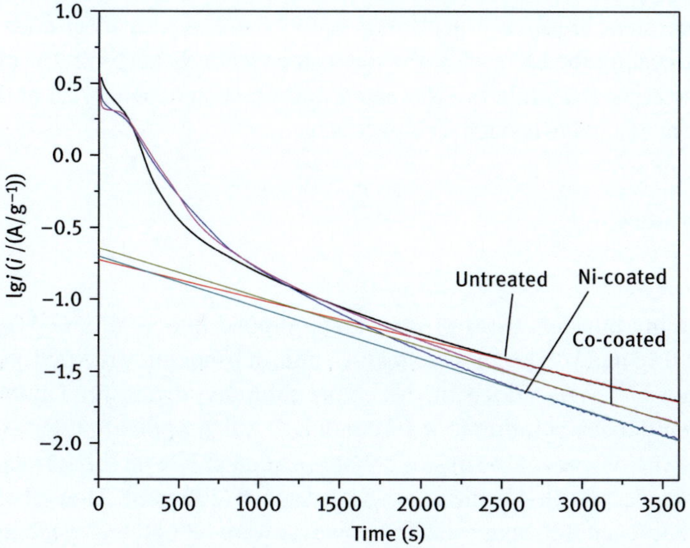

Figure 5.36: Response of discharge current density versus discharge time at a potential step of +500 mV for fully charged alloy electrodes before and after electroless plating.

alloy surface, although they are somewhat destroyed by repeat in and out of hydrogen atoms. For Ni-coated alloy some particles are found and for Co-coated alloy cracks are found in the modification layer. However, for the both modification layer, no serious peeling off is found. The protective layer from Ni- or Co-coating keeps the alloy from forming hydroxides and keeps higher conductivity at the alloy surface, and therefore

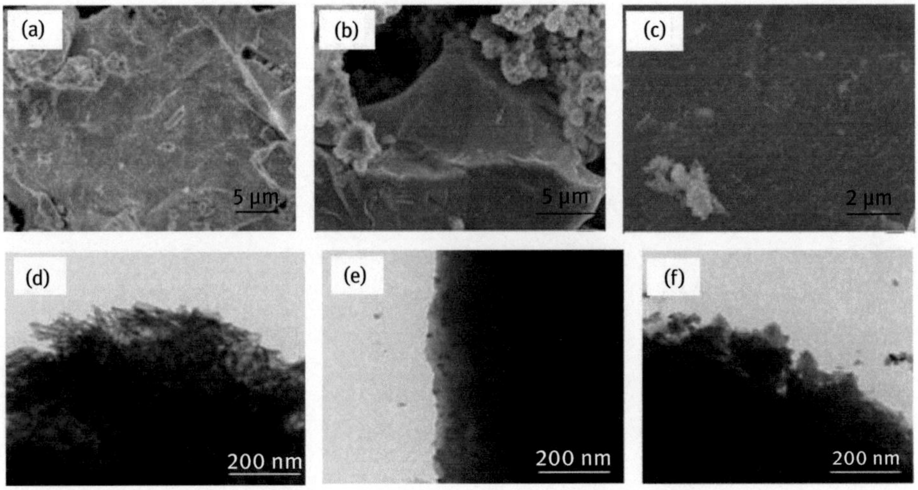

Figure 5.37: Evolution in morphology of alloy before and after electroless plating at the 100th cycle (a) untreated alloy SEM; (b) Ni-coated alloy SEM; (c) Co-coated alloy SEM; (d) untreated alloy TEM; (e) Ni-coated alloy TEM; (f) Co-coated alloy TEM.

the surface coating treatment improves the cycling stability of the alloy electrodes. After charge/discharge cycling for 200 cycles, the untreated electrode keeps 71.5 % of its maximum discharge capacity, while the Ni-coated and Co-coated keeps 79.6 and 76.5 % of their maximum discharge capacity, respectively.

5.3.2 Polymer Modification

5.3.2.1 Experimental

The La–Mg–Ni-based alloy powders (5.00 g) were firstly dipped into 50 mL Na_2SO_4 aqueous solution (c = 10.56 mol L^{-1}) preprocessing for 1 min, subsequently the pyrrole monomers (c = 0.28 mol L^{-1}) were added to the above solution stirring for 1 min. Finally, 50 mL hydrogen dioxide solution (c = 0.19 mol L^{-1}) was dropped to the preceding mixed solution. The process of the in situ polymerization of PPy on the surface of the alloys lasted 8 min along with stirring at a constant speed (200 rpm). After all of the above procedures had been finished, the alloy powders were filtered and washed thoroughly with distilled water. Subsequently, the powders were dried in vacuum at 333 K for 4 h.

5.3.2.2 Microstructure and Electrochemical Properties

The SEM images of the bare and Na_2SO_4-doped PPy-modified alloy powders are shown in Figure 5.38. According to Figure 5.38(a), it can be seen that the untreated alloy powders represent a smooth surface whilst the surface of the modified alloys is attached with massively cotton fibre-shaped substance as shown in Figure 5.38(b) with

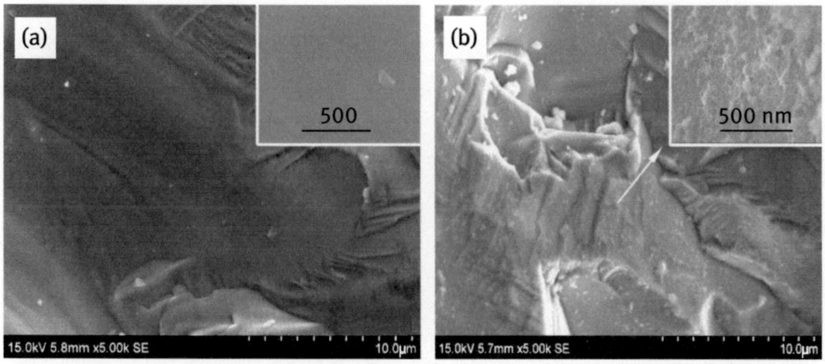

Figure 5.38: SEM images of $La_{0.80}Mg_{0.20}Ni_{2.70}Mn_{0.10}Co_{0.55}Al_{0.10}$ alloys before and after Na_2SO_4-doped PPy modification: (a) SEM before modification; (b) SEM after Na_2SO_4-doped PPy modification.

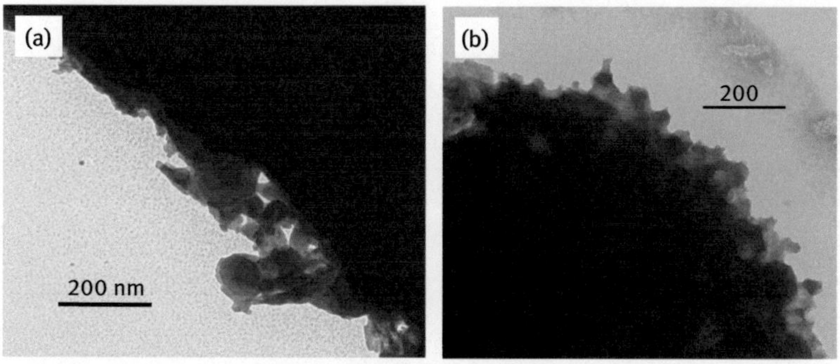

Figure 5.39: TEM micrographs of $La_{0.80}Mg_{0.20}Ni_{2.70}Mn_{0.10}Co_{0.55}Al_{0.10}$ alloys before and after Na_2SO_4-doped PPy modification, (a) TEM before modification; (b) TEM after Na_2SO_4-doped PPy modification.

larger magnification, which is similar to the morphology of PPy [37]. After PPy treatment, the surface becomes rather rough, covered with a relatively uniform layer as shown in Figure 5.38(b) with smaller magnification. The TEM image in Figure 5.39(b) shows that the PPy formed on the alloy surface is nanocrystallized and that the PPy nanoparticles distribute uniformly, combined with Figure 5.38(b) with smaller magnification. Figure 5.40 shows the FTIR spectra of bare and Na_2SO_4-doped PPy-modified alloy powders. For the modified alloys, the characteristic peaks occur at 1,571 and 1,491 cm^{-1} (the fundamental vibrations of the pyrrole rings), 1,288 cm^{-1} (=C–H in-plane vibration), 1,182 cm^{-1} (C–N stretching vibration), 1,100 cm^{-1} (the SO_4^{2-}), 1,039 cm^{-1} (the C–H and N–H in-plane deformation vibrations) and 912 cm^{-1} (C–C out-of-plane ring deformation vibration). The results above reveal that the nano-PPy synthetized by in situ polymerization of pyrrole monomers grew on the surface of the alloy powders.

The capacity retention of bare and treated alloy electrode is plotted in Figure 5.41 as a function of cycle number. As shown in Figure 5.41, it is explicit that there is

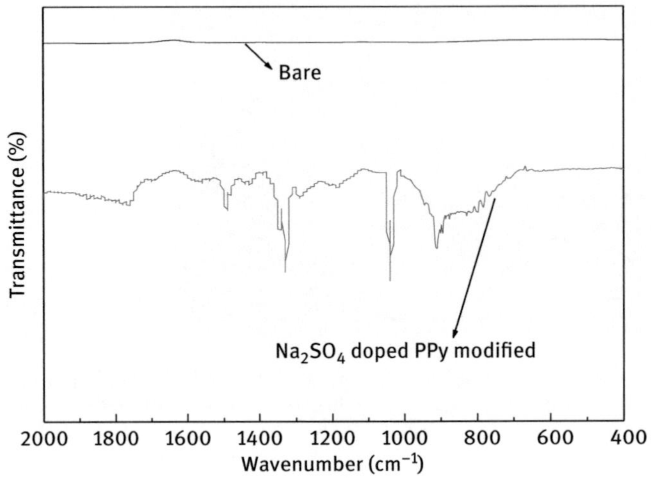

Figure 5.40: FT-IR spectra of $La_{0.80}Mg_{0.20}Ni_{2.70}Mn_{0.10}Co_{0.55}Al_{0.10}$ alloys before and after Na_2SO_4-doped PPy modification.

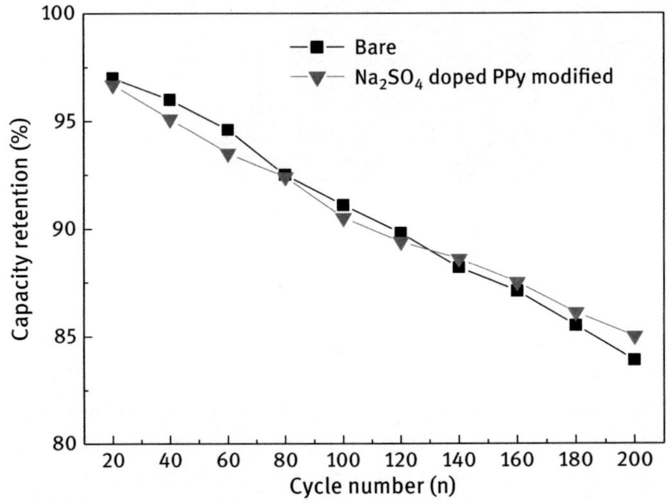

Figure 5.41: Capacity retention of bare and Na_2SO_4-doped PPy-modified $La_{0.80}Mg_{0.20}Ni_{2.70}Mn_{0.10}Co_{0.55}Al_{0.10}$ alloy electrodes versus cycle number.

little difference in both of the maximum discharge capacity and the capacity retention between the bare and the treated alloy electrodes.

Figure 5.42 shows the *HRD* curves of the bare and the treated La–Mg–Ni-based alloy electrodes at different current densities. The *HRD* reflects the discharge capacity of the alloy electrodes at high current densities. As shown in Table 5.6, the *HRD* of the Na_2SO_4-doped PPy-modified alloy electrode at 1,800 mA g^{-1} obviously increases

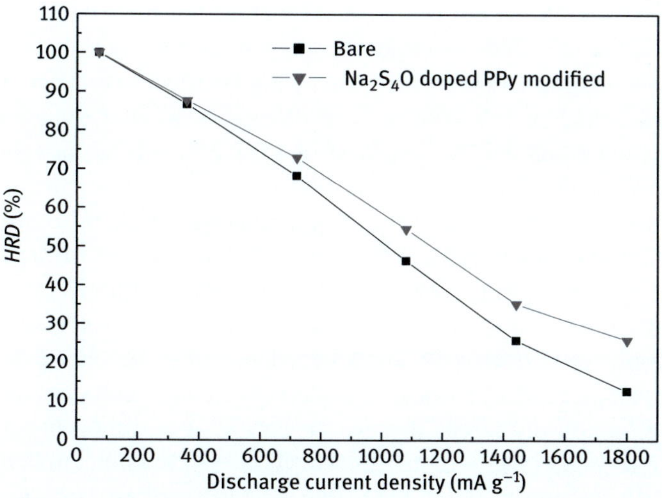

Figure 5.42: *HRD* of bare and Na_2SO_4-doped PPy-modified $La_{0.80}Mg_{0.20}Ni_{2.70}Mn_{0.10}Co_{0.55}Al_{0.10}$ alloy electrodes.

Table 5.6: Electrochemical kinetics parameters of the bare and treated $La_{0.80}Mg_{0.20}Ni_{2.70}Mn_{0.10}Co_{0.55}Al_{0.10}$ alloy electrodes.

Sample	HRD_{1800}[a] (%)	R_p[b] (mΩ)	I_0[c] (mA g^{-1})	R_{ct}[d] (mΩ)	I_L[e] (mA g^{-1})	D[f] ($\times10^{-10}$ cm^2 s^{-1})
Bare alloy	12.5	104.7	245.3	493.0	2,719	2.28
Treated alloy	25.1	99.9	256.9	287.0	2,916	2.88

[a]The capacity retention rate at discharge current density of 1800 mA g^{-1}.
[b]The polarization resistance.
[c]The exchange current density.
[d]The charge–transfer resistance obtained by Z-View program.
[e]The limiting current density.
[f]The hydrogen diffusion coefficient.

from 12.5% to 25.1 % compared with that of the bare alloy electrode. And in Figure 5.42 it can be seen that the modification of nano-PPy on the alloy surface exerts a positive effect on the *HRD* of alloy electrodes. The *HRD* at lower rates improves slightly, and the improvement of *HRD* increases with increasing discharge current density. The HRD_{1800} for the modified alloy reaches two times of that for the bare alloy. This is because Na_2SO_4-doped PPy possesses excellent reversibility between two or more oxidation states [38], which could accelerate the hydrogen absorption/desorption process of the alloy powders. It is the electrocatalytic reaction due to the reversibility of the PPy that accelerates the hydrogen absorption/desorption into/from the alloy. The detail is the released electrons during the reversible process of PPy oxidize hydrogen protons. The above electronic cross reaction accelerates the hydrogen absorption/desorption into/from the alloy. The good reversibility of the PPy redox reaction makes the above electrocatalytic reaction of hydrogen keep circulating. According to Shaktawat et al. [39], the charge carriers of the PPy are generally considered as polarons and bipolarons, which is stabilized by the additional ions called dopants during the preparation. In our system, the Na^+ and $SO4^{2-}$ ions play the role as dopants and can stabilize the synthesized PPy, so that the nano-PPy is unlikely to be seriously damaged during charge/discharge cycles. Moreover, the improvement of *HRD* also corresponds to the microstructural characteristics of the modified alloy powders surface, on which the cotton fibre-shaped PPy doped with Na_2SO_4 distributes uniformly.

The typical charge–discharge curve at 5*C* rate is considered to be a persuasive measurement of high-power performance for battery electrodes. As shown in Figure 5.43, the discharge voltages of the modified alloy electrode at the discharge current density of 1,800 mA g^{-1} obviously exceed those of the bare alloy and show better high-power performance. The regular pattern presented in Figure 5.43 is similar to that of the HRD_{1800} in Figure 5.42. It is reported that the higher discharge voltage confirms higher catalytic activity on the alloy surface and faster hydrogen diffusion on the alloy powder interface [39]. Thus, the polarization of treated alloy electrode possessing higher catalytic activity is restrained well and its discharge capacity is higher than that of the bare alloy electrode at the discharge current density of 1,800 mA g^{-1}.

In order to certify the outstanding electrocatalytic activity of Na_2SO_4-doped PPy nanoparticles attached to alloy surface, the linear polarization curves are tested subsequently. The exchange current density (I_0) is used to characterize

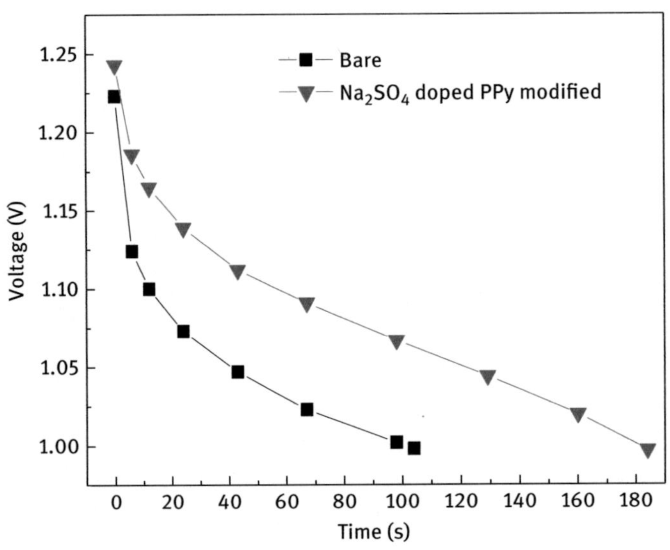

Figure 5.43: Typical charge–discharge curves of bare and Na_2SO_4-doped PPy-modified $La_{0.80}Mg_{0.20}Ni_{2.70}Mn_{0.10}Co_{0.55}Al_{0.10}$ alloy electrodes at a 5*C* rate.

electrocatalytic activity for charge–transfer reaction on the surface of alloy electrodes. In Table 5.6, the exchange current density slightly increases from 245.3 mA g^{-1} to 256.9 mA g^{-1}. This result also corresponds to that of Figure 5.43. They reach an agreement on the view that the PPy doped with Na_2SO_4 possesses better electrocatalytic activity.

Based on the tremendous improvement of *HRD* in Figure 5.42, it can be related to the decrease of the charge transfer resistance of alloy electrode surface. The reason is that the nano-PPy possessing good conductivity forms on the surface of alloy powders, whose catalysis and electrochemical redox reversibility are superior. Hence, the EIS was measured at 50 % DOD at 298 K, and the Nyquist and Bode plots of bare and treated alloy electrodes are presented in Figure 5.44. As shown in Figure 5.44(a), it can be seen that each spectrum is consisted of two semicircles followed by a straight line at the low-frequency range. It is reported by Kuriyama et al. that the semicircles in the high-frequency region and low-frequency region correspond to the contact impendence and the charge–transfer resistance (R_{ct}), respectively, and the sloped straight line corresponds to Warburg impendence [40]. From Figure 5.44(a), it can

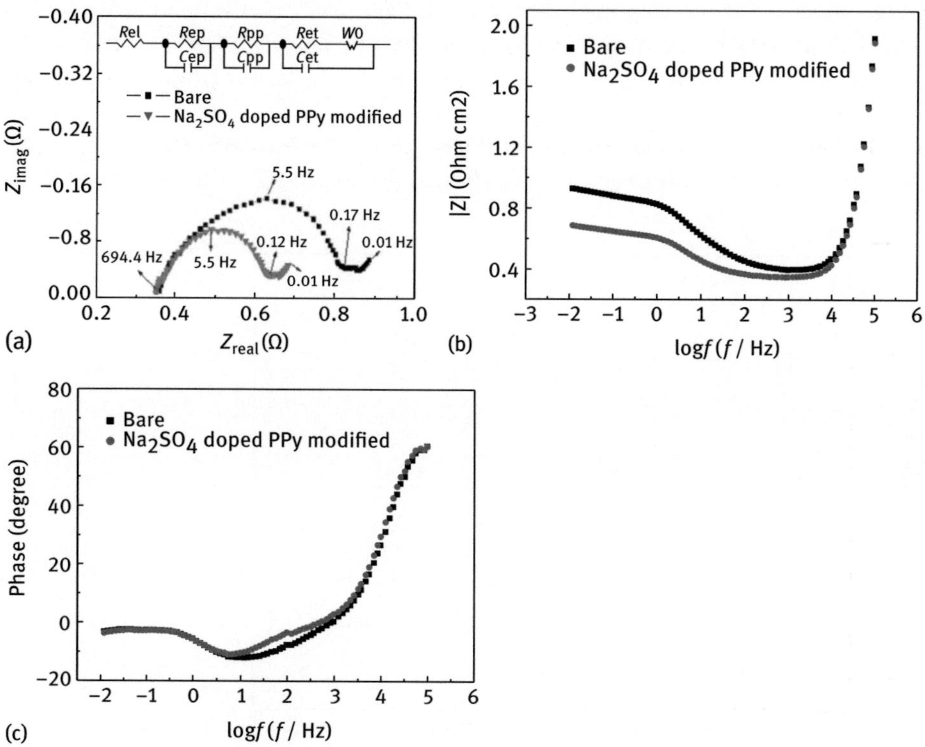

Figure 5.44: Nyquist and Bode plots of bare and Na_2SO_4-doped PPy-modified $La_{0.80}Mg_{0.20}Ni_{2.70}Mn_{0.10}Co_{0.55}Al_{0.10}$ alloy electrodes at 50 % DOD after activation (a) Nyquist plots; (b) modulus of impedance versus log *f*; (c) phase angle versus log *f*.

be seen that there is little change with the radius of the smaller semicircle and obvious decrease in the radius of the larger semicircle of Na_2SO_4-doped PPy-modified alloy electrode in the low-frequency region. This is in line with the change of *HRD* in Figure 5.42 and the R_{ct} values listed in Table 5.6. The data are fitted by Z-view software. The equivalent circuit for fitting is also shown in Figure 5.44(a), in which R_{el} is resistance of electrolyte, R_{cp} corresponds to resistance between current collector and electrode, R_{pp} is attributed to resistance between alloy particles and R_{ct} is charge–transfer resistance. Fitting results show that the charge–transfer resistance for bare alloy electrode is 493.0 mΩ and that for treated alloy electrode is 287.0 mΩ. Thereby, it can be concluded that the nano-PPy possessing good conductivity and electrochemical redox reversibility facilitates the charge transfer rate. In the corresponding Bode plots representing mass transport, the modulus of impedance of modified alloy electrode decreases compared with the bare electrode in Figure 5.44 (b), and the decrease is more obvious with decreasing of frequency. Figure 5.44(c) shows the corresponding to Bode phase plots with respect to the applied frequency. In light of Figure 5.44(c), the phase angle for the modified electrode is more pronounced than that for the bare electrode. Thus, the hydrogen diffusion rate of treated alloy electrode speeds up, which is attributed to the increasing charge–transfer rate. The hydrogen concentration at treated alloy surface is lower with charge–transfer rate increasing, which forms larger gradient with the interior of alloy bulk facilitating the hydrogen diffusion compared with the bare alloy.

The hydrogen diffusion rate in alloy bulk can be characterized by anode polarization curve measurement. Figure 5.45 indicates the anode polarization curves of the

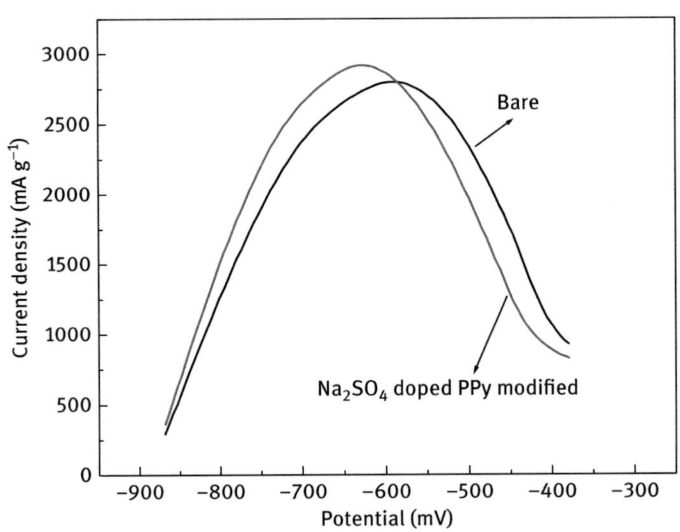

Figure 5.45: Anode polarization curves of bare and Na_2SO_4-doped PPy-modified $La_{0.80}Mg_{0.20}Ni_{2.70}Mn_{0.10}Co_{0.55}Al_{0.10}$ alloy electrodes at 50 % DOD after activation.

bare and the treated alloy electrodes at 50 % DOD and 298 K. During the process of anodic polarization test, as shown in Figure 5.45, the anodic current grows with the enhancement of over potential and reaches a maximum defined as limiting current density I_L. The I_L value figures the hydrogen diffusion rate of the bulk alloy particles [34]. Combined with Table 5.6 and Figure 5.45, it is found that the I_L value rises from 2,719 mA g^{-1} (bare alloy) to 2,916 mA g^{-1} (treated alloy). The improvement of hydrogen diffusion rate in alloy bulk is related to the nano-PPy on the alloy surface for their electrocatalytic activity which brings out the decrease of R_{ct} value. With the increasing charge–transfer rate, the hydrogen concentration at treated alloy surface is lower, which forms larger gradient with the interior of alloy bulk compared with the bare alloy. Thereby, the hydrogen diffusion rate speeds up.

Meanwhile, the hydrogen diffusion rate can also be characterized by potential step method. Figure 5.46 exhibits the correspondence of anodic current density to discharge time of bare and treated alloy electrodes at fully charged state and 298 K. It can be seen that the semilogarithmic value of the anodic current (log i) reveals an approximately linear dependence on time (t). In this situation, the surface concentration of hydrogen atoms get close to zero; hence, the hydrogen diffusion in alloy bulk becomes rate-controlling step of the electrode reaction. The hydrogen diffusion coefficient D listed in Table 5.6 improves from 2.28×10^{-10} cm^2 s^{-1} (bare alloy) to 2.88×10^{-10} cm^2 s^{-1} (Na_2SO_4-doped PPy-modified alloy), that is, the hydrogen diffusion rate of the modified alloy powders is faster than that of the bare alloy. The result of potential step method is in accordance with that of the anode polarization curves in Figure 5.45 for the eminent electrochemical redox reversibility of Na_2SO_4-doped

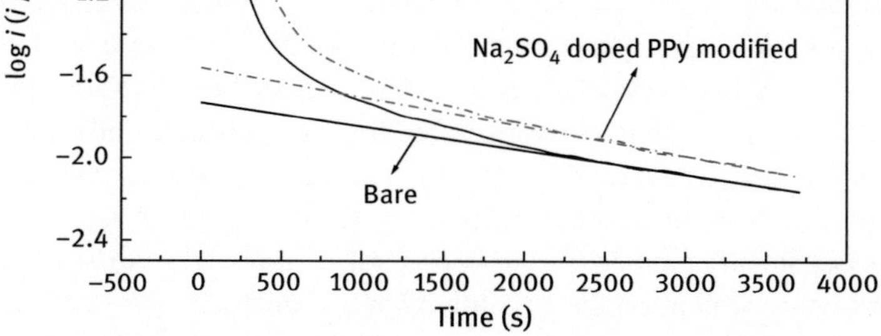

Figure 5.46: Correspondence of anodic current density versus discharge time of bare and Na_2SO_4-doped PPy-modified $La_{0.80}Mg_{0.20}Ni_{2.70}Mn_{0.10}Co_{0.55}Al_{0.10}$ alloy electrodes at fully charged state after activation.

PPy. The hydrogen diffusion rate in the alloy bulk is enhanced by the decrease of charge–transfer resistance during the electrode reaction.

5.4 Conclusions

(1) CuO was added to the AB_3-type La–Mg–Ni-based alloy electrodes in order to improve the electrochemical properties. SEM and cyclic voltammogram show that CuO can be reduced to fine Cu particles during charge processing, depositing at surface of alloy particles. The maximum discharge capacity increases from 314 mAh g^{-1} (blank electrode) to 341 mAh g^{-1} (3.0 wt.% CuO). The HRD increases after CuO is added, and the electrode at 1,440 mA g^{-1} with 0.5 wt.% CuO addition exhibits HRD of 59.9 %. The cycling stability is greatly improved by CuO addition. When the charge/discharge cycle number is more than 40, capacity of electrodes with 1.0 wt.% or more CuO addition decreases more slowly than that of the blank one. And the discharge capacity retention rate at the 200th cycle increases from 71.6 % (blank electrode) to 77.2 % (2.5 wt.% CuO electrode). CuO additive also improves HTD of the electrodes.

(2) 2 wt.% TiO_2, Er_2O_3 and ZnO were added to AB_3-type La–Mg–Ni-based hydrogen storage alloy electrodes and their effect on the electrochemical properties were investigated. The electrochemical measurements showed that HRC, cycling stability and HTD increased in case of using metal oxide additives. The HRC at 1,440 mA g^{-1} increased from 85.1% to 94.1 % (TiO_2), 93.3 % (Er_2O_3) and 90.5 % (ZnO) owing to the catalytic effect of the 3d transition metal oxides (TiO_2 and ZnO) and the heavy rare earth oxides (Er_2O_3) on hydriding of the alloy electrodes. These oxides addition suppressed oxidation of the alloys in alkaline electrolyte during charging/discharging cycles, and therefore cycling stability of the electrodes was ameliorated. The discharge capacity retention rate at the 200th cycle mounted up from 72.9 % (unmodified) to 79.6 % (TiO_2), 87.5 % (Er_2O_3) and 77.9 % (ZnO). The HTD was also improved, increasing from 84.9% to 92.7 % (TiO_2), 91.2 % (Er_2O_3) and 89.5 % (ZnO) at 333 K.

(3) The hydrogen storage alloy electrodes were treated by electroplating Co–Ni film. Morphology and electrochemical properties of the unmodified and modified La–Mg–Ni-based alloys have been investigated. SEM-EDS and XRD results show that Co–Ni alloy layer has been uniformly deposited at the alloy surface. The charge and discharge testing results suggest that the alloy particle treated at 10 mA cm^{-2} has a superior discharge performance compared to the other alloys. After Co–Ni treatment, the active surface area on the surface increases due to the electrocatalysis of metallic layers, which results in a modification of the charge–transfer reaction, therefore the polarization phenomenon of high current densities is suppressed. Consequently, the modified alloy electrode at 10 mA cm^{-2} exhibits the highest exchange current density, and high rate discharge ability of the modified alloy electrodes is up to 57.5 % at 1,875 mA g^{-1} accordingly. The coated

film also acts as a protective film, which prolongs the cycle life of the alloy electrodes.

(4) In order to improve the electrochemical properties of La–Mg–Ni-based alloy electrodes, cobalt or nickel is plated on the alloy particles using an electroplating method. Effect of surface modification on electrode performance was investigated in detail. FESEM images reveal that the metallic layer (spherical cobalt or lamellar nickel) has been deposited at the alloy surface. Electrochemical characteristics have been improved effectively due to the anti-oxidation and superior catalysis of metallic layers. The maximum discharge capacity increases from 316 mAh g^{-1} to 335 mAh g^{-1} (Co-coated) and 336 mAh g^{-1} (Ni-coated), and the discharge capacity retention of alloy plating cobalt after 200 cycles is about 7 % higher. The nickel metallic layer is favourable to preventing self-discharge, the self-discharge rate decreasing from 17% to 7 %. Furthermore, the coated alloy electrodes exhibit higher exchange current densities (I_0) and lower apparent activation enthalpy $\Delta_r H^*$. The HRD controlled by kinetics is improved accordingly, increasing approximately 10 and 17 % for cobalt-coated and nickel-coated alloy electrodes at $I_d = 1,500$ mA g^{-1}, respectively.

(5) Electroless plating Ni and Co was performed on Nd–Mg–Ni-based alloys with a bidentate ligand $N_2H_4{\cdot}HSO_4$ in plating solution. The bidentate ligand makes the coating layer distribute more uniformly at the surface and combine more tightly with the alloy surface. From the SEM results honeycomb-like Ni-coating interlinks at the surface and flake-like Co-coating covers at the surface. Ni and Co shows high electrocatalytic activity during hydrogen anodic electrode process. The catalytic activity facilitates charge transferring at the surface and therefore decreases hydrogen concentration at the surface, thus promoting hydrogen diffusion rate in the bulk owing to larger hydrogen concentration difference. The initial discharge capacity is higher after surface modification. In the discharge process at higher discharge current density, HRD is greatly improved. At the discharge current of 1,800 mA g,$^{-1}$ the HRD increases from 55.3 % (untreated) to 75.7 % (Ni-coated) and 73.6 % (Co-coated), respectively. After charge/ discharge for 100 cycles, the coating layer keeps still complete at the surface and protects the alloy from corrosion in the electrolyte. Consequently, the capacity retention rate at the 200th cycle increases from 71.5 % (untreated) to 79.6 % (Ni-coated) and 76.5 % (Co-coated), respectively.

(6) The La–Mg–Ni-based alloy particles are modified chemically by Na_2SO_4-doped PPy. Micromorphology and electrochemical kinetics of the alloys before and after treatment are studied systematically. In combination with SEM, TEM and FTIR, it is concluded that the cotton fibre-shaped PPy nanoparticles are successfully deposited on the surface of alloy particles. Kinetics of the alloy electrode is signally improved. The charge–transfer resistance R_{ct} decreases; meanwhile, the limiting current density I_L and hydrogen diffusion coefficient D increase owing to

the prominent electrochemical catalytic activity and redox reversibility of nano-PPy. Consequently, *HRD* of the Na_2SO_4-doped PPy-modified alloy electrode at 1,800 mA g^{-1} obviously increases from 12.5% to 25.1 %, as well as the discharge voltages of the treated alloy electrode increase remarkably.

References

[1] Yang SQ, Liu HP, Han SM, Li Y, Shen WZ. Effects of electroless composite plating Ni–Cu–P on the electrochemical properties of La–Mg–Ni-based hydrogen storage alloy. Appl Surf Sci 2013;271:210–15.

[2] Kohno T, Yoshida H, Kawashima F, Inaba T, Sakai I, Yamamoto M, Kanda M. Hydrogen storage properties of new ternary system alloys: La_2MgNi_9, $La_5Mg_2Ni_{23}$, La_3MgNi_{14}. J Alloys Compd 2000;311:L5–7.

[3] Zhang F, Luo Y, Deng A, Tang Z, Kang L, Chen J. A study on structure and electrochemical properties of $(La, Ce, Pr, Nd)_2MgNi_9$ hydrogen storage electrode alloys. Electrochim Acta 2006;52:24–32.

[4] Li R, Wan J, Wang F, Ding C, Yu R. Effect of non-stoichiometry on microstructure and electrochemical performance of $La_{0.8}Gd_xMg_{0.2}Ni_{3.15}Co_{0.25}Al_{0.1}$ (x = 0–0.4) hydrogen storage alloys. J Power Sources 2016;301:229–36.

[5] Ozaki T, Kanemoto M, Kakeya T, Kitano Y, Kuzuhara M, Watada M, Tanase S, Sakai T. Stacking structures and electrode performances of rare earth–Mg–Ni-based alloys for advanced nickel–metal hydride battery. J Alloys Compd 2007;446–447:620–4.

[6] Kitano Y, Ozaki T, Kanemoto M, Komatsu M, Tanase S, Sakai T. Electron diffraction study of layer structures in La-Mg-Ni hydrogen absorption alloys. Mater Trans 2007;48:2123–7.

[7] Zhang L, Li Y, Zhao X, Liu J, Ke D, Du W, Yang S, Han S. Phase transformation and cycling characteristic of Ce_2Ni_7-type single-phase $La_{0.78}Mg_{0.22}Ni_{3.45}$ metal hydride alloy. J Mater Chem A 2015;3:13679–90.

[8] Gao J, Yan XL, Zhao ZY, Chai YJ, Hou DL. Effect of annealed treatment on microstructure and cyclic stability for La–Mg–Ni hydrogen storage alloys. Power J Sources 2012;209:257–61.

[9] Liu J, Li Y, Han D, Yang S, Chen X, Zhang L, Han S. Electrochemical performance and capacity degradation mechanism of single-phase La–Mg–Ni-based hydrogen storage alloys. J Power Sources 2015;300:77–86.

[10] Liu Y, Cao Y, Huang L, Gao M, Pan H. Rare earth–Mg–Ni-based hydrogen storage alloys as negative electrode materials for Ni/MH batteries. J Alloys Compd 2011;509:675–86.

[11] Li MM, Yang CC, Wang CC, Wen Z, Zhu YF, Zhao M, Li JC, Zheng WT, Lian JS, Jiang Q. Design of hydrogen storage alloys/nanoporous metals hybrid electrodes for nickel-metal hydride batteries. Sci Rep 2016;6:27601.

[12] Giza K, Drulis H. Effect of preparation method of metal hydride electrode on efficiency of hydrogen electrosorption process. Int J Mater Res 2016;107:103–8.

[13] Gan H, Yang Y, Shao H. Improvement of the rate performance of hydrogen storage alloys by heat treatments in Ar and H_2/Ar atmosphere for high-power nickel–metal hydride batteries. Electrochim Acta 2015;174:164–71.

[14] Ji Z, Wang Z, Deng J, Yao Q, Zhou H. Effects of surface treatment on electrochemical properties of AB3.8-type hydrogen storage alloy in alkaline electrolyte. Russ J. Electrochem. 2015;51:244–248.

[15] Yuan H, Yang K, Jiang L, Liu X, Wang S. Surface treatment of rare earth-magnesium- nickel based hydrogen storage alloy with lithium hydroxide aqueous solution. Int J Hydrogen Energy 2015;40:4623–9.

[16] Abe H, Tokuhira S, Uchida H, Ohshima T. Surface modifications of hydrogen storage alloy by heavy ion beams with keV to MeV irradiation energies. Nucl Instrum Methods B 2015;365:214–17.

[17] Zeng XX, Wang JM, Wang QL, Kong DS, Shao HB, Zhang JQ, Cao CN. The effects of surface treatment and stannate as an electrolyte additive on the corrosion and electrochemical performances of pure aluminum in an alkaline methanol-water solution. Mater Chem Phys. 2010;121:459–64.

[18] Qu X, Ma L, Jin C, Zhao X, Ding Y. Effects of fluorination treatment on electrochemical properties of $La_{0.67}Mg_{0.33}Ni_{2.25}Co_{0.75}$ hydrogen storage alloy. Rare Met Mater Eng 2011;40:543–6.

[19] Zhao B, Liu L, Ye Y, Hu S, Wu D, Zhang P. Enhanced hydrogen capacity and absorption rate of $LaNi_{4.25}Al_{0.75}$ alloy in impure hydrogen by a combined approach of fluorination and palladium deposition. Int J Hydrogen Energy 2016;41:3465–9.

[20] Ngene P, Westerwaal RJ, Sachdeva S, Haije W, de Smet LC, Dam B. Polymer-induced surface modifications of Pd-based thin films leading to improved kinetics in hydrogen sensing and energy storage applications. Angew Chem-Int Edit 2014;53:12081–5.

[21] Sun L, Lin J, Liang F, Cao Z, Wang L. Ni coated $Ti_{1.4}V_{0.6}Ni$ composite as the negative electrode in Ni-MH battery. Mater Lett 2015;161:685–9.

[22] Li MM, Yang CC, Chen LX, Jiang Q. Hydrogen storage alloys/reduced graphite oxide: an efficient hybrid electrode with enhanced high-rate dischargeability. Electrochim Acta 2016;200:59–65.

[23] Zadorozhnyy MY, Klyamkin SN, Strugova DV, Olifirov LK, Milovzorov GS, Kaloshkin SD, Zadorozhnyy VY, Deposition of polymer coating on metallic powder through ball milling: application to hydrogen storage intermetallics. Int J Energy Res 2016;40:273–9.

[24] Zhang S, Shi P, Deng C. Characteristics of hydrogen storage alloy electrode with cupric oxide additive. Solid State Ionics 2006;177:1193–7.

[25] Strehblow HH, Maurice V, Marcus P. Initial and later stages of anodic oxide formation on Cu, chemical aspects, structure and electronic properties. Electrochim Acta 2001;46:3755–66.

[26] Raju M, Ananth MV, Vijayaraghavan L. Rapid charging characterization of $MmNi_{3.03}Si_{0.85}Co_{0.60}Mn_{0.31}Al_{0.08}$ alloy used as anodes in Ni–MH batteries. Int J Hydrogen Energy 2009;34:3500.

[27] Perez-Alonso FJ, Adan C, Rojas S, Pena MA, Fierro JL. Ni–Co electrodes prepared by electroless-plating deposition. A study of their electrocatalytic activity for the hydrogen and oxygen evolution reactions. Int J Hydrogen Energy 2015;40:51–61.

[28] Durairajan A, Haran BS, Popov BN, White RE. Cycle life and utilization studies on cobalt microencapsulated AB_5 type metal hydride. J Power Sources 1999;83:114–20.

[29] Chen WX, Cyclic voltammetry and electrochemical impedance of $MmNi_{3.6}Co_{0.7}Mn_{0.4}Al_{0.3}$ alloy electrode before and after treatment with a hot alkaline solution containing reducing agent, J Power Sources 2000;90:201–5.

[30] Yang K, Chen S, Wu F. Study on the surface modification of metal hydride electrodes with cobalt. J Power Sources 2008;184:617–21.

[31] Zhang L, Han SM, Li Y, Yang SQ, Zhao X, Liu JJ. Effect of magnesium on the crystal transformation and electrochemical properties of A_2B_7-type metal Hydride alloys, J Electrochem Soc 2014;161:A1844–50.

[32] Zhao XY, Ma LQ, Gao YJ, Ding Y, Shen XD. Effect of surface treatments on microstructure and electrochemical properties of La–Ni–Al hydrogen storage alloy. Int J Hydrogen Energy 2009;34:1904–9.

[33] Geng M. Electrochemical characteristics of Ni-Pd-coated $MmNi_5$-based alloy powder for nickel-metal hydride batteries. J Alloys Compds 1995;217:90–3.

[34] Kuriyama N, Sakai T, Miyamura H, Uehara I, Ishikawa H. Electrochemical impedance and deterioration behavior of metal hydride electrodes. J Alloys Compds 1993;212:183–97.
[35] Notten PH, Hokkeling P. Double-phase hydride forming compounds: a new class of highly electrocatalytic materials. J Electrochem Soc 1991;138:1877–85.
[36] Iwakura C, Oura T, Inoue H, Matsuoka M. Effects of substitution with foreign metals on the crystallographic, themodynamic and electrochemical properties of AB5-type hydrogen storage alloys. Electrochim Acta 1996;41:117–21.
[37] An J, Sun HX, Cui JF, Zhu ZQ, Liang WD, Pei CJ, Yang BP, Li A. Surface modification of polypyrrole-coated foam for the capture of organic solvents and oils. J Mater Sci 2014;49:4575–82.
[38] Akieh MN, Price WE, Bobacka J, Ivaska A. Ion exchange behaviour and charge compensation mechanism of polypyrrole in electrolytes containing mono-, di- and trivalent metal ions. Synth Met 2009;159:2590–8.
[39] Shaktawat V, Jain N, Dixit M, Saxena NS, Sharma K, Sharma TP. Temperature dependence of conductivity of polypyrrole doped with sulphuric acid. Indian J Pure Appl Phys 2008;46:427–30.
[40] Li XF, Wang LZ, Dong HC, Song YH, Shang HD. Electrochemical hydrogen absorbing properties of graphite/AB5 alloy composite electrode. J Alloys Compd 2012;150:114–18.

6 Outlook and Challenges of RE–Mg–Ni-Based Alloys as Negative Electrode Materials for Ni/MH Batteries

Driven by the competitive advantages, tremendous breakthroughs in the research of RE–Mg–Ni-based alloys have been witnessed. First, the composition and structure characteristics of the superlattice phases in RE–Mg–Ni-based alloys have been clearly revealed. Second, the phase transformation mechanisms and conditions of different phases have been understood. Thirdly, the effect of different preparation methods and elemental compositions on the phase transformation has been summarized. Finally, the relationship between phase compositions and electrochemical properties has been clarified. Based on the research, great progresses in developing RE–Mg–Ni-based alloys have been made, which enables this alloy system to become one of the key electrode materials for Ni/MH batteries. Despite the significant improvement in the past few decades, challenges still remain:

1. The significant differences of the melting points of compositional elements make it difficult to obtain alloys with target elemental compositions. Typically, the vapour pressure of Mg element is very high, leading to easy Mg evaporation. Therefore, precisely controlling the amount of constituent elements in alloys is still a significantly technical challenge.
2. Because of the similar elemental compositions of different phases and the strict peritectic reaction temperatures for their formations, most RE–Mg–Ni-based alloys are with multiphase structures. So the disturbance from coexisting mixture phases is confusing the understanding of the impact of a specific phase on the electrochemical properties of alloys. Therefore, revealing the basic characteristics of different superlattice phases is another important challenge.
3. RE–Mg–Ni-based alloys store hydrogen through phase lattice, and the hydrogen absorption/desorption behaviour of different phases is the main factor that determines the electrochemical properties of RE–Mg–Ni-based alloys. But their studies are difficult due to the inaccessibility of observing atomic behaviours. Thus, finding better characterization techniques for studying the interactions between metal and hydrogen atoms and hydrogen distribution in the alloys is another great challenge.
4. From the application perspective, RE–Mg–Ni-based hydrogen storage alloys show great potential as electrode materials for power batteries. But further improvement in the large current discharge capability and cycling stability is still necessary for future applications. For example, it is expected to obtain one battery on the basis of different characteristics and requirements for industrialization by adjusting and controlling the alloy's elemental composition and phase structure, such as the batteries with high capacity, low self-discharge, high power, high-temperature resistance and resistance to overcharge/overdischarge.

DOI 10.1515/9783110501483-006

Index

DOI 10.1515/9783110501483-007

Bei Fragen zur Produktsicherheit wenden Sie sich bitte an:
If you have any questions regarding product safety,
please contact:

Walter de Gruyter GmbH
Genthiner Straße 13
10785 Berlin
productsafety@degruyterbrill.com